CHEMICAL ANALYSIS
IN COMPLEX MATRICES

Ellis Horwood and Prentice Hall

are pleased to announce their collaboration in a new imprint whose list will encompass outstanding works by world-class chemists, aimed at professionals in research, industry and academia. It is intended that the list will become a by-word for quality, and the range of disciplines in chemical science to be covered are:

ANALYTICAL CHEMISTRY
ORGANIC CHEMISTRY
INORGANIC CHEMISTRY
PHYSICAL CHEMISTRY
POLYMER SCIENCE & TECHNOLOGY
ENVIRONMENTAL CHEMISTRY
CHEMICAL COMPUTING & INFORMATION SYSTEMS
BIOCHEMISTRY
BIOTECHNOLOGY

Ellis Horwood PTR Prentice Hall
ANALYTICAL CHEMISTRY SERIES

Series Editors:
Ellis Horwood, M.B.E.
Dr. Mary Masson, University of Aberdeen
Dr. Julian F. Tyson, University of Massachusetts at Amherst
Dr. Frank A. Settle, Virginia Military Institute
Consulting Editors:
Professor J.N. Miller, Loughborough University of Technology
Dr. R.A. Chalmers, University of Aberdeen

Current Ellis Horwood PTR Prentice Hall
Analytical Chemistry Series titles

CHEMICAL ANALYSIS IN COMPLEX MATRICES

Editor

MALCOLM R. SMYTH
School of Chemical Sciences
Dublin City University
Dublin 9, Ireland

ELLIS HORWOOD   **PTR PRENTICE HALL**
NEW YORK   LONDON   TORONTO   SYDNEY   TOKYO   SINGAPORE

First published in 1992 by
ELLIS HORWOOD LIMITED
Market Cross House, Cooper Street,
Chichester, West Sussex, PO19 1EB, England

A division of
Simon & Schuster International Group
A Paramount Communications Company

Printed and bound in Great Britain
by Hartnolls, Bomdin, Cornwall

British Library Cataloguing-in-Publication Data

Chemical analysis in complex matrices. —
(Ellis Horwood series in analytical chemistry)
I. Smyth, M. R. II. Series
660.028
ISBN 0–13–127671–9

Library of Congress Cataloging-in-Publication Data

Chemical analysis in complex matrices / editor, Malcolm R. Smyth.
p. cm. — (Ellis Horwood series in analytical chemistry)
Includes bibliographical references and index.
ISBN 0–13–127671–9
1. Chemistry, Analytic. 2. Chemistry, Technical. I. Smyth, Malcolm R. II. Series
QD75.2.C467   1992
543–dc20   91–42948
                                                                          CIP

Table of contents

1

Chemical analysis in complex matrices — an introduction

Malcolm R. Smyth
School of Chemical Sciences, Dublin City University, Dublin 9, Ireland

1.1 INTRODUCTION

"Analytical Science" can be defined in several ways; depending on the range of disciplines that fuse to create the "academic philosophy" of any particular course. At The University of Manchester Institute of Science and Technology (UMIST), for instance, there is a Master's course in "Instrumentation and Analytical Science" which "is a form of systems engineering concerned with transduction of physical and chemical variables to a convenient quantity (often electrical) and consideration of their subsequent transmission, display, storage and in some cases feedback for control of the original variable" [1]. At Dublin City University, there is a Bachelor's degree course in "Analytical Science" which is defined as "that science which deals with the detection, identification and quantification of chemical, biological and microbiological species in matrices of chemical, biological or environmental importance". While the former is a fusion of mainly chemical, physical and engineering aspects of the subject, the latter course concentrates mainly on the integration of the analytical aspects of chemistry, biochemistry and microbiology. Both courses have their own unifying philosophy, but this book will illustrate the approach taken by the Dublin institution.

During the final year of the course at Dublin City University, the students attend lectures on two modules entitled "The Analytical Approach" and "Industrial Analysis". Both of these modules have the underlying theme of explaining how instrumental methods of analysis can be used to solve problems of chemical, biological, environmental or industrial importance. They stress the importance of first understanding the nature of the matrix to be studied, before going on to show how analytical methods can be developed for the determination of the analyte(s) of interest within that matrix. The choice of the term "The Analytical Approach" has its origins in the section of the American Chemical Society (ACS) Journal *Analytical Chemistry* with the same title, and a book published in 1983 by the ACS, again with the same title [2].

These lecture courses assume a detailed knowledge of instrumental analytical techniques, which are covered during the third year of the course. For the purposes of this text, a good knowledge of instrumental analytical techniques is also assumed, since the emphasis in this volume is on how analytical techniques are used in "problem solving". The choice of the appropriate instrumental method to be used in the "end step" is still obviously an important parameter to be considered, since this will influence the choice of the appropriate sample preparation regime which is required to extract the analyte(s) of interest quantitatively from the matrix with the minimum number/amount of interferences. Some effort is therefore devoted in each chapter to a discussion of the instrumental techniques which are most suitable for particular analyte(s).

1.2   THE ANALYTICAL APPROACH

The "Analytical Approach" is basically a chain of operations which starts by asking questions such as:

"which species do I want to measure, and in which matrix?"

"in which form (speciation) do these species exist in the matrix under investigation?";

"in what concentration ranges are these species likely to be present?";

"what chemical, physical or biological processes operate within the system under investigation which might affect the concentration/stability of the species under investigation?";

"which instrumental analytical technique would be the most suitable to use for this particular application?";

"what are the likely matrix interferences which would interfere with the analysis of the species under investigation?";

"which levels of precision and accuracy does the analytical method require to satisfy the aims and objective of the study?";

"what are the material/equipment needs for the project?";

and ends with obtaining meaningful answers to the problems set in the first place.

If readers look at the example of biopharmaceutical analysis, which is the theme of the second chapter in this volume, they can soon appreciate the importance of asking such questions before embarking on the development of a validated analytical method for the determination of a drug substance and its metabolites in a biological sample. In this case it is necessary to consider: (i) the chemical behaviour and reactivity of the compound of interest (e.g. does the compound hydrolyse and under what conditions?); (ii) the metabolism of the compound in a biological system, including understanding of the conjugation, complexation and binding properties of the compound and its metabolites to endogenous compounds (both small and large biomolecules); (iii) the nature of interferences to the analytical method which are carried through the sample preparation stages of the assay; and (iv) the extent of the validation process requirements. Analytical scientists must therefore have a thorough knowledge of the chemical, physical and biological aspects relating to the problem at hand *before* they set out to develop the analytical method to solve it.

Once the analytical scientist is aware of the complexity of the problem at hand, he/she can then set out to devise an analytical method based on what can be termed "unit processes": i.e. sampling; sample pretreatment; separation/preconcentration; derivatization, determination ("end step"); and statistical treatment of results (Fig. 1.1). Each of these steps can influence the others, and "feedback" is often required

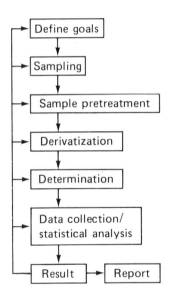

Fig. 1.1 — Unit processes in analytical science.

before an overall analytical method is refined and robust enough to give reliable data. The main aim in any analytical procedure should therefore be to develop a method which involves the minimum number of manipulations to be performed on a minimum size of sample, as carefully and quickly as possible, taking into account the selectivity and sensitivity requirements of the method. This is not as easy to achieve as it sounds, and calls a lot on the experience of the people who are performing the analyses. One of the most important factors to take into account is the fact that any errors introduced at any point in the analytical method are carried throughout the whole procedure, and thus affect the overall "accountability" of the method in terms of the main statistical parameters of "accuracy" and "precision".

Thus, **sampling**, the first link in the chain, assumes a major responsibility in this regard [3].

"No chain is stronger than its weakest link"

It is not the intention of this book to go in detail into the theory either of sampling or statistical evaluation of analytical methods, since these are covered well in other texts [4–6], but rather to highlight practical considerations that are important for decreasing errors that arise in the analysis of compounds in complex matrices. A practical exercise that is set for students at Dublin City University in this regard is to provide them with a map showing part of Dublin Bay. The students are then required to visit the site (bringing their wellington boots with them!) and compile a report on the sampling strategy they would employ to investigate the state of pollution at this particular location. The students can quickly appreciate the complexity of the problem, since they must first ask themselves the question:

"What is meant by the term 'pollution'?"

They must then decide:

"what compound(s)/micro-organisms do I want to determine?"

"how many sampling points should there be?"

"what types of sample are required?"

"how can I take a representative sample of air, sediment or water?"

"how often am I going to take samples?"

"how will I collect and store samples?"

"what environmental factors might influence the sampling process?"

"how am I going to record the sampling process?"

In this way, the idea that sampling has as much to do with common sense as it has to do with statistics is graphically illustrated, and the field trip helps the students to think in an analytically scientific way.

The next link in the chain is **sample pretreatment**, which again is nearly a science in itself. The objectives in sample pretreatment are:

(i) to carry through as high a percentage as possible of the analyte(s) from the original sample to the determination step (i.e. effect a good recovery);

(ii) to ensure that the analyte(s) remain in their original state (for organic analysis and speciation studies), or are fully converted into a stable state (for total metal determinations);

(iii) to carry through as few of the matrix components from the original sample into the determination step.

The sample preparation step often involves a protocol such as:

(i) extract or dissolve analyte(s) into a solution which is compatible for the determination step;

 (ii) remove contaminants and interfering compounds;
(iii) concentrate the analyte(s).

The main means of achieving these aims are to use techniques such as solvent extraction, solid-phase extraction, column chromatography, distillation, evaporation, protein precipitation, etc., and examples of the use of such techniques will be given in each of the subsequent chapters.

If the sensitivity and selectivity requirements of the determination step are not met, it is possible to move directly to the **end step**. In some cases, however, a **derivatization** is required in order to modify the analytes(s) in order to improve conditions for their detection. This inevitably introduces another link into the chain, with its associated error, and should be avoided unless the benefits gained (in terms of increase in sensitivity, lower limits of detection and improved selectivity) far outweigh the disadvantages of increased error, time of analysis, cost, etc. The main points to be considered in the development of any derivatization reaction are:

 (i) efficiency of the derivatization reaction, especially going from the macro- to the micro-scale;
(ii) effect of side reactions;
(iii) decrease of selectivity for related species, e.g. drug metabolites; and
(iv) contamination due to excess of reagents.

If instrumental chromatographic methods are taken as an example, derivatization reactions are carried out in gas chromatography (GC) to:

 (i) increase the volatility of the species of interest;
(ii) produce derivatives which give better responses at more selective and sensitive detectors (e.g. electron-capture detector);
(iii) improve the stability of the species of interest.

Common reagents employed include trimethylsilyl compounds (to improve stability and volatility) and trifluoroacetic acid (for improved detection at an electron capture detector).

In high-performance liquid chromatography (HPLC), derivatization is mainly used to improve the limit of detection of an analytical determination, since polarity and stability of analytes are not as great a problem in HPLC as in GC. These derivatization reactions can be carried out in a pre-column, on-column or post-column fashion depending on the analysis. Examples of such derivatization reactions will be given in subsequent chapters.

For the **determination step**, there is a wide range of instrumental analytical techniques available, depending on the analyte(s), and the application. The most commonly used methods are those based on chromatographic, spectroscopic, electrochemical and bioanalytical methodology. As mentioned previously, it is beyond the scope of this book to discuss the theory of such techniques, but many examples will be given of their application for particular analytes in particular matrices. From this, the reader should gain an important insight into the importance of understanding the chemical, physical and biological behaviour of particular analytes before deciding on the appropriate choice of instrumental techniques to be used in the determination step.

Any analytical method used in a real-life analysis also needs to be properly validated. The importance of **statistical evaluation** of results should not be understated, and too many methods in the literature are often found wanting in this regard when put to the test. It is again beyond the scope of this book to discuss the subject in great detail, and the reader is referred to standard texts such as that by Miller and Miller in this series [6]. The role of statistics can have a profound effect on all the "links in the chain" and provides an important method of feedback to help improve the precision and accuracy of each individual step in the process (Fig. 1.1).

It has not been possible in this volume to deal with every matrix which the analytical scientist might meet during his or her career. The subjects covered do however include a wide range of matrices from body fluids to atmospheric, food and beverage samples and industrial sealants. It is hoped that the reader finds the individual accounts both stimulating and enlightening.

REFERENCES

[1] *Booklet for M.Sc. course in Instrumentation and Analytical Science*, University of Manchester Institute of Science and Technology, Manchester, U.K.
[2] J. G. Grasselli, (ed.), *The Analytical Approach*, American Chemical Society, Washington D.C., 1983.
[3] B. Kratochvil and J. K. Taylor, *Anal. Chem.*, 1981, **53**, 924A.
[4] R. Smith and G. V. James, *The Sampling of Bulk Materials*, Royal Society of Chemistry, London, 1981.
[5] G. Kateman, and F. W. Pijpers, *Quality Control in Analytical Chemistry*, Wiley, New York, 1981.
[6] J. C. Miller and J. N. Miller, *Statistics for Analytical Chemistry*, Second Ed., Ellis Horwood, Chichester, 1988.

2

Drug analysis in biological fluids

Mary T. Kelly
School of Chemical Sciences, Dublin City University, Dublin 9, Ireland

2.1 INTRODUCTION

Chemists and biochemists have been analysing biological fluids since it was discovered that a relationship existed between the concentration of various endogenous substances, e.g. glucose, hormones, enzymes; and the cause of a disease [1]. Interest in the fate of exogenous compounds following their introduction into the body can be considered to extend back to last century when investigations were first carried out on the metabolism of quinine. During the 1920s, with the seminal work by Widmark on the detoxification of alcohol [2], scientific attention became more focused on the study of the metabolism and elimination of drugs. It was not until after the end of the Second World War, however, that biopharmaceutical analysis, as the analysis of drugs in biological media is known [3], began to be used routinely in a systematic manner. This was due, at least in part, to the growing availability of analytical instrumentation capable of determining the low drug concentrations (micro- or nanograms per ml) which tend to occur in biological media [4].

2.2 THE RATIONALE FOR BIOPHARMACEUTICAL ANALYSIS

There are a variety of circumstances in which it is either necessary or desirable to measure the concentration in the body of an ingested or administered drug substance. These situations extend from the areas of forensic science to the clinical laboratory and will be discussed individually below.

Therapeutic drug monitoring
Developments in drug measurement techniques have given rise to the popularity of individual therapeutic monitoring programmes as an adjunct to routine patient care, and the extensive application of drug assays over the past 60–70 years has led to

significant improvements in the therapeutic use of drugs which are characterized by wide individual variations in dose–response relationships. A drug frequently monitored in clinical practice is digoxin which has a very narrow therapeutic window (or range), where small variations in plasma levels produce large oscillations between sub-therapeutic and toxic responses. It has been reported that after the implementation of a drug monitoring programme for this drug, and subsequent tailoring of dosage to individual requirements, the incidence of toxic side effects was significantly reduced [5]. Furthermore, since sub- and supra-therapeutic effects produce similar symptoms [6], a suitable assay is a useful tool in determining which course of action to take in an emergency. In fact, so routine is the monitoring of this drug, that it is usually determined by immunoassay, with one of the many commercial kits designed for this purpose. Other drugs which are commonly dose-adjusted to suit individual subjects are theophylline, which also has a narrow therapeutic index [3,7], and the hydantoins, which exhibit dose-dependent kinetics [8].

Therapeutic drug monitoring may also be required in disease states, especially in relation to renal or hepatic failure, where the ability of an individual to metabolize and eliminate drugs is severely compromised. It may also be warranted where drug interactions are likely to occur; for instance, when the co-administration of an enzyme-inducing or inhibiting drug can lead to an intensification or reduction in response to the point where the patient's health is put at risk. It is, perhaps, important to stress that therapeutic monitoring is not necessary for the majority of administered drugs: the establishment and implementation of a monitoring programme is both time-consuming and expensive, and it is usually necessary to evaluate critically the need for such a programme in terms of the ultimate benefit for the patient.

Pharmacokinetics

In order to secure registration for non-experimental use, a new drug substance must be subjected to a series of toxicity and metabolic tests in both animals and humans. These tests serve to establish that the proposed compound is (i) not toxic in either its original or metabolized forms, (ii) produces the desired pharmacological response in man, and (iii) is well tolerated at high concentrations without unacceptable side effects. These types of tests in humans are often referred to as 'Phase 1 of Clinical Testing', and it is after the drug has passed this stage that the pharmacokinetic profile of the drug must be established. Pharmacokinetics is the study of absorption, distribution, metabolism and elimination of drugs [3], and biopharmaceutical analysis is central to the generation of the raw data used to develop that profile. The concentration of the administered drug in various body fluids (usually plasma, serum and urine) is typically determined after single-dose and multiple-dose regimens, given both intravenously and by the intended route of administration. The data obtained from such measurements are then used to plot drug concentration–time profiles from which various pharmacokinetic parameters (such as the half-life of the drug in the body) can be derived. A typical concentration–time profile of a drug after the administration of a single dose is shown in Fig. 2.1.

After the basic pharmacokinetic profile of the drug has been established, it is then necessary to show that the intended dosage form delivers sufficient drug to elicit the

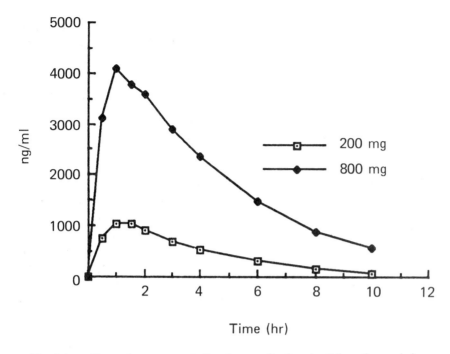

Fig. 2.1 — Mean plasma concentration–time profile for cimetidine after oral doseage.
Reproduced with permission of the copyright holder, J. Bloomfield, Trinity College, Dublin.

desired pharmacological response. This is usually achieved by comparison of the
area under the concentration–time curve (AUC) for the most easily absorbed
formulation with the intended dosage form. For instance, the plasma concentrations
for an oral formulation would be compared with those achieved by an intravenous
injection. When a drug company is seeking registration of a generic form of an
established product it is necessary to demonstrate bioequivalence between their
formulation and the existing product. The importance of establishing bioequivalence
or otherwise between formulations was vividly demonstrated in relation to digoxin
[9], when a generic drug was found to produce lower plasma levels than the patented
brand of digoxin. Though original formulations are largely interchangeable with
their generic equivalents, it is still common practice for pharmacists to ensure that
patients receive the same type of digoxin unless otherwise specified by the prescriber.

Other applications
Drug analysis in biological fluids also has a role to play in providing evidence of
patient compliance with the dosage regimen. Since subjects do not, in general,
comply with even the most explicit instructions, research continues to design novel
sustained delivery systems, and to find long-acting analogues and alternatives to
existing drugs. The field of analytical toxicology has grown in scope and importance
with respect to the increasing number of athletes taking not only a

range of steroid drugs, but also agents designed to mask their detection by standard analytical means. Biopharmaceutical analysis is widely employed in establishing the abuse of therapeutic drugs (such as barbiturates), the ingestion of illegal hedonistic drugs, and it may be useful in proving compliance with a drug treatment programme. Drug analysis is also widely used in the forensic laboratory, and the outcome of such tests can often provide critical evidence in a criminal trial. It may, in cases of overdosage, be necessary to carry out biopharmaceutical analysis in order to select the most appropriate remedial action, and in some instances, such determinations form part of a series of tests to establish the nature of the ingested substance.

2.3   ANALYTICAL METHODS IN DRUG ANALYSIS

The basic requirements for a biopharmaceutical assay are that it is applicable to the compounds of interest, is reproducible, selective for the analyte or analytes concerned, and is of sufficient sensitivity to detect the levels demanded by the application for which it was designed. Additional requirements may include factors such as cost, speed and technical simplicity, i.e. it should be relatively simple to perform and execute, and ideally should not involve lengthy and tedious sample-preparation stages. The relative importance of these criteria may vary from one scenario to the next. In an emergency situation, for instance, speed may be more important than sensitivity, whereas during clinical trials, technical simplicity, or the ability to measure a range of metabolites, may take precedence over speed and cost.

The first step in the development of a biopharmaceutical assay is the selection of a suitable technique to measure the drug of interest, and its metabolites, when required. Most biopharmaceutical assays currently involve either chromatographic separations or competitive binding assays. Less widely used, but nonetheless of considerable significance, are the spectroscopic and electrochemical techniques, which, because of their relative simplicity, continue to find use in certain applications.

2.3.1   Spectroscopic methods

Absorption spectrophotometry and spectrofluorimetry are techniques which do not involve a separation step, and although convenient, they frequently lack the selectivity and sensitivity associated with chromatographic and competitive-binding assay techniques. Some drugs do, however, lend themselves satisfactorily to these relatively simple procedures, particularly if they are present at relatively high concentrations in the body. Such drugs include the salicylates, which may be determined by colorimetry after solvent extraction [10], or protein precipitation [11]. Some publications describing simple spectrophotometric and fluorometric assays for drugs are listed in Table 2.1. Though the use of selective extraction schemes, supplemented by colorogenic and fluorogenic reactions, can considerably enhance the sensitivity and selectivity of spectrophotometric methods [22], much attention has recently focused on the development of special techniques and more sophisticated, computer-supported instrumentation. The latter include diode-array spectrophotometry, [23–25], or luminescence spectrophotometry, which permits the capture of the entire luminescence excitation–emission matrix [26,27].

Table 2.1 — Some spectrophotometric and spectrofluorometric assays for drugs in biological fluids

Drug	Method	Reference
Cephalosporins	Drugs were subjected to Cu-catalysed degradation at 75°C in aqueous buffer. Absorption maximum was at 326 nm. Limit of detection 4.3 µg/ml.	12
Paracetamol	Proteins were precipitated from plasma with methanol and the supernatant incubated with aryl acrylamidase for 2 min. After addition of o-cresol and ammoniacal copper sulphate, absorbance was measured at 615 nm. Linear range 0.15–6 µg/ml.	13
Piroxicam	Proteins were precipitated with acetonitrile and the supernatant mixed with perchloric acid. Absorbance was measured at 330 nm. Linear range 0.5–8 µg/ml.	14
Chloroquine	Drug was extracted from basified plasma into ether which was then back-extracted with 0.05M sulphuric acid. Absorbance was measured at 345 nm.	15
NSAIDS	Blood was mixed with acidified sodium tungstate. After centrifugation, the supernatant was mixed with 3,6-di(dimethylamino)xanthylium chloride in chloroform. The organic layer was removed, evaporated, and the residue reconstituted in water. Fluorescence emission was measured at 566 nm (excitation at 542 nm). Linear range 50–90 ng/ml.	16
Acrylonitrile	Alkaline permanganate followed by bromine solution added to plasma samples. After removal of excess of bromine solution, pyridine and phloroglucinol were added and the pH was raised to 9. Absorbance was measured at 540 nm. Linear range 10–70 µg/ml.	17
Rifampicin	Plasma was extracted with chloroform–hexane. The organic layer was removed, evaporated, and the residue reconstituted in methanol. Absorbance measured at 343 nm. Linear range 1–435 µg/ml.	18
Pethidine	Plasma (adjusted to pH 10.5) was extracted with chloroform, back extracted into phthalate buffer, pH 3, and re-extracted into chloroform. Absorbance measured at 412 nm. Linear range 1.0–4.0 µg/ml.	19
Haloperidol and primozide	Plasma proteins were precipitated and supernatant treated with acetate buffer, pH 2.8, and 10 µM methyl orange followed by extraction into chloroform. Absorbance of dried organic layer was measured at 425 nm vs. a reagent blank. Limit of detection 50 ng/ml.	20
Heptacaine	Plasma was extracted with chloroform followed by TLC separation. Eluted bands were extracted with methanol or ethanol. Absorbance was measured at 236 nm. Analytical range for heptacaine was 7–16 µg/ml.	21

A widely used special routine is 'difference spectrophotometry', where the spectrum of a sample under one set of conditions (e.g. in acidic solution) is subtracted from a the spectrum of a sample under a different set of conditions (e.g. in alkaline solution). For example, the spectral differences between pH 6.0 and pH 9.5 for two basic drugs, nifedipine and tyramine, are clearly illustrated in Fig. 2.2, and difference spectrophotometry was recently used in the determination of drotaverine in biological tissues [28]. Derivative spectrophotometry is commonly used in biopharmaceutical analysis, particularly in view of its ability to confer added discrimination between the analyte and matrix interferences. First, or higher-order, derivative spectra can considerably sharpen absorption bands, and may be used to resolve

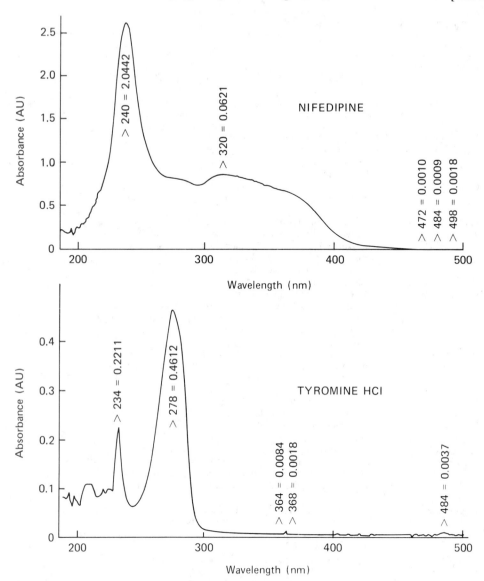

Fig. 2.2(a) — Ultraviolet absorption spectra of nifedipine and tyramine in 0.025M ammonium
nitrate–methanol [1:4], pH 9.5.

overlapping peaks. The theoretical aspects of this method were first considered by
Hammond and Price [29] and Geise and French [30] in the 1950s, and since then a
considerable body of literature on the subject has been accumulated [31–35]. Some
biopharmaceutical applications of this method include the determination of benzodiazepines in blood [36], atracurium [37] and tetrahydropalimitine [38] in plasma,
and a rapid assay for paraquat in plasma [39].

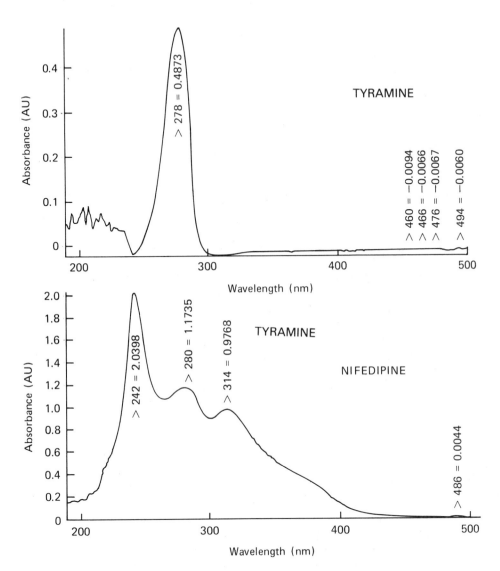

Fig. 2.2(b) — Ultraviolet absorption spectra of nifedipine and tyramine in $0.025M$ ammonium nitrate–methanol [1:4], pH 6.0.

The derivative method may also be applied to fluorescence measurements and it becomes a particularly useful tool when combined with the synchronous scanning technique [40]. Here, both the excitation and emission monochromators are scanned simultaneously, separated by a spectral interval which may be tailored to the compound of interest, and adjusted to produce the optimum resolution. This method has been used in the determination of salicylates in urine, where the monochromators were scanned at a 90 nm difference, with first-derivative measurements at 416 and 392 nm for salicyclic and salicyluric acids respectively [41]. Detection limits were 20 ng/ml for each compound with rectilinear calibrations up to levels greater than 0.22 µg/ml. Synchronous scanning of excitation and emission wavelengths has also been used with room-temperature phosphorescence spectra for the determination of warfarin in blood [42].

2.3.2 Immunoassay techniques
The introduction of competitive binding assays by Yalow and Berson in 1959 [43] proved to be a significant development in the fields of both clinical and analytical chemistry. Initially applied only to biopolymers and proteins, these techniques have been extended to the realm of drug analysis by the covalent coupling of drug molecules to a protein, in order to confer immunogenic properties on them [44]. The absolute numbers of analyses carried out by immunoassay are legion, and no attempt will be made to reference them in this text, but as an indication of their importance, it has been estimated that over 90% of laboratories participating in the College of American Pathologists Therapeutic Monitoring Survey used some form of immunoassay for the determination of phenytoin and theophylline [45]. However, for less routine applications, the time and expense involved in the development of assay kits is rarely justified, and one-off analysis is ideally performed by using some kind of chromatographic method.

Immunoassays are competitive binding techniques in which an unknown concentration of unlabelled drug in a sample competes with a known concentration of labelled drug for binding sites on an antibody raised towards that drug. The basic principle of competitive immunoassays is shown in the reaction below:

$$Ab + Ag + Ag^* \rightleftharpoons AbAg + AbAg^*$$

where Ag represents the free (unlabelled) drug-antigen, Ag^* the labelled drug-antigen and Ab, the antibody. In this case, the drug-antigen is created by linking the drug to a protein (usually bovine serum albumin), as drugs do not usually possess inherent immunogenic properties. In the past, drugs were usually tagged with a radioisotopic label such as tritium or Iodine-125, hence the term 'radioimmunoassay'. These labels are low energy beta and gamma emitters, respectively, and they may readily be detected in a liquid scintillation counter. Extremely low levels of analyte (<1 ng/ml) may be determined in radioimmunoassay, and this is one of the factors which have contributed to its continued widespread use in spite of the difficulties associated with the handling and disposal of radioactive materials. Furthermore, the incorporation of a radioactive element into an analyte has little or no effect on the chemical behaviour of the analyte or the environment in which it occurs. Nowadays, however, non-isotopic labels, such as enzymes, fluorophores, particles and cells are gaining broader acceptance because of enhanced user safety, extended reagent shelf life, and ready adaptability to conventional instrumentation [44]. An exceptional case is the immunoassay for digoxin, where, since plasma concentrations of this drug lie in the sub-nanogram range, the more sensitive radioimmunoassay remains the method of choice in many laboratories [26,45].

In the original so-called 'heterogeneous' immunoassays, the bound and unbound fractions are separated after reaction between the labelled drug, the unlabelled drug and the antibody. This is usually accomplished by extraction, electrophoresis or simple centrifugation of the drug antigen–antibody complex. The labelled drug in the bound or unbound fractions, or both, is measured by using a detection scheme appropriate to the type of label used, and the amount of drug present in the initial sample is found by interpolation on a standard curve of fraction bound *vs.* concentration. The typical form of such a calibration curve is shown in Fig. 2.3.

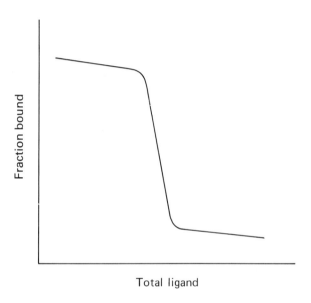

Fig. 2.3 — Typical form of a competitive binding assay calibration curve. Reproduced with permission of the copyright holder, R. O'Kennedy, Dublin City University.

2.3.2.1 *Enzyme immunoassays*

Heterogeneous enzyme immunoassays (EIA) originated some 15 years ago in the work of Van Weeman and Schuurs [46], and the principles underlying the various assay protocols are identical to those in which other labels are employed. EIA is analogous to classical radioimmunoassay in that there is a competition between labelled and unlabelled drug-antigen for a specific amount of antibody. A widely used derivative of the original heterogeneous EIA is the Enzyme Linked Immunosorbent Assay (ELISA) technique, where the antibody is labelled with an enzyme, reacted with drug-antigen from the sample, and then added to excess of solid-phase antigen. The activity of antibody–enzyme complex which remains bound to the solid phase after washing is inversely proportional to the concentration of drug-antigen originally present. This procedure is depicted in Fig. 2.4. The solid support most

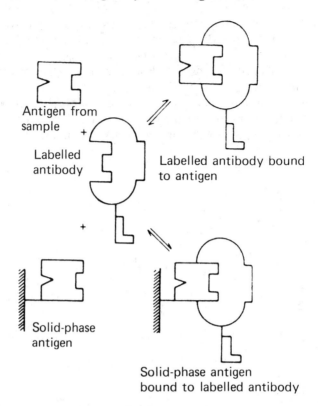

Fig. 2.4 — Schematic representation of ELISA system. Enzyme-labelled antibody reacts specifically with antigen in the sample and is then added to excess of solid-phase antigen. After washing, the enzyme label still attached to the solid phase is measured.

frequently used in large-scale screening work required in the study of infectious diseases has been the microtitration plate [47]. These plates are cheap, and require only small volumes of reagent.

Homogeneous immunoassays are so-called because they do not require separation of bound and free fractions prior to measurement of the percentage binding. The most extensively used homogeneous EIA system is the Enzyme Multiplied Immunoassay Technique (EMIT). This technique, developed by Rubenstein *et al.* [48], depends on a change in the specific enzyme activity when an antibody binds to a drug-antigen labelled with that enzyme. The more unlabelled drug-antigen present, the less the modification in enzyme activity, and therefore, the calibration curve for this technique is the reciprocal of that obtained with ELISA. The principle of the EMIT procedure is depicted in Fig. 2.5. The earliest EMIT systems employed lysozyme as the enzyme label, but modern drug assays use glucose-6-phosphate dehydrogenase and malate dehydrogenase almost exclusively. The latter enzyme is used in the assay for thyroxine, in a rare instance where binding of the enzyme-labelled antigen actually results in increased enzyme activity. This is because binding of malate dehydrogenase to thyroxine substantially inhibits enzyme activity, an inhibition which is at least partially overcome by antibody binding [49].

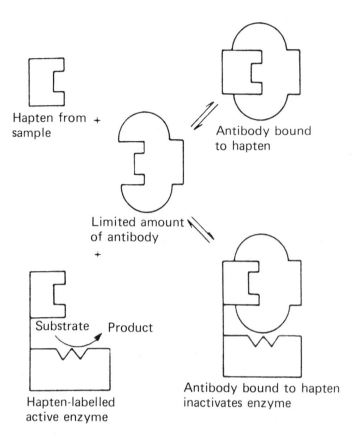

Fig. 2.5 — Schematic representation of EMIT system. Conjugation of hapten to enzyme does not destroy enzyme activity, but combination with hapten-specific antibody causes marked inhibition of enzyme activity. The measured enzyme activity is dependent on the relative amounts of free hapten and hapten-labelled enzyme.

2.3.2.2 Other immunoassays

Substrate-labelled fluorescence, and fluorescence-polarization immunoassay are other homogeneous immunoassay techniques. In the former case, the label is a fluorophore and the antigen is a fluorophore–drug (F–D) conjugate which does not fluoresce until it is reacted with a fluorescence-inducing enzyme. Only the fraction of F–D which is not bound to the antibody (Ab) will react with the fluorescence-inducing enzyme. Upon mixing of the labelled drug (F–D), the unlabelled drug (D) and the antibody, the labelled and unlabelled versions of the drug species compete for the antibody, and as a result, the amount of fluorescence produced is proportional to the amount of unlabelled drug present.

In fluorescence polarization immunoassay, the label is again a fluorophore and the antigen is a drug–fluorophore conjugate. The unlabelled drug, labelled drug and the antibody are irradiated with polarized light. Only the bound species fluoresces with light polarized in the same plane as the incident beam, since the unbound species will lose polarization through Brownian motion. The more (unlabelled) drug present in a sample or standard, the less bound fluorescent species produced, and the extent of polarization will decrease.

2.3.2.3 *Antibody reagents*

An important limitation of immunoassays has been the restricted availability of analytically useful antibodies. Antisera prepared by conventional animal immunization techniques contain populations of different antibodies varying both in antigenic specificity and affinity [50]. Large numbers of animals must be immunized, and careful screening is required to identify animals that produce analytically useful antisera. With the advent of monoclonal antibody technology, the problems of lack of uniformity and unavailability of antibodies are more readily solved.

Monoclonal antibodies are produced by fusing an antibody-producing spleen cell with a tumour cell [51]. By cloning the progeny of a single hybridoma, the resulting cells, which are clones of the parent, produce antibodies which are identical in chemical structure and homogeneous in binding affinity. Monoclonal antibodies are highly specific since all the binding sites are directed at the same antigenic determinant. Large quantities of monoclonal antibodies can be produced, since hybridomas, by virtue of their tumour origin, will propagate readily.

2.3.3 Chromatographic techniques in biopharmaceutical analysis

Chromatography is a physical method of separation in which the compounds to be separated are distributed between two phases, one stationary and one mobile. In gas chromatography (GC), the mobile phase is a gas, and in high-performance liquid chromatography (HPLC), thin-layer chromatography (TLC) and paper chromatography (PC), the mobile phase is a liquid. In all cases, the stationary phase is a solid, or a liquid coated onto a solid surface. The chromatographic process occurs as a result of repeated sorption onto and desorption from the stationary phase as the sample components traverse a column or plate under the influence of the flowing mobile phase. Separation betweeen individual components arises from differences in the distribution coefficients between the two phases.

The word 'chromatography' was first coined by the Russian botanist, Tswett, who in 1903, produced coloured bands by separating concentrated plant extracts on a column of adsorbent material [52]. In 1941, Martin and Synge described the method of liquid–liquid partition chromatography, and in the same work, laid the foundation of gas–liquid chromatography, which was to enjoy enormous popularity during the following years [53]. The practical application of GC was first realized by James and Martin some 10 years later in 1952 [54] and the technique soon evolved into a far more sophisticated form of separation than liquid chromatography (as it existed then), since greater efficiencies could be achieved by the employment of smaller particles with a gaseous mobile phase. In the early 1960s, Giddings showed that the theoretical principles developed to accommodate separation processes in gas chromatography could readily be applied to liquid chromatographic systems [55], and the successful implementation of practical liquid chromatography is largely attributed to Horvath, Lipsky, and Preiss [56], Huber [57] and Kirkland [58], whose new high-pressure systems (operating up to 5000 psi) were able to overcome the problem of the high viscosities of liquids relative to gases. Thus, the way was paved for efficient and rapid separations in liquid chromatography, comparable with those then being achieved by gas chromatography. As a result of the use of high-pressure systems and the achievement of separation efficiencies which were vastly superior to those pertaining in older low-pressure LC systems, the technique soon become known as 'High-Performance' or 'High-Pressure Liquid Chromatography' (HPLC), and to this day the same acronym is generally taken to refer to either or both of these meanings.

Chromatographic techniques, which combine selectivity with highly sensitive detection schemes, today represent possibly the most useful tools at the disposal of analytical scientists engaged in biopharmaceutical analysis. HPLC and GC have been developed to an advanced level of technical sophistication and are now used in more instances than other methods such as PC or TLC, although the latter still finds some usage in quantitative and qualitative screening applications. In addition, derivatives of these chromatographic modes have been developed over the years and the various inter-relationships may be seen in Table 2.2 which classifies chromatographic methods according to mobile and stationary phases.

Table 2.2 — Classification of chromatographic modes according to mobile phase

Mobile phase	Stationary phase	Mechanism	Technique
Gas	Solid	Adsorption	Column
Gas	Liquid	Partition	Column
Liquid	Solid	Adsorption	Column, paper Thin layer
Liquid	Solid	Ion exchange	Column
Liquid	Solid	Exclusion	Column
Liquid	Liquid	Partition	Column, paper Thin layer
Supercritical fluid	Solid	Adsorption	Column
Supercritical fluid	Liquid	Partition	Column

2.3.3.1 Gas chromatography

A schematic diagram of a typical gas chromatograph is shown in Fig 2.6. As in liquid chromatography, the column forms the core component of the system and for GC, columns may be classified as either packed or non-packed. In the former case, the columns are made of stainless steel, copper, or more usually, glass; they may be up to 2 m in length, and usually have a diameter of about 5 mm. They are packed with microparticulate material which is either uncoated, or coated with a liquid which then acts as the stationary phase on an inert solid support. Uncoated materials form the basis for the original GC technique, gas–solid chromatography (GSC), while stationary phases which are a liquid on a solid support involve the relatively newer and now more common, gas–liquid chromatography (GLC). Although GSC does find application in certain areas of chemical analysis, it is little used in drug separations since it suffers a number of drawbacks which are beyond the scope of the present text to discuss. Stationary supports and liquid phases are available separately

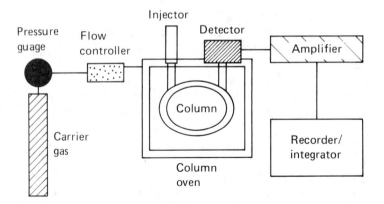

Fig. 2.6 — Schematic diagram of the essential components of a gas chromatograph.

or with the liquid already coated onto the solid support, and it is the enormous variety of solid support–stationary phase combinations available for gas chromatography that have contributed to its widespread use as an analytical technique over the years. Examples of the most widely used solid supports and liquid materials are presented in Table 2.3. GC stationary phases may also be formed by chemical bonding (rather than simple coating) of the liquid phase onto a solid material. These generally feature greater hydrolytic and thermal stability than their coated counterparts. However, they do tend to be more expensive, and cannot, as a rule be prepared in-house and tailored to individual requirements.

Non-packed columns are known almost universally as 'capillary columns' by virtue of the fact that they have much narrower diameters than packed columns (i.e. 0.2–0.6 mm as compared to about 5 mm). A more accurate description sometimes used is 'open-tubular', a term that reflects the nature of these columns; that is, they are open tubes, the inner walls of which are coated with stationary phase. Apart from the thickness of the layer (1–5 microns), there are a few possible options in terms of stationary phase coating, the best known of which are wall-coated and support-coated open tubular columns (WCOT and SCOT, respectively). WCOT columns have the liquid stationary phase coated directly onto the inner walls, whereas in SCOT columns, the walls are treated with a solid support onto which the stationary phase is then coated. The stationary phases in capillary GC are more-or-less the same as those used in packed GC, which allows for easy transfer of a method developed on a packed column to capillary GC.

Since non-packed columns are configured in an open tubular design, it is not necessary for the gas to negotiate a bed of packed material and it can traverse the column virtually unimpeded at very high velocities, though the actual flow-rates are substantially lower than in packed GC on account of the smaller column diameters. Capillary GC therefore allows for rapid analyses, and the use of very long columns (100 m or more) can provide resolving power unparalelled by other separation techniques. Chromatograms of some drugs separated by capillary GC are shown in Fig. 2.7.

Table 2.3

Name	Type of material	Comment
	STATIONARY SUPPORTS FOR GC	
Chromosorb P	Flux calcined diatomaceous earth	Prepared from Johns-Manville firebrick. Surface area (SA) 4 m²/g. Preparative GC.
Chromosorb W	lfux calcined diatomaceous earth	SA 1 m²/g. Medium capacity Analytical GC.
Chromosorb G	Flux calcined diatomaceous earth	SA 0.5 m²/g. More inert than Chromosorb W. Useful where low percentage loadings are required.
Unisorb	Flux calcined diatomaceous earth	Higher density than Chromosorb W.
	SOLID STATIONARY PHASES FOR GC	
Porasils	Spherical silica particles of controlled particle size and pore diameter	Available in grades A–F in order of decreasing SA and increasing pore diameter. May also be used as stationary supports.
Chromosorb 101	Styrene–divinylbenzene	SA 50 m²/g.
Chromosorb 102	Styrene–divinylbenzene	SA 400 m²/g.
Chromosorb 103	Polystyrene	SA 20 m²/g.
Chromosorb 106	Polystyrene	SA 800 m²/g.
Chromosorbs 105, 107, 108	Acrylic esters	Varying surface areas
Porapaks	Cross-linked polystyrene beads	Available as Porapaks P, PS, Q, QS, R, S, N, T in order of increasing polarity. Porapak P is STY–DVB co-polymer, Q is ethylvinylbenzene–DVB co-polymer. Other Porapaks are variations of these with different monomers providing a range of polarities.
Carbopak B	Graphitized carbon black	SA 12 m²/g.
Carbopak C	Graphitized carbon black	SA 100 m²/g.
Molecular sieves	Sodium and calcium aluminium silicates.	
	LIQUID STATIONARY PHASES FOR GC	
OV1, OV101, DB1, SE30, BP1	Polysiloxanes	Dimethylsilicone
OV17		Phenylmethylsilicone, 50% phenyl.
OV73, CPSIL8, BP5		Diphenyldimethylsilicone gum
OV210, DB210		Trifluoropropylmethylsilicone
OV1701, CPSIL9CB DB1701, AP10		Dimethylphenylcyano substituted polymer
Carbowax series	Polyethylene glycols	Prepared by polymerization of ethylene oxide and fractionated according to molecular weight. Carbowax 20M: MWt=15000
Apiezon	Greases	Prepared by high-temperature treatment and distillation of lubricating oils. Very hydrophobic.
	CHEMICALLY BONDED MATERIALS FOR GC	
Durapak	Organic functionality bonded to spherical totally porous silica	Examples include Carbowax 400, n-octane or phenyl isocyanate on Porasil C.
Ultrabond	Organic functionality bonded to usual GC supports	Support is coated, subjected to high temperature treatment and then extracted. About 0.25% of organic material remains bound.

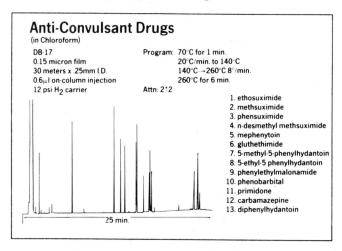

Fig. 2.7 — Separation of anti-convulsant drugs by capillary GC. Reproduced with permission of the copyright holders, Jones Chromatography, Mid Glamorgan, UK.

A disadvantage of capillary GC is that the sample loading capacity is substantially lower than that of packed columns and, as a result, it is usually necessary to split the sample in what is known as a 'split injection technique'. Allied to this is the fact that capillary systems are highly prone to contamination by matrix interferents, so more elaborate sample clean-up is required and higher purity gases must be used. Therefore, while capillary GC is useful, and indeed often essential for the separation of extremely complex mixtures of environmental samples (such as pesticide residues and polychlorinated biphenyls), in biopharmaceutical applications, where it is sought to separate the analyte, internal standard and possibly a small number of metabolites, the resolving capacity provided by a well-prepared packed column is usually adequate.

As outlined above, a packed GC method may readily be transferred to a capillary system, and while gas chromatographs designed for packed columns cannot accommodate the narrow diameters (0.25 mm) of capillary columns, wide- or mega-bore open-tubular columns (diameters of approximately 0.5 mm) may be fitted to these machines by using the appropriate conversion kit for the inlet and detector connections. Mega-bore columns are well suited to bioanalytical applications since they offer superior resolution to packed GC, and as they often feature thicker stationary phase films than capillary GC, they can tolerate larger quantities of sample and splitless injection techniques. Examples of some drug separations on a megabore column are shown in Fig. 2.8, which illustrates the fact that column efficiencies are comparable with those on narrower-bore columns.

An important reason for the popularity of gas chromatography is its ability to incorporate the flame-ionization detector (FID) for detection purposes. This is an almost universal detection mode, with the additional benefits of being cheap, sensitive (5×10^{-12} g/sec), of wide linear range (10^7), and simple to use. Gas chromatography is also compatible with other sensitive detectors which are more selective than FID. These include the nitrogen–phosphorus detector (NPD), a variant of the FID, which is selective for nitrogen- and phosphorus-containing

BASIC DRUG SCREEN

Column: DB– 1301
                     30m ✕ 0.53mm I.D.
                     J&W P/N 125– 1332
Film thickness: 1.0 micron
Carrier: Helium @ 10 mL/min
Oven: 150°C to 250°C (20°C/min)
Injection: 1.0 μL direct flash vaporization
Detector: FID

1. Methamphetamine
2. Phendimetrazine
3. Meperidine
4. Phencyclidine
5. Mepivicaine (IS)
6. Amitriptyline
7. Methaqualone
8. Pentazocine
9. Codeine
10. Diazepam
11. Nordiazepam

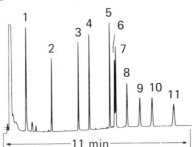

Fig. 2.8—Separation of basic drugs on a megabore column. Reproduced with permission of the copyright holders, Jones Chromatography, Mid Glamorgan, UK.

compounds; and the electron-capture detector (ECD), which is sensitive to electron-rich substances, for example organohalogens. Another detection mode which can be used with GC is mass spectrometry (MS). Because it combines high sensitivity with unique selectivity, GC–MS finds wide usage in forensic applications for its capacity to provide definitive identification of trace concentrations [59].

Compounds which are to be analysed by GC must be volatile or capable of being rendered so by appropriate derivatization procedures. Gas chromatography is ideally suited to compounds which may be volatilized at 200–300°C without decomposition or the need for further derivatization. Drugs which fall into this category include the benzodiazepines and the tricyclic antidepressants. Many drug compounds, however, for reasons of instability, intermolecular hydrogen bonding or high molecular weights, require derivatization to increase their volatility and, frequently, their stability. Derivatives, since they are less polar, are less likely to become adsorbed onto polar sites in the column (adsorption produces skewed peaks, long retention times and may promote decomposition or rearrangement of the analyte), and furthermore, derivatization frequently introduces an added degree of selectivity into the extraction and separation stages.

Derivatization may be achieved in many ways, arguably the most popular of which is silylation, since most polar and active functional groups are readily converted into their alkylsilyl esters and ethers. Drugs are most commonly converted into their trimethylsilyl derivatives by reaction with agents such as hexamethyldisilazane (HMDS), N,O-bis(trimethylsilyl)-trifluoroacetamide (BSTFA) or trimethyl-chlorosilane (TMCS), shown along with other reagents in Table 2.4. Reactions usually proceed smoothly at ambient or elevated temperatures in anhydrous conditions, and although these TMS derivatives are highly volatile, they may suffer the drawback of limited stability. This problem may be circumvented by the formation of derivatives, or alternatively, alkyldimethyl anologues, which combine the volatility of the trimethyl derivatives with the stability of the trialkyl derivatives. The formation of higher alkyl derivatives may also serve to distinguish between drugs and metabolites which differ in their degree of steric hinderance arround the active centre. A wide variety of functional geroups will readily undergo silylation reactions, including phenols, carboxylic acids, thiols, primary and secondary amines and amides.

Table 2.4 — Some silylating agents used in gas chromatography

1. Hexamethyldisilazane (HMDS)	$(CH_3)_3Si-N-Si(CH_3)_3$
	        |
	        H

2. Trimethylchlorosilane (TMCS)	$(CH_3)_3Si-Cl$
3. N-Trimethylsilyldiethylamine (TMSDEA)	$(CH_3)_3Si-N-(C_2H_5)_2$
4. N-methyl-N-(trimethylsilyl)acetamide (MSTA)	$CH_3-\underset{\underset{O}{\parallel}}{C}-\underset{CH_3}{\overset{\mid}{N}}-Si(CH_3)_3$
5. N-methyl-N-(trimethylsilyl)trifluoroacetamide	$CF_3-\underset{\underset{O}{\parallel}}{C}-\underset{CH_3}{\overset{\mid}{N}}-Si(CH_3)_3$
6. N,O-bis(trimethylsilyl)acetamide	$CH_3-C\begin{matrix} \nearrow O-Si(CH_3)_3 \\ \searrow N-Si(CH_3)_3 \end{matrix}$

An active hydrogen on a drug may be replaced by reaction with an appropriate alkyl halide. Groups which may be converted in this way include alcohols, phenols and carboxylic acids. A catalyst such as silver oxide, or if necessary, sodium hydride, may be added to accelerate the reaction. Many alkyl derivatives may be prepared by reaction betweeen the substrate and a diazoalkane, the most popular of which is diazomethane. Because the rate of reaction varies greatly according to the reactivity of the functional group, this approach to alkylation permits differentiation of drugs with functional groups differing in their degrees of reactivity, for example carboxylic acids and alcohols. An example of this approach is the use of diazomethane to methylate the carboxylic acid function of the anti-inflammatory drugs (NSAIDS). However, since the diazoalkanes are toxic and explosive under some circumstances, a safer route to methylation is to use a reagent consisting of boron trifluoride and methanol. The acidic nitrogen function of the barbiturate and hydantoin moieties are readily converted to their N-methyl derivatives by reaction with trimethylphenylammonium bromide or hydroxide [60]. A method has been described where the latter reagent and a mixture of anticonvulsants are introduced directly into a heated (200°C) injection port, where the derivatization reaction is allowed to take place just before the start of chromatography [61]. This is known as the 'on-column' technique, and considerably reduces the inconvenience traditionally associated with GC derivatization steps. Quaternary ammonium reagents may also be used for the technique of extractive alkylation of acidic drugs. In this case, the drug is ionized and extracted as an ion-pair into an immiscible organic solvent. In the poorly solvating environment of the organic solvent the drug, which forms the anion of the ion pair, will react readily with an alkyl halide, yielding a derivative which has been simultaneously extracted and alkylated.

Drugs which contain amine, alcohol, and other moieties with replaceable hydrogens are suitable for acylation, and reagents which permit the introduction of different acyl groups, for instance acetyl- or isobutyl- groups are commercially available. Another useful acylation reaction is the introduction of halogen groups to enhance detection by electron capture. Such derivatizations may be accomplished by the reaction between the substrate with haloacyl reagents, yielding, for instance, monochloroacetyl, chlorodifluoroacetyl and heptafluorobutyryl derivatives. These reagents not only increase response in the ECD, but increase the volatility of the substrate (especially the fluoride-containing derivatives) and are equally suitable for flame-ionization detection. Halogen groups may also be introduced by silylation reactions using, for example, a halomethyldimethylsilyl reagent in place of the trimethylsilyl reagent.

2.3.3.2   *High-performance liquid chromatography*
The configuration of a typical liquid chromatograph is shown in Fig. 2.9. The advantage of this chromatographic mode is that, unlike GC, it is suitable for the analysis of compounds that are thermally labile and non-volatile. The efficiency and sensitivity of HPLC has been greatly improved with recent advances in column technology in terms of the production of microparticulate packings of smaller particle sizes and particle size ranges (3–5 microns), as well as improved column structures of shorter lengths and diameters (1-mm and less). Liquid chromatography has the further advantages of almost never requiring derivatization, and of being well suited to simple methods of sample preparation. For example, in some instances, where plasma drug levels are in the high ng/ml range and above, a simple protein precipitation step will suffice as sample pre-treatment, and in addition the rapid, efficient technique of column switching for trace enrichment and sample clean-up is readily applicable to liquid chromatographic methods in both automated and manual versions.

Considerable advancement in the field of HPLC was made with the development of silica supports modified by covalent bonding of the stationary phase. These bonded-phase materials gave a significant improvement in terms of column stability and durability, and are largely responsible for the enormous popularity enjoyed by HPLC today. It is now possible to modify silica chemically with a variety of reagents to generate stationary phases that may be polar, e.g. aminopropyl, non-polar, e.g. octadecyl, or ionic, e.g. quaternary ammonium. Bonded-phase chromatography may be operated in the normal-phase mode (where the stationary phase is more

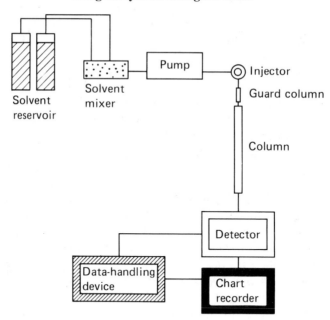

Fig. 2.9 — Schematic diagram of the typical components of a HPLC chromatograph.

polar than the mobile phase), in the reversed-phase mode (where the eluent is more polar than the stationary phase) and in the ion-exchange mode (where separations proceed through electrostatic interactions). The properties and applications of a selection of modified and unmodified stationary phases for HPLC are listed in Table 2.5

Of the various possible types of HPLC separations, the reversed-phase mode is undoubtedly the most widely used and its importance cannot be over-emphasized. There have been many reports describing the theoretical and practical aspects of the technique [62], and it is estimated that somewhere between 75–90% of all HPLC separations are carried out on reversed-phase columns [63]. The near-universal application of reversed-phase chromatography stems from the fact that practically all organic molecules possess hydrophobic regions in their structure that are capable of interacting with the non-polar stationary phase, and most drugs are sufficiently hydrophobic in character to be retained by the non-polar mechanisms involved in reversed-phase chromatography. On the other hand, since the mobile phase is polar and contains water, reversed-phase chromatography is well suited to the separation of more polar molecules which are either insoluble in non-polar organic solvents or bind too strongly to solid adsorbents for normal-phase chromatography. Retention and selectivity are easily controlled by changing the composition of the mobile phase, for example by altering the type and percentage of organic modifier (e.g methanol, acetonitrile), by changing the pH, or by the addition of ion-pairing reagents for the separation of ionic species.

The most favoured reversed-phase material at present is silica bonded with octadecyl silane to form a highly hydrophobic C18 surface; and a similar, but slightly

Table 2.5 — Properties of some stationary phases used in HPLC

SILICA-BASED MATERIALS

Product	Particle shape	Particle size μm	Specific surface area, m²	Mean pore diameter nm	Speciific pore volume ml/g	pH of a 1% suspension
Apex	Sph	3,5,10	170	10	0.7	8.0
Apex WP	Sph	7	100	30	0.8	7.0
Hypersil	Irr	3,5,10	170	11.5	0.7	9.0
Hypersil WP	Irr	5,10	60	30	0.6	8.0
Lichrosorb 60	Irr	5,10	490	60	0.8	6.5
Lichrosorb 100	Irr	5	320	11.5	0.7	7.0
Nucleosil 50	Sph	5,10	450	5	0.8	7.0
Nucleosil 100	Sph	3,5,10	350	10	1.0	7.0
Partisil 5	Irr	5	500	6	0.8	7.0
Partisil 10	Irr	10	500	6	0.8	7.0
Porasil	Irr	5,10	350	10	1.1	7.2
Spherisorb	Sph	3,5,10	190	8.1	1.6	9.5
Spherisorb WP	Sph	5,10	190	30	1.5	7.0
Zorbax BPSil	Sph	5	300	5.6	0.5	3.9

Chemically modified forms of the above silicas are available, bonded with C18, C8, C6, C2, C1, cyano, cyclohexyl, phenyl, amino, diol, sulphonic acid and quaternary ammonium groups for reversed-phase, normal phase and ion-exchange chromatography. The range of materials available varies from manufacturer to manufacturer.

OTHER MATERIALS

Name	Type of material	Applications
Hypercarb	Porous graphitized carbon	Drugs, amino acids, peptides
Alumina	Hydrated aluminium oxide. Available as α, β or γ alumina; the latter form is mainly used in HPLC	Strongly basic compounds which bind too strongly to silica
PRP-1	Fully porous reversed-phase styrene–divinylbenzene co-polymer	Suitable for separations at extremes of pH
PRP-X100	PRP1 resin with quaternary amine bonded to surface	anion-exchange chromatography
PRP-X2000	PRP1 resin with sulphonic acid bonded to surface	cation-exchange chromatography
PL GEL	Spherical styrene–divinylbenzene polymer beads of controlled cross-linking and pore size	Gel permeation chromatography
PLRP-S	Similar to PRP-1. Available in pore sizes of 10, 30, and 100 nm.	Ion suppression at pH values not possible on silica
PL-SAX	Same as PLRP-S with strong anion-exchange group bonded	Anion-exchange chromatography

more polar stationary phase consists of silica modified with octyl carbon, i.e. C8. The alkylnitrile- (cyano-) substituted phase is of intermediate polarity and provides good selectivity for the separation of double-bond isomers and ring compounds differing in either the number or position of double bonds [64]. This type of packing may be used in either the normal or reversed-phase modes and has actually been proposed by Massart and Buydens as an almost universal stationary support [65]. Cyano columns present an attractive alternative to C18 columns for the separation of basic drugs, which tend to interact with residual silanol groups in C18 materials. A high proportion of unmodified silanol groups is present in C18-modified silica because during reaction, they are physically inaccessible to the sterically hindered octadecyl chains. Ionic interactions between these acidic silanols and basic analytes operate to the detriment of peak shape and retention times. The silanol groups are, however, more readily accessible to the smaller cyano groups with the result that a greater degree of surface modification is achieved and fewer unreacted moieties are available to take part in secondary ionic interactions. Good separations and peak shapes have been obtained in the determination of tricyclic antidepressants on a cyano column after solid-phase extraction from human plasma [66,67] and for the analysis of tripelennamine in bovine plasma and milk [68]. Sample chromatograms showing the separation of tricyclic antidepressants on a cyano column are shown in Fig. 2.10. These were obtained after on-line solid-phase extraction, which will be discussed in a later section.

A disadvantage of HPLC is that the detectors used routinely are no match for the flame-ionization detector in GC with regard to sensitivity and universal applicability. Ultraviolet absorption (UV) is the most widely used detection scheme in HPLC; adequate detection depends on the presence of a chromophore within the molecule and the limit of detection will depend on the molar absorptivity of the drug at the detection wavelength. When the therapeutic compound absorbs strongly at a reasonably high wavelength, and is present in µg/ml concentrations in plasma, HPLC–UV is an ideal technique for analysis. Problems can arise, however, where the compound has a weak chromophore and/or when it occurs in the low ng/ml range in the relevant biological matrix: in these cases, if the detection wavelength is in the low UV region, selectivity may be severely compromised and the analyst may have to use an alternative detection mode such as electrochemical or fluorescence detection with pre-or post-column derivatization.

2.3.3.3 Thin-layer chromatography

Although less widely applied than the 'closed-bed' techniques of gas and column liquid chromatography, thin-layer chromatography, because of its technical simplicity, remains a useful procedure in drug analysis, particularly for clean-up and qualitative methods. Indeed, the development of high-performance stationary phases and the refinement of detection schemes designed for TLC are proof in themselves that the technique continues to have a vital role to play in a number of applications in clinical, organic and analytical chemistry laboratories.

Thin layer chromatography is, as the name implies, a technique in which the stationary phase is spread as thin layer on a flat plate. The sample mixtures are dotted along one end of the plate, which is then placed in a closed tank in which the mobile phase has been allowed to come into equilibrium with the local atmosphere. Separation is effected by the capillary movement of the liquid up the column bed, whereby the individal spots spread out into a series of bands corresponding to the components in the mixture, as illustrated by the schematic diagram of a thin layer chromatogram shown in Fig. 2.11. Under a given set of chromatographic conditions, components will appear at a characteristic position known as the R_f (Retardation factor) value, analogous to the retention time in column chromatographic separations. If the compounds are not coloured, detection may be achieved by spraying the plate with a reagent to react with the analytes and/or by inspection under UV

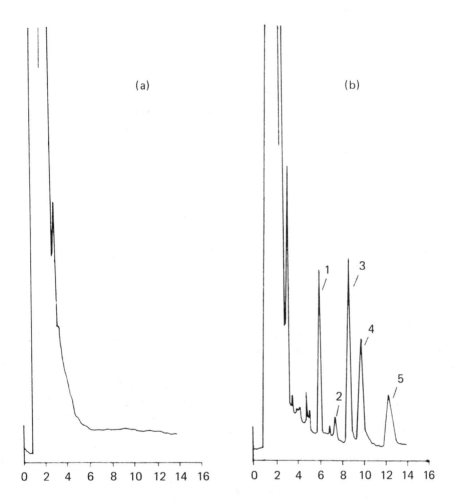

Fig. 2.10 — Chromatograms of (A) drug-free human plasma and (B) plasma from a patient receiving a 50-mg daily oral dose of amitriptyline. Peaks: 1: 12-hydroxynortriptyline; 2: 10-hydroxyamitriptyline; 3: desmethyldoxepin (internal standard); 4: nortriptyline; 5: amitriptyline. Chromatographic conditions: Column: Techsphere 3 μm cyano (10-cm). Mobile phase: acetonitrile–acetate buffer, 0.05M, pH 7.0 [6:4]. Reproduced with permission of A. Power.

light. Quantification may be effected by measuring the area of each band, or alternatively, the bands may be removed for further processing, testing or more accurate analysis.

The original TLC technology involved the use of unmodified silica or alumina coated onto glass plates, so the early methods were confined to the adsorption mode of chromatography. Reproducibility was critically dependent on the removal of adsorbed moisture, which would otherwise introduce an element of partition into the separation processs. Another problem with early TLC plates was lack of uniformity

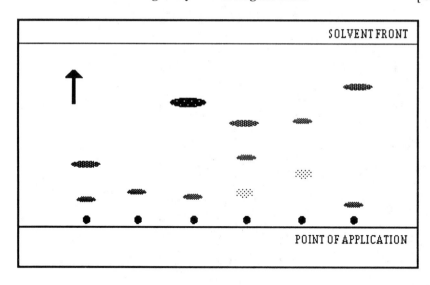

Fig. 2.11 — Diagram of a developed TLC plate. Arrow indicates direction of solvent flow.

in the thickness of the thin layer, and this also contributed to irreproducible separations. This problem was largely overcome with the invention by Stahl of a device to produce a uniform thin layer on the glass plates [69]. More recently, the production by commercial suppliers of plastic and foil-backed plates to which the particles are fixed by special binders, has contributed to a significant improvement in the reproducibility of TLC. With the added advantage that the band of interest may be cut out, these commercially produced plates have made the home-produced glass-backed plates less important, despite their much lower cost.

High-efficiency plates, where the average particle size is smaller, with a narrower size distribution, and which are coated with a thinner, more uniform layer [70], minimize band broadening and thus generate greater resolving power for the technique and more sensitive detection of the separated fraction. The use of these plates has given rise to the method known as High-Performance Thin Layer Chromatography (HPTLC). A drawback associated with HPTLC plates is that they have a much smaller sample capacity (ng rather than µg) than conventional TLC plates, and as a result, automatic spotting devices (such as nanomats, contact spotters and microsyringes) are required to apply nanolitre volumes of sample with precision. Such devices deliver small uniform sample aliquots and cause minimal damage to the plate surface at the point of application. Accurate quantification is necessary to match the high efficiency separations achievable in HPTLC, and this demands that measurement of the eluted bands be carried out *in situ* with some kind of automated detection device. This is because of the subjective nature of visual scanning, and the fact that less than µg quantities per spot could not be detected by the human eye. In addition, it is difficult to determine accurately the edges of a band, and a generous margin of error around the area of interest may not be possible if the plate contains several bands. Removal of the band followed by extraction of the analyte into a suitable solvent is, therefore, a risky process, as not all the band may be removed, extraction may be incomplete, and results may be obscured by co-extracted interferents originating from the plate binders.

Quantification of the analyte bands *in situ* is accomplished by densitometric measurements. These involve shining light of suitable wavelengths onto the plates and measuring the amount of light transmitted or reflected by the band. Measurements may be made by monitoring at the wavelength of the incident light, or at the emission wavelength of fluorescence generated by the analyte. Drug determinations are usually based on reflectance measurements, since transmission measurements are limited to longer wavelengths, owing to the opacity of the plate itself at the shorter UV wavelengths at which most drugs absorb or fluoresce. The available optical configurations (Fig. 2.12) range from a simple single-beam arrangement to more

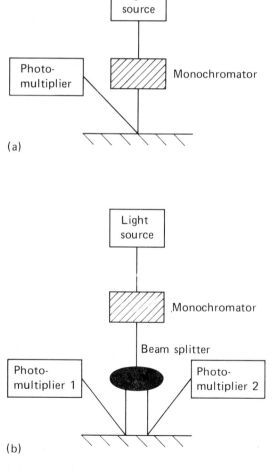

Fig. 2.12 — Optical configurations for densitometry. (A): Single-beam arrangement. (B): Dual-beam arrangement: monochromatic light is split into two beams of the same wavelength; one beam scans the sample band and the other between the lanes to allow for background correction.

sophisticated dual-beam instruments which allow for compensation of background interference and plate irregularities. The diffusely reflected or scattered light is then impinged on a photomultiplier, converted into an electronic signal and stored by an appropriate data-collection device.

In scanning densitometry the plate is scanned with a beam of monochromatic light by moving the plate rather than the light source. This operation is controlled mechanically and the scan length may be adjusted to accommodate the size and number of bands. Chromatograms from a scanning densitometric scan resemble those of a column chromatographic trace, and although the concentrations from such reflectance–absorbance scans are not linearly related to concentration, quantification is achieved by including appropriate internal standards on each plate.

Other variations, applicable in both conventional and high-performance TLC, are based on the method of plate development. In two-dimensional TLC, the plate is run conventionally with a first mobile phase which is evaporated after development. The plate is then rotated through 90° and run in a perpendicular direction with another mobile phase. The principle of two-dimensional TLC is depicted in Fig. 2.13, which shows the emergence of an extra band as a result of the second development. This method is ideal for resolution of complex mixtures that could not be separated with a single eluent, and is useful for generation of a metabolic profile where bands may be removed for structural identification and determination. Another advance includes the development of circular and anticircular chromatography. In the former case, the samples are spotted as a small circle in the centre of the plate, and development occurs in a radial direction from the centre to the periphery of the plate. Anticircular chromatography is the opposite; samples are dotted in a large circle around the edge of the plate and development occurs inwards. These approaches have the advantages that plate development occurs rapidly and a large number of samples may be run simultaneously.

A very recent innovation dispenses with the backing material altogether. In this arrangement, 8 μm silica particles are tightly bound in a web of PTFE fibrils in such a way that they are individually suspended and are free to interact fully with the solvent. The silica–PTFE matrix is formed into flexible sheets which obviate the need for any kind of backing materials or binders, thus eliminating a not inconsiderable source of interference. The commercial suppliers of these sheets claim that they combine the high efficiency and resolving power of HPTLC with the high capacity of ordinary TLC.

Since TLC is a rapid procedure, capable of the simultaneous analysis of large numbers of samples, it is ideally suited to drug screening methods when there is a requirement to establish the presence of a particular compound or group of compounds. An example would be the testing for drugs of abuse in plasma or, more usually, in urine, in which drug concentrations tend to be higher. In a drug screening exercise, an unknown sample is run alongside a series of mixed standards, for example, opioids, cannabinoids and barbiturates, and in this way, the analyst can form a preliminary opinion as to what group of compounds the suspected material belongs to. By then running the unknown against a series of individual standards from the suspected group, a tentative identification may be made, and this can be confirmed by GC–MS or LC–MS. TLC has also found use in the elucidation of

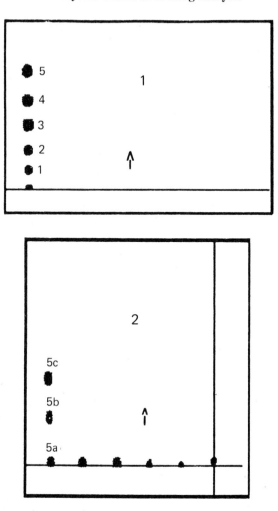

Fig. 2.13 — Schematic diagram of 2-dimensional TLC. After development in one direction, the plate is rotated through 90°, and the plate developed again with a different solvent. In this example, the second development reveals second and third components not separated in the initial development. Arrow indicates direction of solvent flow.

metabolic pathways by using radiolabelled drugs; the various zones of radioactivity correspond to different metabolites which may be quantified by scintillation counting or by photographing the plate on an X-ray film.

In order to increase sensitivity, it is common to use a densito-fluorimetric technique for quantification. This may be based on the native fluorescence of the compound of interest, derivatization prior to application to the TLC plate, or may involve spraying with a suitable reagent following development. In one example, the diuretic drug xipamide was extracted from plasma, derivatized with a specially prepared fluorescent reagent, and spotted on a silica plate, which was then developed with a mixture of toluene and ethyl acetate. Fluorescence of the derivative was measured at 404 nm (excitation at 313 nm) [71]. With continuing improvements in terms of sample-application methods, higher efficiency plates, advanced detection schemes and the flexibility offered by digitized manipulation of data, it is likely that TLC will continue as part of the armoury of many of those engaged in biopharmaceutical analysis.

2.3.3.4 Chiral chromatography

Many pharmacologically active compounds exist in two enantiomeric forms, and it is becoming increasingly important in biopharmaceutical analysis to be able to recognize, and distinguish analytically, between enantiomeric entities. Enantiomers are stereoisomers with mirror images that are superimposable on one another and which have optical activity in that they rotate the plane of polarized light to the left or right. This is because they possess an asymmetric or chiral centre, and chirality usually occurs in a molecule because it has a carbon atom to which four different substituents are attached. Depending on the type of nomenclature used and rules applied, the isomer which rotates polarized light to the left is known as the 'L' (levorotatory), '−' or 'S' (sinister) isomer, and the dextrorotatory ('D') form is known as '+' or 'R' (rectus) isomer. Diastereoisomers possess two chiral centres; they are stereoisomers with mirror images that are not superimposable, and unlike enantiomers, they differ in their physical properties. The distinction between enantiomers and diastereomers may be seen from the structures in Fig. 2.14.

Owing to the stereoselective nature of drug–enzyme and drug–receptor interactions and drug transport systems, enantiomeric forms of a drug may exhibit different bioavailabilities, have differing activities, and may even have opposite activities. For example, the R and S forms of quinalbarbitone exhibit equipotent anticonvulsant activity, but the S enantiomer is more toxic and is a potent anaesthetic [72]. The importance of enantiomeric elucidation was illustrated no more tragically than in the well documented case of thalidomide (N-phthalylglutamic acid imide) in which the S form is teratogenic whereas the R form is not. Women treated with a formulation containing a racemic mixture of this drug during the first trimester of pregnancy subsequently gave birth to children with severely debilitating deformities.

Nowadays, such is the emphasis on chiral discrimination that regulatory authorities frequently require evidence that an analytical method is capable of chiral recognition as part of the submissions for drug registration. In this respect, it is normally necessary not only to determine the optical purity or the relative concentrations of the R and S forms of the pure drug substance, but also to establish the eudismic ratio (the ratio of activities of different enantiomers in biological media) which recognizes that different enantiomers can have different rates and extents of absorption into the body. This can present an additional problem for the biopharmaceutical analyst since enantiomers do not differ in their distribution ratios, and therefore are not separable by simple chromatographic means.

In both GC and HPLC, enantiomers may be separated by use of chiral stationary phases, and in HPLC, an alternative strategy is the use of chiral mobile phases. Alternatively enantiomers may be converted into diastereomeric derivatives prior to chromatography.

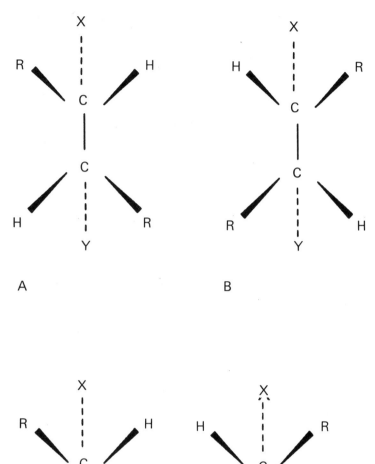

Fig. 2.14 — Enantiomers and diastereomers. Enantiomers are superimposable mirror images of one another. The mirror images of diastereomers are non-superimposable. In the diagram, A and B are enantiomers; C and D are enantiomers. Both A and B are diastereomers of both C and D, and vice versa.

Chiral mobile phases

Chiral mobile phases may be used in normal- or reversed-phase chromatography, and are traditional organic or aqueous-organic eluents containing additives which react with the enantiomers in such a way that their distribution coefficients are no longer equal. In many cases these additives react with enantiomers to form diastereoisomers which then elute with different retention times. For example, *N*-(1-phenylethyl)phthalamic acid enantiomers may be separated by using quinine in the mobile phase for diastereomeric complex formation [72]. Other types of additives are cyclodextrins which are molecules composed of several cyclic-bonded glucose units. They are composed of a hydrophobic central cavity and a hydrophilic exterior made up of hydroxyl groups arranged around the perimeter of the molecule. α-, β- and γ-cyclodextrin contain, respectively, six, seven and eight glucose units, with β-cyclodextrin (Fig. 2.15) being the most widely used.

Cyclodextrin will interact with an enantiomer if the hydrophobic portion of the guest molecule can be accommodated in the hydrophobic cavity of the cyclodextrin, while the hydrophilic portions interact with the exterior hydroxyl groups. Chiral discrimination can be achieved in this way because of the stereospecific nature of these associations, and the interaction of a number of substances, including drug compounds, with cyclodextrin-modified eluents has been studied by Han and Purdie [73]. Cyclodextrin-modified mobile phases may be used on conventional silica-based columns or on more unusual stationary phases; for instance, superior chiral discrimination has been claimed for enantiomers when separated on a porous graphitic carbon stationary phase as compared to separations generated on C18-modified silica [74].

Chiral stationary phases

The other approach to chiral separation, applicable to both gas and liquid chromatography, is to employ a chiral stationary phase. There are many types of phases currently available, some of the more common of which are as follows:

(*a*) Ligand-exchange phases are composed of an immobilized chiral ligand forming a co-ordination complex unit with a transition metal. During passage of an enantiometric mixture through the column, diastereomeric mixed-ligand sorption complexes will form by a displacement exchange mechanism. An L-aminoacid such as L-proline is an example of a ligand bonded by an alkyl chain to silica gel [75], and it would typically be equilibrated on the column with an aqueous mobile phase containing about 5 m*M* copper sulphate [76]. Although selectivities are high on this kind of material, the mobile phase is relatively complex and the exchange mechanisms can be critically affected by pH [75].

(*b*) Protein phases involve the use of silica gel upon which proteins, usually bovine serum albumin or alpha acid glycoprotein, have been immobilized. Although these materials have high selectivities, they are characterized by low efficiencies and low sample capacities. Furthermore, separation is again dependent on pH, temperature and ionic strength [77].

(*c*) Cyclodextrins, instead of being used as mobile phase additives, may instead be bonded to silica to form chiral cavity phases. This technique evolved from the

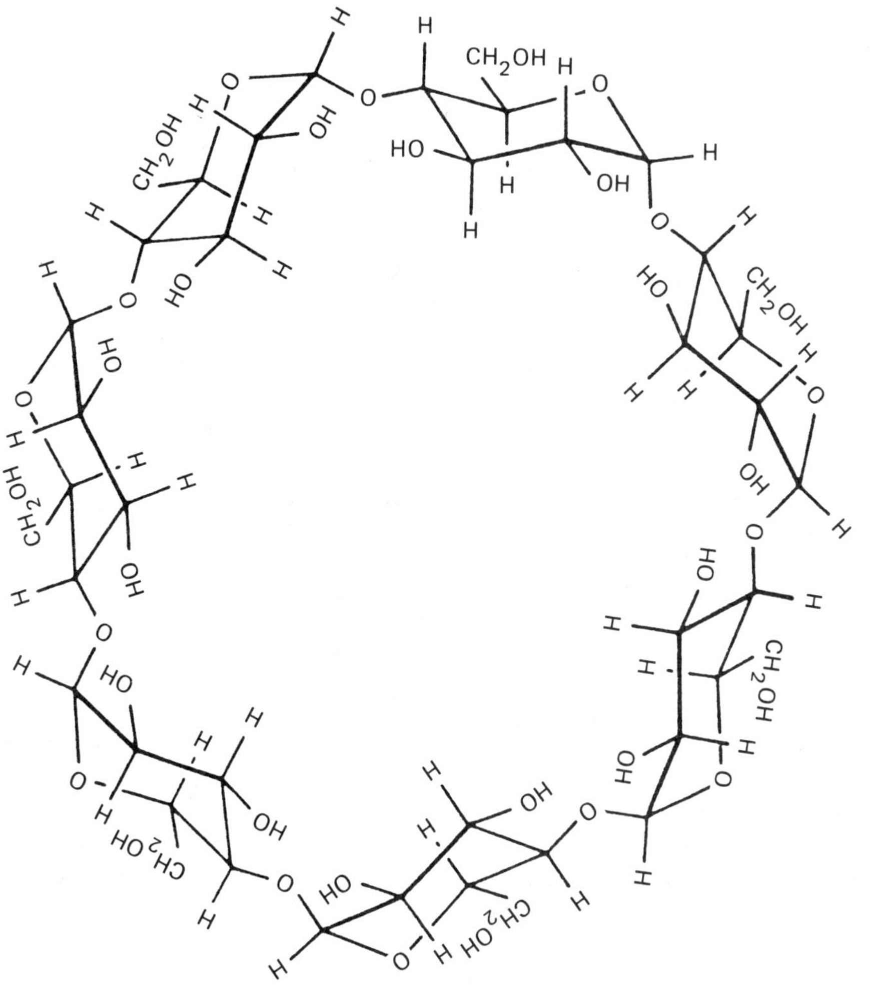

Fig. 2.15 — Structure of β-cyclodextrin.

work of Cram and colleagues who originally used a chiral crown ether bound to silica gel [78]. Nowadays, it is possible to generate dynamically modified chiral stationary phases by passing a mobile phase containing cyclodextrins through a reversed-phase C18 column [79]. Cyclodextrin-bound phases constitute one of the more popular types of materials, since by virtue of their relatively polar exteriors, they are compatible with aqueous solutions and generally have excellent selectivities [77]. Like most other kinds of chiral recognition packings, however, they suffer the drawback of reduced efficiency, though new materials are rapidly and continually being developed which offer efficiencies approaching those of achiral separations.

(d) Packing materials based on derivatives of D-phenylglycine were first developed by W. H. Pirkle [80], and though for that reason they have become known as

'Pirkle phases', the different materials are commonly identified as CSP (Chiral Stationary Phase) 1, 2, 3, __, etc. Analytes associate with these packings by a combination of dipole–dipole, hydrogen bonding, charge transfer and steric interactions [75]. Originally applicable only in the normal-phase mode, their use has been extended to reversed-phase systems with newer materials offering increasingly impressive separation factors. Pirkle stationary phases are applicable to a variety of drugs containing chiral centres, and a common example is the separation of R and S propranolol on the dinitrobenzoylphenyl CSP [81].

Enantiomeric separations may also be achieved in GC through the employment of chiral stationary phases. One type of material depends on the formation of hydrogen bonds between the solute and the stationary phase, the most successful of which are those formed by the covalent attachment of diamide-type moieties (for instance, L-valine t-butylamide) to a silicone polymer backbone. This type of stationary phase is amenable to the high temperatures commonly used in GC.

Another type of packing makes use of metal coordination chemistry. A chiral β-dicarbonyl ligand is complexed to a transition metal, the solution dissolved in squalane, and the mixture is then coated on the inner walls of the column or on an inert support, thus forming the stationary phase. This approach, by virtue of the hydrophobicity of squalane, is well suited to non-polar compounds and solute retention depends on electron donation to the metal.

2.3.3.5 Supercritical fluid chromatography

Although it was first reported in 1962 [82], supercritical chromatography (SFC) has evolved as a viable analytical technique principally in the last decade. It may be regarded as a chromatographic mode which is complementary to both the gas and liquid methods, since it can offer the resolving power, rapid analysis times, and the range of detection schemes available in capillary gas chromatography, and the ability to separate the thermally labile and involatile compounds traditionally determined by HPLC.

In SFC, the mobile phase is a dense fluid, the properties of which are intermediate between those of a gas and a liquid: the viscosities and diffusion coefficients of supercritical fluids are similar to those of gases, while their solubilizing powers approach those of liquids [63]. A supercritical fluid is obtained by heating a gas or a liquid to a temperature above its critical temperature (T_c), while simultaneously compressing it to a pressure above its critical pressure (P_c). The most widely used supercritical eluent is carbon dioxide because of its low cost, relative lack of toxicity and the fact that it has a conveniently accessible critical temperature and pressure of 30°C and 72.9 atm, respectively. Pentane is also quite widely used, but suffers the drawback of having a higher critical temperature (196.9°C, P_c=33.3 atm) and is thus less applicable to thermolabile compounds. Alternative eluents such as nitrous oxide, ammonia and benzene are less suitable because of the hazards associated with their use, and the elaborate precautions required to prevent their leakage into the atmosphere.

The column in SFC may be either a packed GC- or HPLC-type column, or a capillary-type GC column which offers relatively unhindered solvent flow. Since the latter columns are capable of high resolution, they are suitable for the separation of complex samples which do not lend themselves well to traditional capillary GC. For instance, a high-molecular-weight thermolabile sample may be separated, provided that an eluent with a reasonably low critical temperature is employed. In the case of packed columns, less costly analysis and higher sample loadability are achieved at the expense of reduced efficiency and resolving power. However, the instrumental requirements for packed SFC are less stringent, since these systems are less sensitive to extracolumn band broadening. For capillary SFC, the mobile phase is delivered by a syringe-type pump, while for packed SFC a high-pressure reciprocating HPLC-type pump is used, with the pump heads temperature-controlled to maintain the supercritical fluid. In both cases, the eluent stream may be passed into an LC detector (e.g. UV), or after provision for gas expansion, it may be fed into a gas-phase detector such as an FID or mass spectrometer.

One major advantage of SFC in biopharmaceutical analysis, particularly in relation to trace determinations and forensic applications, is that drugs separated on a packed column may readily be detected by mass spectrometry. This provides the benefits of accurate identification and quantification without the need for lengthy sample preparation associated with capillary GC or the complicated interfacing technology needed in LC–MS. Detection schemes for SFC have beeen reviewed by Hill *et al.* [83,84] and several articles covering the general area of instrumentation in SFC have been published [85–90].

Retention characteristics in SFC are more complicated than those in gas or liquid chromatographic systems. They depend on, among other factors, the density of the eluent and on how well it dissolves the analyte. The polarity of the mobile phase is determined by its density: the more dense the eluent the more polar its characteristics, and because of the relatively polar nature of many drugs, high densities are frequently required to achieve their separation in SFC. Solvent density can be increased by raising either the temperature or pressure; the latter approach is more favoured and it tends to reduce the sensitivity of the system to pressure fluctuations. Similarly, gradient elution may be accomplished by increasing the solvent density over a suitable time frame, though it must be recognized that a linear increase in pressure will not produce a linear increase in density owing to the non-linear nature of pressure–density isotherms, especially at lower temperatures near the critical point. The density of the mobile phase may additionally be increased in an isocratic or gradient fashion, by the inclusion of polar organic modifiers such as methanol, an approach which has been reviewed in a number of detailed articles [91–95].

Both capillary and packed versions of SFC have been applied to the analysis of drugs in biological fluids, and some application examples are presented in Table 2.6. The instrumental aspects and applications of packed and capillary columns in SFC have been reviewed by Petersen [101] and Janssen [102], respectively.

Further developments in the field of SFC include its extension to ion-pair chromatography [100,103], separation of enantiomers by using chiral stationary phases [94,104–106], and the investigation of novel phases involving the inclusion of micelles [107].

SFC has by no means usurped the position of GC and HPLC in the field of biopharmaceutical analysis, and most workers tend to continue with separations

Table 2.6 — Applications of supercritical fluid separations in drug analysis

Drug	Stationary phase	Reference
Propranolol and analogues	Aminopropylsilica	94
Oxytetracycline	Methylpolysiloxane	95
Alkaloids	Aminopropylsilica	96
Alkaloids	Unmodified silica	96
Barbiturates	Polystyrene–divinylbenzene co-polymer	97
Caffeine, diphenylamine	Cyanopropyl silica	98
Purine and pyrimidine drugs	Cross-linked cyanopropyl silica	99
Isradipin, bopindolol, salicyclic acid	Silica, diol, cyano	100

which can successfully be executed by either of these two techniques. This is due to a number of disadvantages associated with SFC which include technical complexity of the instrumentation requirements, lack of availability of reasonably priced commercial instruments, and limited selectivity owing to the restricted number of mobile phases. Nevertheless, SFC accounts for a large area of interest in the field of analytical chemistry, and to date, the theory and practice of SFC have generated a considerable body of literature. For further information, the reader is directed to the literature [88,98,108–116]. Biomedical applications of SFC have been reviewed by Niessen *et al.* [117].

2.3.4 Voltammetric techniques

Voltammetric methods are current–voltage techniques which are applied to the determination of substances that can be oxidized or reduced at an electrode. The basic principle involves applying a potential to a solution containing the electroactive substance and measuring how the resulting current changes as a function of applied potential. The current obtained is proportional to the concentration of the species in solution. A voltammetric system consists of a cell containing a working electrode, an auxiliary electrode and a reference electrode, usually a standard calomel electrode (SCE), which are connected to a potentiostat which provides the input waveform and permits measurement of the current produced.

Selectivity in electrochemistry is achieved by adjusting the applied potential in a similar fashion to wavelength selection in spectrophotometry. Electroactive analytes will be oxidized or reduced above a certain potential, and the higher this potential, the lower the selectivity of the measurement. The potential range over which voltammetric measurements can be made depends on the solvent, the supporting electrolyte, the pH, and the material of the working electrode. Since oxygen is readily reduced, it constitutes a major source of interference, and it is generally necessary to purge the electrochemical cell with nitrogen to ensure an oxygen-free environment. The most common electrode materials used in analytical voltammetry are based on mercury and carbon. Solid carbon electrodes which may be glassy (vitreous) or made of carbon paste, are used for the measurement of oxidizable species at reasonably positive potentials, but they suffer some drawbacks in terms of reproducibility and electrode contamination. However, new work in this area [118] has led to the design of a carbon fibre micro-electrode which is so inexpensive as to be virtually disposable, and could seriously be considered as a possible method to circumvent the problem of electrode contamination.

When the working electrode is a dropping mercury electrode (DME), the voltammetric measurement technique is called 'polarography'. The DME is preferred for drugs containing reducible moieties which cannot be measured using solid electrodes. Since the surface of the droppping mercury electrode is constantly renewed, the method is highly reproducible and surface contamination is rarely a problem. Because potentials more negative than $-0.1V$ can be used, oxygen removal using a high grade of nitrogen is again a necessary pre-requisite. Drugs which contain reducible moieties and which are commonly measured by polarography include the benzodiazepines, the tricyclic antidepressants, antineoplastic agents and certain antibiotics.

A major problem with conventional direct-current (DC) polarographic measurements is the high background current which renders measurements at low concentrations (below about $10^{-4}M$) almost impossible. The major component of the background signal is the charging current, which is the current required to charge the electrode–solution interface as the potential of the electrode changes. Pulse polarography was developed to minimize the charging current, and can, as a result, improve limits of detection by several orders of magnitude. In pulse polarography, rather than applying a linear voltage ramp over the lifetime of the mercury drop, a series of potential pulses are applied near the end of the drop lifetime, when the size of the surface changes little with time. In normal pulse polarography, the amplitude of the potential pulses is increased with each successive drop and the current is measured during the second half of the potential pulse. In differential-pulse polarography, the potential pulses have a constant amplitude and are superimposed on a slowly increasing linear DC potential ramp. The current is measured before and after application of the pulse and this difference is plotted as a function of potential. Differential-pulse polarography gives superior resolution of adjacent signals, and is therefore well-suited to the analysis of mixtures. It has been widely applied to the measurement of drugs in biological fluids. One well known example is the determination of chlorodiazepoxide and its two metabolites, desmethylchlorodiazepoxide, and demoxepam in plasma after separation by thin-layer chromatography [119]. Lower limits of detection have been realized for the determination of bromazepam in plasma (down to 10 ng/ml) by the use of a polarographic microcell and a miniature electrode [120]. Since no metabolites of bromazepam occur in blood, no separation is necessary prior to analysis.

The differential-pulse technique has been used for determination of methadone in urine [121]. After liquid–liquid extraction of the drug into cyclohexane, the organic phase was evaporated and the residue reconstituted in Britton–Robinson buffer. The analyte was oxidized at a carbon paste working electrode by using both linear sweep and differential-pulse voltammetry from 0.3 to 1.3 and 0.2 to 1.0V respectively. Methadone could be measured down to 1 μM by using this technique. The literature is replete with examples of drugs determined by differential-pulse polarography, e.g. diazepam [122], streptomycin and chloramphenicol [123], trimethoprim [124] and digoxin [125].

Improved sensitivity is gained by use of stripping voltammetry. This method may conveniently be carried out at a mercury electrode (either hanging drop or mercury film on glassy carbon [126]), or at a carbon-paste electrode. In its more usual form, i.e. anodic-stripping voltammetry, (ASV) a constant negative potential is applied first with stirring, during which time the analyte is pre-concentrated at the drop surface. This is known as the 'accumulation time' and its duration is determined by a number of factors, principally the concentration of analyte to be measured. After an interval without stirring to allow the contents of reaction vessel to become quiescent, the potential is scanned anodically, so that the reduced analyte is re-oxidized and stripped off the electrode generating an anodic wave. Because the concentration of analyte on the electrode is much greater than that in the original solution, the size of the anodic wave is much greater than the corresponding polarographic reduction wave. In cathodic stripping voltammetry (CSV) the reverse reaction occurs: the analyte is oxidized on the surface of the electrode then stripped off by application of a cathodic scan. Anodic stripping voltammetry is popular for the determination of metal ions that may be reduced to form amalgams on the electrode surface. Cathodic stripping voltammetry has been used for determination of anions such as chloride, and sulphur-containing species such as thiols.

For most organic species such as drug substances, pre-concentration may be achieved by adsorptive accumulation, where the analytes are concentrated by adsorption instead of by amalgam formation on the electrode surface. This technnique, known as adsorptive stripping voltammetry, again involves the application of a fixed potential under the influence of which the drug is accumulated at the electrode. Adsorptive stripping procedures may be either anodic or cathodic with a linear, or more usually a differential potential pulse, during the stripping step. Adsorptive stripping voltammetry has been applied to the determination of pipemidic acid in urine [127]. Different scan modes were investigated, i.e. linear-sweep, differential-pulse and square-wave, and it was found that the linear-sweep mode yielded the lowest limit of detection. The drug was extracted from urine by using a solid-phase C-18 extraction cartridge, and eluted with methanol which was then removed by evaporation; the residue was subsequently dissolved in 5 ml of $0.1M$ $HClO_4$ and transferred to the polarographic cell. The accumulation potential was -0.5 V, the scan rate 1 V/sec with an accumulation time of 20 sec. The dynamic range for the determination of the drug in urine was 2.0×10^{-7} to $1.0 \times 10^{-5}M$ in the cell, with a correlation coefficient of 0.999. A linear sweep voltammogram for pipemidic acid in urine is shown in Fig. 2.16.

Other drugs which have been determined by using similar approaches include methotrexate [128] and mitomycin C [129], digoxin and digitoxin [130], diazepam and nitrazepam [131], reserpine [132] and diltiazem [133]. In addition, procedures have been published describing adsorptive stripping voltammetry at carbon paste rather than mercury electrodes, e.g. for adriamycin [134]. Cathodic stripping procedures have occasionally been described in biopharmaceutical analysis. These include a method for penicillins which requires prior conversion to the electroactive penicilloic acid [135] and a method which can detect down to 5 nmoles of nifedipine [136]. A method has been described for the determination of methotrexate in serum by alternating (as opposed to direct) current polarography [137]. A voltage of 30 mV

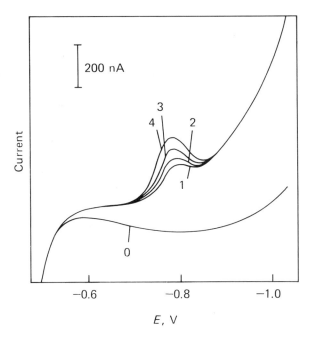

Fig. 2.16—Linear-sweep voltammetric curves obtained for extracts of a urine sample. 0: Drug-free urine, 1: Urine containing 30.35 µg/ml of pipemedic acid, 2–4: Successive standard addition of $1.5 \times 10^{-7} M$ drug. For analytical conditions, see text. Reproduced with permission of the copyright holder, M. R. Smyth, Dublin City University.

was applied at a fixed frequency of 75 Hz and a scan rate of 10 mV/sec. Results compared favourably with those of liquid chromatography and the method was linear from 0.5 to 5µM with a detection limit of 0.3µM.

2.4 SAMPLE PREPARATION

2.4.1 Introduction

It is worth noting that with some analytical methods, it is possible to carry out drug determinations without any sample pre-treatment. This is clearly the case with immunoassays, and it is one of the obvious advantages that these methods enjoy over other techniques. In some cases, the drug may be present in high concentrations and a diluted sample may be measured directly by some of the more sensitive voltammetric techniques. Since there is no separation or selective extraction, this approach depends, of course, on the drug having a redox potential different from that of any endogeneous compounds or metabolites likely to interfere in the analysis.

In HPLC, direct injection is again only feasible when the liquid sample (urine, bile, serum, plasma) contains a high concentration of analyte. It is important to ensure that the proteins are soluble in the mobile phase to prevent precipitate formation, especially for serum or plasma samples. When a pre-column and a column-switching valve are used for direct injection, protein and other interferences may be diverted to waste before HPLC separation of the analyte. This approach will be discussed later under the heading of On-line Solid-phase Extraction. Some workers prefer not to use a pre-column, but rather to carry out direct injection and then clean the analytical column by washing with an appropriate solvent at regular intervals, or replace the top 1–2 mm of the column when there are signs of contamination, as shown by an increase in column back-pressure or peak tailing. There are, however, obvious limitations to this procedure in terms of limit of detection and column lifetime.

A technique which is particularly advantageous for direct injection is micellar chromatography. This chromatographic mode was first demonstrated by Armstrong and Henry in 1975 [138], who showed that eluents containing surfactant solutions (for example sodium dodecyl sulphate) at concentrations above the critical micelle concentration (CMC) could be used to separate phenols and polynuclear aromatic hydrocarbons. Because of the unique solubilizing power of micelles, they can form soluble protein–surfactant complexes, thus preventing precipitation on the analytical column. This process also displaces drugs from serum components, freeing protein-bound drugs to partition onto the stationary phase [139–141]. Several papers have been published describing drug monitoring methods involving direct serum injection into micellar systems [139–145]. For instance, Granneman and Senello have shown that surfactants, in competitively binding proteins, will release antibiotics from serum components [145]. In the direct serum injection technique described by DeLuccia et al., [139] 20-µl serum samples were introduced onto the column, and limits of detection were of the order of µg/ml. Arunyanart and Cline-Love [140] achieved lower limits of detection by a similar procedure with fluorescence rather than UV detection. Micellar chromatography has also been used for the screening of illegal drugs in sport [146]. The drugs were separated on a C18 column by using a mobile phase of 50mM SDS with 3% butanol added to increase efficiency.

Columns have been specially designed for direct injection of serum in, for example, internal surface reversed-phase (ISRP) chromatography, a concept which was developed by Hagestam and Pinkerton [147]. These columns are prepared by reacting a hydrophobic peptide, such as glycine–phenylalanine–phenylalanine with a silica support. The support is then incubated with an enzyme which cleaves the peptide from the external surface only, since the enzyme, by virtue of its large size, cannot penetrate the internal pore structure. The internal phase consequently retains a hydrophobic surface which small molecules can access and interact with, while proteins are excluded and elute rapidly off the column since they do not interact with the hydrophilic external surface. This process is depicted diagramatically in Fig. 2.17. The retention properties of IRSP columns have also been investigated by Nakagawa et al. [148] where the hydrophobic internal surface consisted of a glycine–phenylalanine–phenylalanine tripeptide, and it was found that the addition of SDS could increase the retention of certain drugs at low pH. Furthermore, IRSP columns may be used in protein binding studies. For example, Shibukawa et al. [149] demonstrated the separation of the fraction of warfarin which is bound to the strong binding site on bovine serum albumin (BSA) from the fraction which binds to the weak binding site on the same protein.

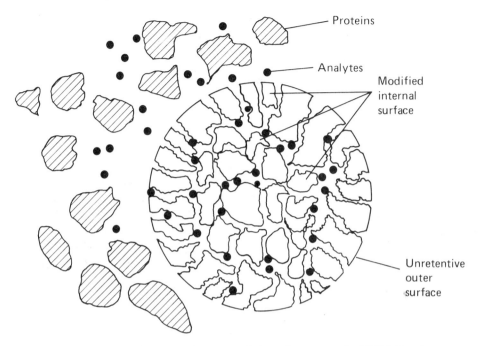

Fig. 2.17 — Principle of internal-surface reversed-phase chromatography (IRSP). Proteins are excluded from the modified internal pore structure and are rapidly eluted, as they do not react with the unretentive outer surface.

Untreated body fluids may also be injected onto columns packed with wide-pore (200–300 Å) particles. Such columns are readily available commercially and are widely used for the separation of proteins and other macromolecules which could not permeate the smaller (60–100 Å) pore of traditional packings. They are characterized by lower plate numbers, though if adequate separation can be achieved, the lower inlet pressures which they permit can prove useful for direct injection in some biopharmaceutical applications [150]. Micellar chromatography has also been carried out on wide-pore columns [151]. It was observed that a low hydrophobicity (C-1) column was more easily wetted by the Brij-35 mobile phase, while protein precipitation on the column was further minimized by the employment of a wide-pore packing material.

Direct injection is limited by the fact that there is no pre-concentration of the drug prior to analysis, so large volumes of untreated biological fluid would need to be introduced onto the column to detect drugs with therapeutic concentrations in the mid to low ng/ml ranges. Injecting large volumes of serum or plasma onto an HPLC column creates the further problem of blockage and build-up of strongly retained endogenous components, and because there is no sample clean-up, the biological matrix will produce a large signal eclipsing the peaks of early eluting compounds. On the other hand, direct injection is a time-saving and cost-effective exercise, and should be considered whenever detection and interference are not problems, for example in the determination of theophylline in serum.

In most instances, a biological sample containing a compound of interest requires some kind of sample pretreatment. Such procedures are executed principally to isolate the drug from interfering matrix substances, but also to:

(i) liberate the drug from protein binding sites;
(ii) concentrate the drug for more sensitive analysis;
(iii) separate the drug from other drugs or metabolites.

As outlined above, there are certain criteria that must be considered in selecting an analytical method for a drug assay. The same criteria apply to the selection of a sample preparation procedure, with the additional proviso that the type of sample pre-treatment is appropriate to the intended methodology. For instance, a simple protein precipitation step would not provide adequate clean-up of a viscous biological sample prior to GC analysis. The amount of sample preparation required also depends on the chemical nature and concentration of the drug and metabolites, the type of sample, and on the nature of the interfering substances. Furthermore, it will be affected by instrumental factors such as the susceptibility of the detector to contamination, (particularly with electrochemical methods and with electrochemical detection in HPLC) and in the case of UV absorbance detection, the detection wavelength. If very low concentrations of the drug are to be measured (as is increasingly the case) then the analyst may need to consider some kind of pre-concentration step in addition to sample clean-up. If the drug is present in a highly contaminated matrix, then a very selective or multiple-step extraction procedure may be necessary. Due consideration must be given to the co-extraction of metabolites (which frequently are more polar conjugates of the parent compound), to the extent of protein binding, and whether or not it is desired to measure bound or unbound fractions (or both) of the drug. The sample preparation step should therefore be capable of reducing the concentration of endogenous compounds and of concentrating the sample where this is required. In HPLC, removal of endogenous components is particularly important when these interferents could become irreversibly adsorbed onto the packing material of the column, as is the case with lipids, or precipitated in the chromatographic system, as is the case with proteins. To achieve this, either the interfering substances may be removed from the sample while the compounds of interest are retained in the original aqueous phase (e.g. protein precipitation), or alternatively, the drugs may be selectively removed from the biological specimens leaving most of the gross contamination behind (e.g. liquid extraction).

2.4.2 Protein precipitation

The preparation of protein-free solutions is especially important for analysis of blood and tissue extracts. Protein precipitation steps include the addition of an acid, a solution containing a heavy metal ion, or a water-miscible organic solvent, to the biological fluid. Following admixture of the precipitant, the sample is centrifuged to produce a clear supernatant containing the compound of interest. The protein-free solution may then be subjected to further processing, such as liquid extraction with an immiscible organic solvent, or it might be introduced directly into the analytical system. There are a number of protein precipitants available, the more common of which will be described in the following section.

2.4.2.1 *Precipitation by addition of organic solvents*
On addition of a water-miscible organic solvent, such as methanol, ethanol, acetonitrile or acetone to a biological sample such as plasma, the solubility of proteins is lowered and they are thus precipitated out. In addition, drugs are released from protein binding sites. It is important to use a protein precipitating agent in which the analyte is highly soluble, otherwise it may adsorb onto, or co-precipitate with, the protein. It is sometimes necessary to use a mixture of two solvents to obtain quantitative recovery. For example, methanol with dimethyl sulphoxide has been employed to extract porphyrins from liver tissue [152]. In this case, methanol was added to precipitate the protein, while dimethyl sulphoxide released the otherwise adsorbed porphyrins from the protein.

Diluting serum by up to threefold with the organic solvent will effect removal of 99% of the proteins [153,154]. This degree of protein removal gives a cleaner extract for injection onto a HPLC column, but, owing to the dilution factor, it effectively decreases the sensitivity of the method. This drawback can be counteracted to some extent by increasing the volume of injection, but this can, however, adversely affect chromatographic efficiency and peak shape. Alternatively, the supernatant may be evaporated, but this measure also serves to concentrate any remaining interfering compounds.

The relative precipitating powers of various solvents and other agents have been evaluated by Blanchard [154], who found that acetonitrile gave the highest percentage precipitation in the lowest volume-ratio to plasma. Acetonitrile is particularly suitable for sample preparation prior to HPLC since it is a common component in many mobile phases. In some cases, salt–organic solvent combinations are used for protein removal, the salt helping to promote salting out effects which encourage protein precipitation (owing to their reduced solubility in a more strongly ionic environment). The salt also causes the organic solvent to form a separate layer from the aqueous biological phase, which facilitates separation after centrifugation. This approach has been applied to the assay of anticonvulsants and analgesics in plasma [155]. In this case, after the addition of acetonitrile, the solution was saturated with sodium sulphate–sodium chloride and the upper layer was injected without further treatment onto the column.

2.4.2.2 *Precipitation by the addition of inorganic compounds*
Proteins are positively charged in strongly acidic and negatively charged in strongly basic solutions, owing to their zwitterionic nature. Acidic or anionic precipitants such as trichloroacetic acid or perchloric acid form insoluble protein salts with the cationic form of proteins at low pH. These two acids are widely used to extract compounds from tissue and blood samples, and a 10–20% (w/v) solution is usually used. After centrifugation, an aliquot of the supernatant may be injected onto the analytical column in HPLC, and in this case, the mobile phase should contain a high

molar concentration of buffer to protect the column from damage by the strongly acidic solution [156]. More commonly, the acid solution is neutralized with an alkali or removed by extraction into ether prior to analysis. Acidic protein precipitants are obviously unsuitable for compounds which are prone to acid hydrolysis, and in addition, perchloric acid can sometimes constitute an explosion hazard. A combination of an anionic precipitant with an organic solvent may also be useful where the latter can aid the solubilization and extraction of the analytes from the proteins. An example of this method is the combination of tungstic acid and ethanol, which is suitable for compounds that are unstable at low pH. The drawback, however, is that two volumes of ethanol are needed for complete denaturation, and this reduces the sensitivity of analysis [157]. Proteins may also be precipitated by forming insoluble salts with cationic precipitants such as copper or zinc in alkaline solution [158], e.g. zinc sulphate in sodium hydroxide or barium hydroxide, or copper sulphate in potassium hydroxide. Cationic protein precipitants may also be used in conjunction with organic precipitants: a method has been described for several therapeutic drugs which involves the addition of 10 ml of 10% $ZnSO_4$ to 100 ml of sample and 100 ml of methanol or acetonitrile [159]. This approach to protein precipitation trends to produce clean extracts, but it should be bourne in mind that precipitation by insoluble salt formation is unsuitable for compounds having a tendency to form metal complexes.

The supernatant liquid contains other constituents which remain after protein precipitation, and frequently the drug signal can be accompanied and complicated by interference from other components. Thus, in HPLC, the likelihood of interfering peaks is about the same as for direct injection, and it may be necessary to manipulate the chromatographic conditions carefully to obtain adequate separation and detection of the drug signal. Moreover, small and late-eluting peaks (including late-eluting peaks from previous samples) may co-elute with the drug, thereby distorting peak measurements and adversely affecting assay precision. The potential for late-eluting peaks to undermine assay precision has been addressed in detail by Van der Wal and Snyder [160]. Another potential problem with protein precipitations is that if the supernatant is allowed to stand for any period of time (for instance in an autosampler vial), further precipitation may occur which would clog the chromatographic column. This may be avoided by refrigerating the sample precipitant mixture prior to centrifugation or by re-centrifugation, having allowed the supernatant to stand (preferably under refrigeration) for about 15 minutes. The deproteinization procedure can sometimes give low recoveries for drugs which are strongly protein bound; and as mentioned above, certain precipitants can co-precipitate or degrade drugs or their metabolites.

Where dilution of the sample is not of major importance, organic precipitants should be considered as they are less aggressive than the ionic precipitants, with the added advantage of not requiring neutralization prior to analysis. Although methanol is approximately half as effective as acetonitrile as a protein precipitant [154, 161], it is preferred because of its low cost and low toxicity. The protein precipitation method is particularly useful for highly polar drugs such as antibiotics, and amphoteric drugs like the sulphonamides, which are difficult to extract from plasma with organic solvents. Analyses with good reproducibilities are often possible, even without internal standardization. Limits of detection for this technique are usually limited to the μg/ml range, because of the level of interferences remaining, the significant dilution of the analyte, and the fact that only small aliquots can be injected in HPLC without overloading the column.

2.4.3 Ultrafiltration

A protein-free solution may be obtained by filtration through a size-selective semi-permeable membrane under pressure or by centrifugation in a membrane cone. The basic process differs from ordinary filtration only in the size of the particles which are separated, and in the hydrostatic pressure (1–10 atmospheres) applied as a driving force for the separation process. Ultrafiltration membranes are microporous in their structure, and all molecules greater than the largest pore diameter will be retained, while all molecules smaller than the smallest pore completely pass through the membrane. All molecules smaller than the largest pores, but greater than the smallest pores will be filtered or retained in accordance with the pore size distribution. Compounds which are protein-bound will remain behind with the proteins unless they can be displaced from the binding sites before filtration. Hence, ultrafiltration is a useful technique when it is sought to measure the free (and not protein-bound) fraction of a drug, though consideration must be given to the fact that the dynamic nature of the filtration process can disturb protein-binding equilibria. Commercially available filters have, for instance, been used to separate the bound and unbound fractions of anti-epileptics [162–164].

A major factor limiting the speed and effectiveness of an ultrafiltration process is the build-up on the upstream surface of the membrane of a layer of protein molecules which cannot traverse the membrane. This is called 'concentration polarization'. Since the layer serves to reduce the rate of ultrafiltration, control of the thickness of the layer is of major importance. If a high shear is not maintained at the membrane surface, the layer will increase in thickness until the flux rate drops to a very low level. Higher shear can be obtained by using either high flow rates across the membrane surface, or rapid stirring. Another problem associated with ultrafiltration is the fact that non-ideal membranes may allow a higher filtration rate for water rather than solute molecules, thus resulting in reduction of the drug concentration. The pH and temperature must also be carefully controlled throughout the procedure, and the chances of an analyte adsorbing into the inert membrane must be taken into account, particularly if it is present in trace quantities.

2.4.4 Liquid–liquid extraction

Liquid–liquid extraction is an extremely versatile method of sample preparation; convenience, ease of use and the ready availability of highly purified organic solvents have contributed to its continued widespread use in spite of the attractions offered by newer and alternative extraction modes. The separations are usually quite clean because the relatively small interfacial area between the two phases helps to avoid events analogous to the undesirable co-precipitation phenomena which can adversely affect precipitation separations. Its popularity may also be ascribed to the formation of well-defined liquid bi-layers, and the ease with which analyte distribution between the two phases may be altered by changing the phase ratio, the nature of the extraction solvent or the chemical equilibria, the latter by the simple expedient of pH adjustment. In addition, liquid–liquid extraction, because it is more selective than protein precipitation, permits a pre-concentration of the analyte.

Liquid–liquid partition is based on the extraction of an analyte depending on its partition between an aqueous and an immiscible organic phase. The partition coefficient (K_D) of the total analyte concentration is given by:

$$K_D = [A]_o/[A]_a$$

where $[A]_o$ and $[A]_a$ are the concentrations of analyte in the organic and aqueous phases respectively. Hence, the degree of extraction is dependent on K_D, and since K_D is dependent on the type of organic solvent as well as on pH and ionic strength of the sample, it is obvious that these must be optimized for maximum analyte recovery and minimum co-extraction of interferents [165]. It is well known that repeated extractions with small portions of solvent can recover much more analyte than a single batch extraction (the principle underlying countercurrent extraction), but for practical purposes, a biological sample would be extracted no more than twice in most applications.

2.4.4.1 *Solvent selection*
The solvent is selected to provide maximum extraction efficiency (large K_D) with minimum carry-over of contaminants. Polarity is usually the most important factor in the choice of the extraction solvent, and generally as this increases, the range of compounds extracted also increases [166]. Hence, the solvent should be selected with the minimum polarity consistent with quantitative recovery of the drug. This means, however, that drugs of high polarity are difficult to extract and require polar, and hence non-selective, solvents. The influence of extraction solvent on the level of recovered interferents is illustrated in Fig. 2.18 and the physical properties of some common extracting solvents are listed in Table 2.7. The most popular solvents for extraction are diethyl ether, ethyl acetate, hexane, dichloromethane and chloroform. Solvents such as benzene and carbon tetrachloride should be avoided because of their toxicity (carbon tetrachloride is ten times more toxic than chloroform); ether or dichloromethane should be used in preference.

Although ether is flammable, it has the advantage that it is volatile, which permits rapid evaporation following phase separation; phase transfer is facilitated by its low density and the fact that it forms the upper layer when mixed with aqueous solutions. In addition, ether emulsions are easier to break. Chin and Fastlich [167] have listed a number of drugs that are extractable in the presence of sodium dihydrogen phosphate with diethyl ether, but not with hexane. However, although ether extracts a wide range of drugs, it will also extract a wide range of interfering substances from the biological matrix; this is also true of ethyl acetate which has a further drawback of tending to dissolve many of the components found in plastic laboratory consumables. Bailey and Kelner [168] investigated the extraction recoveries of acidic drugs from water and plasma with hexane, ether, toluene, n-butyl chloride and chloroform. The barbiturates, sulphonamides and diuretics were optimally extracted into diethyl ether in acidic conditions, while the tricyclic antidepressants will dissolve in hexane, which gives cleaner extracts owing to its lower polarity and hence greater selectivity.

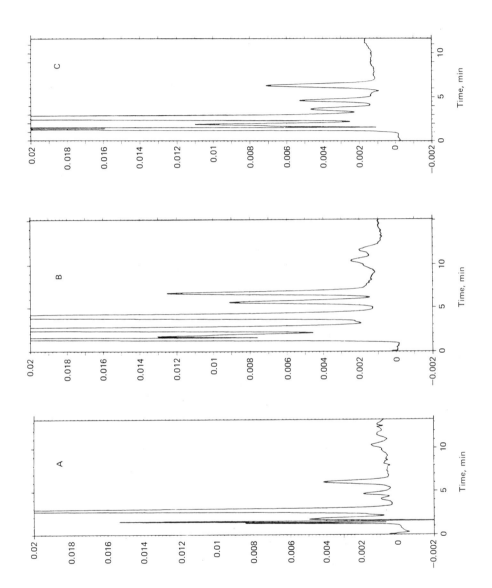

Fig. 2.18 — Influence of solvent on the level of interferences recovered from drug-free plasma
during liquid extraction. (A): Diethyl ether; (B): Ethyl acetate; (C): Methyl isobutyl ketone.
Chromatography by HPLC. Conditions: Column: Pierce C8, 10-μm (22-cm), mobile phase:
0.02*M* acetate buffer, pH 7–methanol [65:35].

Table 2.7 — Properties of some solvents used in extractions

Solvent	ε_0 Alumina	Viscosity cP, 25°C	Ultraviolet cut-off (nm)
Cyclohexane	<0.001	0.90	200
Hexane	0.01	0.30	190
Carbon tetrachloride	0.18	0.90	265
Benzene	0.32	0.65	278
Diethyl ether	0.38	0.24	218
Chloroform	0.40	0.53	245
Dichloromethane	0.42	0.41	233
Tetrahydrofuran	0.45	0.46	212
Acetone	0.56	0.30	330
Dioxan	0.56	1.20	215
Ethyl acetate	0.58	0.43	256
Acetonitrile	0.65	0.34	190
Dimethylsulphoxide	0.75	2.00	268
Propan-1-ol	0.82	1.90	240
Ethanol	0.88	1.08	210
Methanol	0.95	0.54	205
Water		large	

The solvent strength parameter, ε^0, is defined as the adsorption energy per unit area of standard adsorbent and is given as an index of polarity.

There is frequently a trade-off between the amount of drug recovered and the levels of extracted interferences which can be tolerated. It is usually possible to arrive at an acceptable compromise through the selection of a mixture of solvents which, although not effecting complete extraction of the drug itself, does give cleaner extracts. For example, ether is not useful for the extraction of alkaloids, but a chloroform–propan-2-ol mixture can be used to extract amphoteric alkaloids such as morphine [169]. Barbiturates can be extracted from blood at pH 7.5 with a mixture containing equal volumes of hexane and ether, while the use of a more polar solvent or lower extraction pH substantially increases the co-extraction of interferences [170]. Hexane mixed with chloroform or ethyl acetate, and dichloromethane or chloroform mixed with an alcohol are common extraction mixtures, and Van Damme et al. [171] have demonstrated increased recoveries of tricyclic drugs with hexane and dichloromethane if 5% of propan-2-ol is added to the extracting solvent. Although this may be due to the increased polarity of the binary mixture, it is likely that the propan-2-ol helps reduce drug loss through adsorption onto glassware.

As well as having the correct polarity the solvent should be of the highest purity; it should not be toxic or highly inflammable, and it should have a suitable volatility to permit evaporation and concentration of the analyte. Thus, all other factors being equal, diethyl ether, by virtue of its greater volatility, could be a more useful solvent than ethyl acetate for a particular extraction. The solvent should be redistilled if a preservative or a trace impurity is found to react with the drug or produce a signal which interferes with that of the drug itself. This is a more common problem than might initially be imagined, and such is the potential for solvent impurities to undermine assay integrity that significant literature attention has been devoted to the problem [172]. It must be remembered that trace impurities will also be concentrated to higher than trace levels during the solvent evaporation stage, and if the analyst is considering using a particular solvent, or has a range of solvent options available, then a useful exercise would be to evaporate a large volume (10–20 ml) of the solvent(s) under nitrogen, reconstitute in a small volume of the mobile phase (for chromatographic analysis) and thereby test for the presence of possible interfering peaks.

Solvent impurities may not only be responsible for unwanted signals in the analytical system, but they may also promote chemical reactions leading to degradation or transformation of the drug itself. In a commonly cited example [173], the presence of phosgene in chloroform promoted the formation of carbamate derivatives of the tricyclic antidepressants during extraction from urine. For this reason, ethanol is often included in chloroform as a preservative, but this measure can give rise to another set of problems, including an interfering signal from the ethanol itself or the promotion of unwanted chemical reactions such as the conversion of narcodeine to narcodeine carbamate [174]. Artefacts such as these are exacerbated when concentration of the analyte is involved.

Impurity problems can be reduced by extracting the drug into a very small volume of organic solvent, though obviously this approach precludes subsequent concentration of the analyte and it is usually more appropriate to purify the solvent if artefact-promoting contaminants are thought to be present. Solvents may readily be purified by distillation, and passage through an alumina column may be used to remove ethanol from chloroform or peroxides from ether [70].

Since it is only the un-ionized form of a drug that is extracted into the organic solvent, suppression of ionization is used to increase the hydrophobicity of a compound and promote its partition into the organic phase. Thus, acidic drugs are extracted under acidic conditions and basic drugs are extracted under basic conditions. The optimum extraction pH for acidic species is 1–2 pH units (or more) below their pK_a values, and for basic species it is 1–2 pH units (or more) above their pK_a values. The extraction of neutral drugs is independent of pH; i.e. they are extracted over a wide pH range, but they remain in the organic phase if a back-extraction into an alkaline or acidic aqueous medium is carried out. Where possible, a higher pH is often desirable to ensure cleaner extracts since many endogenous compounds are acidic and favour extraction at acidic pH. It is rarely possible to extract acidic and basic drugs together in a simple operation which avoids the use of complexing agents, etc. The required procedure usually involves adjustment of the sample to high pH and extraction of the basic drug, followed by low pH adjustment and extraction of the acidic component, or vice-versa. Typical examples would include the determination of trimethoprim and a sulphonamide-type drug or the co-analysis of propranolol, which is basic, and a diuretic such as frusemide, which is acidic. These pairs of drugs are often administered in combined therapy, frequently as a combined formulation, so their co-determination would present a not unusual problem in clinical analysis.

A method to limit the carry-over of interferents is to adjust the sample pH such that the drug is ionized, and to pre-extract with an organic phase which is then discarded. This is then followed by an extraction tailored to remove the drug, possibly with a different solvent. For example, urine contains many endogenous compounds, and a preliminary extraction from acidic urine improves the purity of a subsequent basic extract. Sometimes, where K_D or the amount of drug present is large, the drug may be extracted into a very small volume of organic solvent (100 µl or less), and after centrifugation, an aliquot of the organic phase is taken directly for analysis. By eliminating the solvent evaporation step the possibility of drug loss by evaporation, decomposition or adsorption onto glassware is reduced, although such methods limit the sensitivity of the procedure. While this approach is quick and simple, it has obvious shortcomings in reversed-phase HPLC where only a limited amount of a non-polar solvent can be injected into an aqueous eluent stream.

After the initial extraction into an organic solvent, an analyte can be back-extracted into a new aqueous phase with pH adjusted so that the analyte ionizes again. This should reduce the amount of neutral contaminants, because they will remain in the organic phase. For back-extraction of basic drugs into an acidic aqueous phase, sulphuric and phosphoric acids are preferred to hydrochloric acid, because many hydrochlorides are soluble in organic solvents. In addition, oxygen-containing solvents with a strong co-ordinating ability such as methyl isobutyl ketone, and especially diethyl ether, form oxonium cations with protons under acidic conditions, and these oxonium systems can complicate the extraction process by the formation of interfering complexes with anions in the sample. The aqueous solution is then used for separation, or more commonly, the pH is again adjusted to suppress ionization, and the analyte is re-extracted into a fresh aliquot of organic phase. This solvent is then evaporated and the residue reconstituted in a suitable solvent. Back extraction is more or less mandatory in GC since extraneous substances in the initial extract tend to contaminate GC detectors, particularly the nitrogen–phosphorus and electron-capture detectors. This approach is also widely used prior to HPLC analysis as it greatly improves the selectivity of the solvent partition technique, though overall recovery of the drug can be reduced and it adds considerably to analysis time, not a trivial consideration where large numbers of samples have to be analysed.

A useful method for dealing with a highly polar ionic drug is to convert it into a neutral ion-pair complex by the addition of an excess of suitable ions of opposite charge, and to extract this into an organic solvent. Common ion-pairing agents for acids are tetra-alkylammonium salts (for example, tetrabutylammonium hydroxide or chloride), and for bases, alkyl sulphonates (for example, heptanesulphonic acid) may be used. Formation of the complex depends on factors such as the pH of the aqueous phase, the type of organic solvent, and the nature and concentration of the counter-ion. The ion-pair extraction technique is useful for a variety of ionizable compounds, but it offers particular advantages for compounds that are difficult to extract in the uncharged form, such as penicillins, amino acids and conjugated metabolites, and quaternary ammonium compounds such as tubocuarine that are ionized at all pH values. The application of ion-pair extraction has been discusssed in detail by Schill [175] and Tomlinson [176].

Extraction can be difficult for compounds which are soluble in water at all pH values, for example water-soluble amphoteric and neutral drugs. In some cases, a 'salting out' procedure may prove useful. The addition of a salt to an aqueous solution increases its ionic strength and hence its polarity, and this tends to decrease affinity of polar compounds for the aqueous phase, thus shifting the partition equilibriun in favour of extraction. Such an approach has been used by Horning *et al*.

[177] in the screening of plasma and urine for acidic, basic and neutral drugs. By using ammonium carbonate as the salt and ethyl acetate as the extracting solvent, they were able to extract drugs with recoveries better than 80%. Sodium chloride and sodium bicarbonate are popular choices for the salting-out technique and both have been used in the extraction of drugs from urine into a relatively polar mixture of dichloromethane–acetone (2:1) [178]. This method was also applied to the extraction of the diuretic, xipamide, from plasma [179]. It was found that the addition of lithium chloride to the sample increased the extraction of the drug into ether, while interference from co-extracted endogenous compounds was reduced (Fig. 2.19).

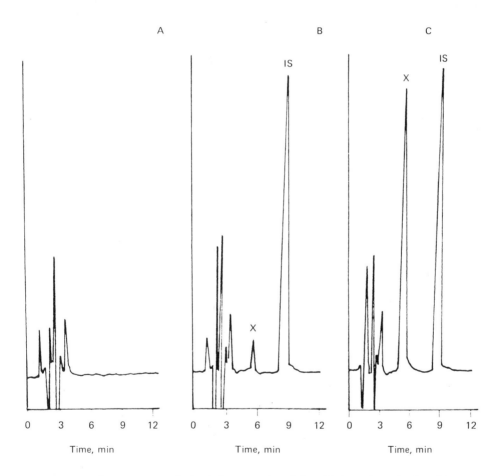

Fig. 2.19 — Chromatograms obtained after extraction of xipamide from plasma by the salting out technique. After adjustment of sample to pH 5, 100 mg of lithium chloride was added before extraction into diethyl ether. (A): Blank human plasma; (B): Plasma spiked with 250 ng/ml of xipamide and 10 µg/ml of mephenesin (IS); (C): Plasma spiked with 2000 ng/ml of xipamide and 10 µg/ml of IS. Chromatography by HPLC. Conditions: Columns: Column: Pierce C8, 10 µm (22-cm), Mobile phase: 0.02M acetate buffer, pH 6.7 acetonitrile [3:1]. Reproduced from D. Dadgar and M. Kelly, *Analyst*, 1988, **113**, 1223, by permission of the copyright holders, the Royal Society of Chemistry.

This procedure was also used in the extraction of cimetidine from plasma into ethyl acetate. As shown in Fig. 2.20, the addition of potassium carbonate to the plasma

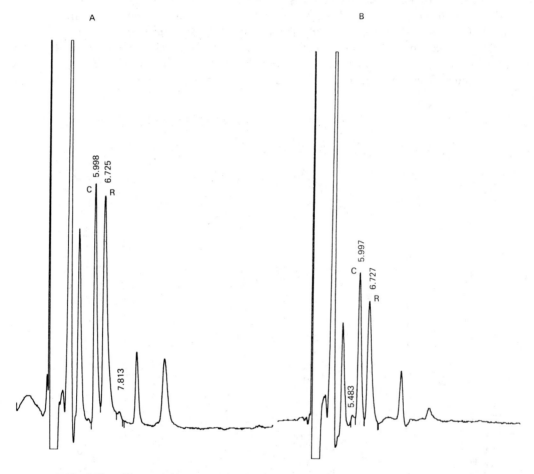

Fig. 2.20 — Plasma extracts containing 250 mg/ml of cimetidine (C) and ranitidine (R). (B): sample pH was adjusted with sodium hydroxide followed by extraction into ethyl acetate; (A): As in (B), except that 100 μg of saturated potassium carbonate solution was added to each 250-μl plasma sample before extraction. Chromatography by HPLC. Conditions: Column: Spherisorb ODS1 10-μm (25-cm); Mobile phase: 0.01M phosphate buffer, containing 0.25 g/l of heptanesulphonic acid–acetonitrile [3:1], pH of the final solution adjusted to 6.2.

effected a significant increase in the recovery of cimetidine, a relatively polar compound which is particularly difficult to extract. Salting out agents may also be used to promote ion-pair formation. For example, bi- and tervalent metal nitrates, which have a pronounced capacity for hydration, bind large numbers of water molecules, thus lowering the dielectric susceptibility of the aqueous phase and increasing the formation of ion-association complexes.

An advantage of salting out is that emulsion formation is minimized, because the combination of salt and solvent helps to form better phase boundaries and reduce the water content of the organic phase. It is clearly important that the chosen solvent and the biological fluid should be easy to separate after mixing. Polar solvents which are water-miscible (for example n-propanol) can be forced to form a separate layer by using this salting out technique, resulting in a two-phase system with the upper layer consisting mainly of the organic solvent.

Compounds insufficiently soluble in organic solvents can be extracted after preparation of derivatives. Derivatization, however, is more commonly used to improve detectability rather than solubility of the analytes, though in certain cases derivatization is necessary to release analytes from their protein binding sites.

2.4.4.2 *Extraction of metabolites*
Metabolites of a drug are usually more polar than the drug itself, and if drug metabolites are to be measured, then the chosen solvent should be sufficiently polar to extract all the compounds of interest so that they can subsequently be separated and determined chromatographically. However, selective separation of a drug from its major metabolites can be achieved by extraction with a number of solvents of increasing polarity, and/or by extraction at different pH values. Another method for extracting the more polar metabolites is to extract the drug, then add a high concentration of a salt such as sodium chloride to the aqueous phase, which shifts partition in favour of extraction and forces the metabolites into the organic phase. This salting out procedure has been used in the extraction of hydroxy metabolites of barbiturates from urine [44]. Occasionally, the parent drug and metabolites are of similar polarity and are co-extracted. Separation in these cases may be obtained by partitioning between the organic solvent and a buffer of appropriate pH, though good specificity is difficult to achieve, and some co-extraction of metabolites is inevitable. Separation of a parent drug and its metabolites is usually achieved by chromatographic means, and HPLC is generally the method of choice [181,182]. Drug metabolism is discussed in more detail in a later section, and more in-depth reading is available in the review by Martin and Reid on the isolation of metabolites by solvent extraction and other methods [183].

2.4.4.3 *Practical considerations with liquid–liquid extraction*
Unlike solid-phase extraction, liquid–liquid extraction is not a technique which lends itself well to automation, though Snyder [184] has described a system which is suitable for continuous sample handling and on-line coupling with the actual analysis stage. In this system, the analyte is introduced into a flowing stream which is then segmented with the extraction solvent and delivered to a mixing coil, where under the influence of a variety of mixing phenomena, rapid extraction takes place. After extraction, the stream is passed to a phase separator whence the phase containing the analyte is diverted to the analytical system or on for further processing. Some analyte loss must be suffered since 10–20% of this phase must be discarded to avoid contamination by the other phase. If it is desired to evaporate the solvent, either to concentrate the analyte or eventually to re-dissolve it in another solvent, this may also be accomplished using a commercially available module, developed specifically

for this purpose [185]. Automated on-line liquid extraction and evaporation has been described for the determination of theophylline and other drugs in serum [186].

Solvent extraction is relatively time-consuming; it often requires the removal of solvents by evaporation, and this lengthy step may lead to loss through evaporation or decomposition of unstable compounds, particularly when heating is required. Another problem associated with solvent extraction is adsorption of analytes onto glassware, which can occur at the extraction or solvent-evaporation stage, and is much more noticeable when a drug is present at low concentrations. This problem may be partially overcome by silanization of glassware or by the inclusion of 1–2% of ethanol, propan-2-ol or isoamyl alcohol into the extracting solvent. A method similar to, but more expedient than, silylation is to treat the glassware with 1% of butylamine in hexane, an approach which may be accomplished by simply adding the mixture to the glassware and allowing to stand for a few minutes before decanting. These measures are, however, not always effective, and others may have to be devised [153]. Emulsion formation is another drawback, particularly if a solvent of intermediate polarity (e.g. isoamyl alcohol) is included in the extracting solvent, or if the volume ratio of organic solvent to aqueous phase is not sufficiently large. Emulsions may be avoided by changing the phase ratio, or may be broken by filtration over glass wool, centrifugation, or refrigeration. However, these approaches tend to be unsatisfactory in that they add substantially to the analysis time, and full recovery of the two separate phases is rarely achieved.

One way in which the problem of emulsion formation in liquid–liquid systems may be avoided is to carry out the extraction on a solid support. In this arrangement, a hydrophilic packing material such as inert particles of diatomaceous earth (kieselguhr), is used to adsorb the sample, including water, over a large surface area, the sample forming a thin aqueous film over the surface of each particle. A small volume of water-immiscible organic solvent is then passed through the column, and this effectively extracts the drug from the aqueous film of sample. Water and endogenous materials, such as pigments and other polar compounds, are retained in the adsorbed phase. Such methods may also minimize the problem of adsorption onto glassware, especially if the eluate can be collected in a polyethylene or polypropylene test-tube. Furthermore, they can provide for very efficient extractions owing to the high surface area of the film. Disposable cartridges packed with diatomaceous earth (Chem Elut and Tox Elut) are now commercially available and are popular for the extraction of steroids from plasma and, more particularly, from urine.

Although liquid extraction on a solid support can circumvent the problem of emulsion formation, the eluate must still be collected and evaporated if it is desired to concentrate the analyte. In addition to the problems of loss though evaporation and degradation mentioned earlier, recovery may be reduced by incomplete dissolution of the drug in the reconstituting solvent. However, a strong solvent used for reconstitution of the residue may undermine the integrity of the analytical step: for instance, Van Damme *et al.* [171], when investigating the effect of organic modifier on the chromatographic characteristics of a group of barbiturates, found that as the percentage of acetonitrile in the dissolving solvent was increased, there was a deterioration in both peak shape and peak area. These findings are consistent with the fact that analytes introduced onto an HPLC column in a solvent which is significantly stronger in its elution strength than the mobile phase, generates poor peak shapes and lower peak areas. It is, therefore, generally desirable to reconstitute the residue in either the mobile phase or a solvent which is weaker than it, though, as outlined above, complete dissolution of the drug may not be obtained.

2.4.5 Solid-phase extraction (SPE)

Solid-phase extraction (SPE) is a technique which can circumvent many of the problems associated with liquid–liquid extraction. The general approach to solid-phase (liquid–solid) extraction is the adsorption of the drug from a liquid onto a solid adsorbent or stationary phase immobilized on a solid support. Solid-phase extraction using silica, alumina, celite, talc, charcoal, ion-exchange or hydrophobic resins has long been common practice in the clinical laboratory, and the manufacture of modern HPLC stationary phases has led to new methods in the solid-phase extraction technique. Nowadays, the materials available for SPE are myriad and silica gels bonded with a variety of functional groups, e.g. alkyl-, phenyl-, cyano- and diol- moieties, are commonly employed to provide specific interaction with analytes. Simple procedures involve the addition of the solid to the liquid, agitation and separation by centrifugation. However, the most satisfactory approach, which permits facile and reproducible recovery of the analyte, is to pass the liquid through a short, packed bed of sorbent material.

Solid-phase extraction is usually carried out in small columns packed with a material similar to those used for analytical separations. These short columns have a number of applications apart from simple extraction purposes, including trace enrichment and stabilization of otherwise labile compounds when storage of samples is required. Short columns may also be used to carry out derivatization reactions, and lately they have been employed in immunoassay techniques where the antigen–antibody complexation reaction takes place on an adsorbent within a column.

Solid-phase extraction methodology is based on chromatographic principles, differing only in that compounds in an analytically useful situation would have capacity factors between 1 and 10, whereas in an efficient extraction scheme, they would be either greater than 1000, where the analyte is totally retained, or less than 0.001 where the analyte is eluted with the void volume. This is the principle behind the concept of 'digital chromatography' meaning that the analyte is either totally retained on, or completely eluted off the column, and the two opposing situations should ideally pertain during the loading/washing and analyte elution stages, respectively.

After the selection of an appropriate sorbent on which to carry out the extraction, conditions are optimized to maximize the retention of the compounds of interest, while minimizing retention of interfering substances. The simultaneous achievement of these two objectives is rarely possible, but selectivity can be maximized by judicious choice of pH and eluting solvents. Solid-phase extraction may be performed off-line or on-line by using a switching valve to channel the eluate from the pre-column either to waste or onto the analytical column. These two methods of solid-phase extraction will be discussed separately, since although the underlying principles governing retention mechanisms are identical, the technologies involved in carrying out extractions in the two types of systems are quite different.

2.4.4.1 Off-line solid-phase extraction

Extractions of this type may be conducted with home-made columns such as a
Pasteur pipette packed with adsorbent. However, disposable cartridges, generally
configured as luer-tipped syringes (Fig. 2.21) in various sizes (500 µl up to 20 ml) are

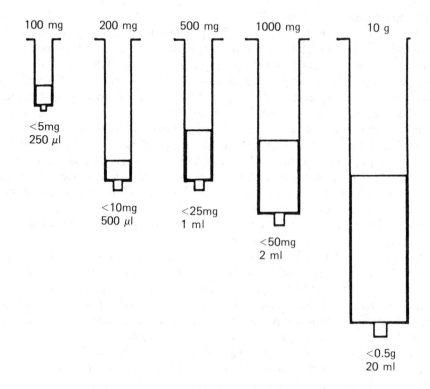

Fig. 2.21 — Examples of disposable cartridges for solid-phase extraction. The capacity of these
cartridges is approximattely 5% of the sorbent weight. The volumes shown are the minimum
elution volumes for quantitative recovery. Reproduced with permission of the copyright
holders, Jones Chromatograpphy, Mid Glamorgan, UK.

available commercially. An important consideration in selection of the size of
cartridge to be used is the amount of drug to be extracted; it is generally desirable to
use the smallest amount of sorbent that will retain that amount of analyte. The
capacities of sorbents depend on a number of factors and generally differ from one
manufacturer to another. For the examples shown in Fig. 2.21, the sorbents can
retain approximately 5% of their own weight and feature a minimum elution volume
of approximately 2.0 ml per g of sorbent. Using the smallest cartridge appropriate to
a particular application means that the drug can be eluted in a small volume of
solvent, thus minimizing the dilution factor, possibly eliminating the need for solvent
evaporation, and certainly cutting down on the consumption of expensive organic
solvents. One of the restrictions with this approach is that if the biological matrix is

very viscous or highly contaminated, a larger sorbent bed may be required to avoid blockage of the cartridge. The above considerations would also apply to the on-line method of sample preparation, though in such cases, the analyst is generally more limited in terms of choice of column size.

Extraction cartridges are generally for single use only, since components retained on the column from a previous sample could affect performance in subsequent use. However, some workers [187,188] have shown that the C18-bonded phase disposable columns can be regenerated and used again without any loss of performance. Since samples for biopharmaceutical analysis are generally aqueous, the most popular and useful bonded materials for these applications are the reversed phases, commonly, octyl- (C8) or octadecyl- (C18-) silica. This is due, at least in part, to their versatility; all the methods described for the liquid–liquid partition of an analyte from an aqueous phase into an organic phase are also applicable to solid-phase extraction with C8- or C18-bonded silica. As in liquid–liquid systems, the general approach in reversed-phase solid-phase extraction is based on the fact that the majority of drugs, though they may contain one or more ionic moieties, are usually less polar than the components of the biological matrix. In optimizing isolation selectivity, the analyst can exploit not only the different affinities of the drug and matrix components for various sorbents, but also differences in pK_a values between the drug and other species in the mixture. The usual strategy is to adjust the pH of the sample so that the analyte is uncharged and its interaction with the stationary phase will be maximal, and ideally, such that the endogenous interferents will be ionized and not retained. After the sample is loaded onto the SPE column, undesirable compounds may be removed by washing with an aqueous buffer (pH adjusted to maximize retention of the drug on the column), or simply with water. Any interferents which are still adsorbed at this stage may be removed by washing with a specific solvent or aqueous–organic mixture. The compounds of interest are then eluted by washing the column with an appropriate solvent, and this may be pH-adjusted to ionize the analyte and effect rapid removal from the column. Selectivity on this basis is most easily achieved for basic drugs since, as mentioned earlier, many endogenous interferents are acidic and will be ionized at high pH values. The strongest wash solvent which does not elute the drug should be used to remove endogeneous interferents. The weakest solvent to give complete recovery of the drug should be used in the elution stage in case more strongly retained interferences are removed also.

A typical liquid–solid extraction procedure with a reversed-phase cartridge is as follows.

(i)   Activate the cartridge with methanol.
(ii) Equilibrate with water or a buffer of controlled molarity and pH, with or without an ion-pairing reagent.
(iii) Load and sample onto the cartridge directly, or after centrifugation, dilution with water or buffer, pH adjustment, ion-pair formation, derivatization or complexation.
(iv) Wash the cartridge with an appropriate eluent to remove early eluting impurities.

(v) Elute the analyte selectively with a solvent mixture optimized for its recovery, leaving late-eluting compounds on the solid phase.

The eluate may then be further processed, or introduced directly into the analytical system. The elution solvent may be miscible or immiscible with water because (where it is used) little sample water remains on the column, though if it is desired to remove all traces of residual water, this may be accomplished by centrifugation or by passing a stream of nitrogen over the sorbent bed.

The cartridge may be treated with a solution containing a masking agent (for example triethylamine or ammonium acetate) after activation with methanol. This suppresses the ion-exchange properties of residual silanol moieties on reversed-phase packings, and so produces more reproducible extractions. Unreacted silanols can have a significant impact on reversed-phase extractions in that they interact ionically with basic analytes and can render elution with simple aqueous-organic eluents very difficult. However, this is not to say that drug–silanol interactions are undesirable. In fact, in quite a number of nominally reversed-phase applications, the secondary ion-exchange reaction is a vital component in a mixed retention mechanism. The role of unreacted silanols is so significant that batch-to-batch and between-brand variations in the number of residual silanols can seriously undermine assay reproducibility, a problem which has been reviewed by Ruane and Wilson [189].

Drug retention by ion exchange has been applied to highly polar, ionizable drugs such as gentamycin, which are not extracted by hydrophobic packings [190]. This drug has also been extracted on unmodified silica [191], where the retention mechanism is also one of ion-exchange involving the positively charged gentamycin and the negatively charged silanol groups. Carbon-based materials and synthetic resins such as styrene–divinyl benzene are also used for the extraction of drugs from biological fluids; chloramphenicol has been extracted from serum on a 50-mg cartridge of graphitized carbon black [192], and in a comparative study, the same material compared favourably with C-18 for the extraction of beta-blocking drugs [193]. Co-polymeric columns have been used for the extraction of drugs of abuse from urine prior to GC analysis [194], and XAD resins have been used as extraction media for various drugs in biological fluids, including diazepam, chlorpromazine [195,196], codeine [197], oxycodone [198], and prior to the determination of methaqualone in plasma by HPLC [199].

Although disposable cartridges may be processed by hand, the difficulty in manually forcing the viscous sample followed by a succession of solvents through the tightly packed sorbent bed is impractical for any kind of routine work. For small-scale operations, a vacuum assembly with space for 10–20 cartridges may be used. Such assemblies are commercially available and are a necessary pre-requisite for reproducible separations. The repetitive nature of the operations involved in solid-phase sample preparation means it is suitable for robotic automation. Automation procedures involving laboratory robots have been described [200,201] and the type of instrument configurations available on the market have been reviewed [202,203]. The technology involved is complicated, requires dedicated instruments and relies on extensive computer support. However, it has advanced to the stage where all the analyst has to do is filter or centrifuge the sample prior to placing it in the autosampler rack, and succcessive operations are carried out in a timed sequence which may be programmed directly or via a system controller. As peak heights or areas are measured electronically and subjected to on-line mathematical manipulation, re-analysis may be carried out if necessary; and thus the system is capable of continuous unattended operation.

2.4.5.2   *On-line solid-phase extraction*

On-line solid-phase extraction is a clean-up method which is used in conjunction with chromatographic procedures, principally HPLC, and the following discussion will treat the subject from that standpoint only. The small column on which extraction takes place is mounted on-line between the injector and the analytical column, which is normally separated by a switching valve to control the direction of solvent flow. The extraction column is usually made of stainless steel and may readily be re-packed. Unlike the disposable cartridges, it is retained for a number of samples which means that adequate steps must be taken to ensure thorough regeneration of the packing material and to avoid analyte and matrix carry-over or other 'shadow effects'. The incorporation of a switching valve into the instrument arrangement for on-line solid-phase extraction is why these type of operations are frequently termed 'column switching' operations.

An example of a scheme for an on-line sample clean-up procedure incorporating two pumps and a six-port switching valve is shown in Fig. 2.22. This assembly has been used for a number of applications including the determination of tricyclic anti-depressants in plasma [66,67], separation of basic drugs on an unmodified silica column [204], and on an alumina column [205], chromatograms for which are shown in Figs. 2.23 and 2.24.

The typical operation of this instrument arrangement is as follows: Pump A is used to deliver the washing solvent. The mobile phase eluent is delivered by pump B. When in position 1, washing solution is passed by pump A via the injector and valve onto the concentration column. Meanwhile, mobile phase is pumped by pump B via the valve onto the analytical column, which is thus maintained in a state of constant equilibration. Injections are made when the valve is in this position. An injected biological sample, for example, is swept onto the pre-column by the washing eluent, whereupon the matrix components are eluted to waste and the compounds of interest are selectively enriched on top of a judiciously chosen sorbent. After a pre-determined wash time, the valve is switched to position 2, which causes the mobile phase to be re-routed onto the concentration column where it desorbs the retained compounds and flushes them onto the analytical column. While this is happening, the washing eluent is passing through the valve and hence to waste. In this arrangement, the drugs are backflushed off the concentration column, which helps to reduce broadening of the analyte bands during the washing step. Band broadening may also be minimized by the employment of small (e.g. 5–10 μm) totally porous particles rather than larger (35–50 μm) pellicular ones; however, direct injection of untreated plasma demands the use of larger particles, otherwise the concentration column would become blocked after one or two injections. In these cases, the analyst is obliged to devise other methods to reduce band broadening such as the employment of short, narrow-bore tubing connections, use of a suitable washing eluent of low

COLUMN-SWITCHING ASSEMBLY

POSITION 1

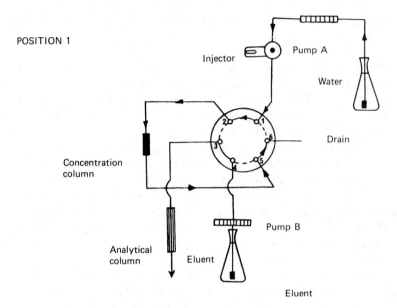

POSITION 2

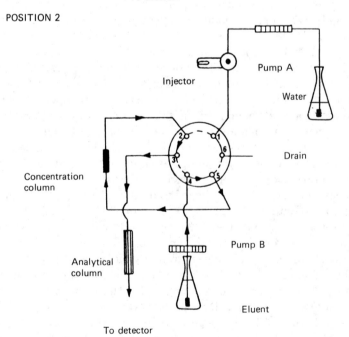

Fig. 2.22 — Instrument arrangement for on-line solid phase extraction by column switching with a single concentration column. Operation is explained in text.

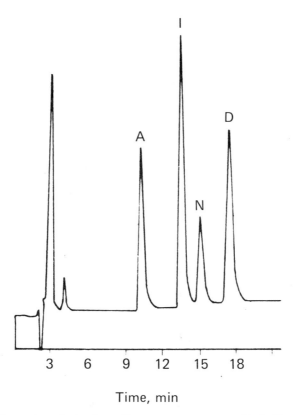

Time, min

Fig. 2.23 — Separation of tricyclic antidepressants on an unmodified silica column after enrichment on C18 concentration column, by using the instrument arrangement in Fig. 2.22. Washing eluent: water; Analytical mobile phase: $0.025M$ ammonium nitrate buffer, pH 9.5–methanol [1:4]. Peaks: (A): amitriptyline; (B): imipramine; (C): nortriptyline; (D): desipramine.

elution strength (for example a totally aqueous buffer at a pH which maximizes drug retention on the column), optimizing of the flow rate, or by keeping the duration of the wash stage as short as is consistent with adequate removal of proteins and other interferents.

A paper by Roth *et al.* [206] reported on a system which contained two pumps, one for the washing solvent and one for the mobile phase, in conjunction with two switching valves and two pre-columns. The instrument assembly for this pre-column system is shown in Fig. 2.25. An attractive feature of this arrangement was the so-called 'alternating pre-column sample enrichment technique' which involved efficient and time-saving operation of the analytical column. The sample was injected and loaded onto the first precolumn, and after a purge phase, the analytes were backflushed onto the analytical column. Meanwhile, a new sample was injected onto the second pre-column, and while the first sample was eluted on the analytical column, the second sample was purged and backflushed into the mobile phase

b

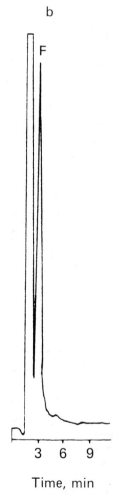

Time, min

Fig. 2.24 — Separation on an alumina column after on-line solid-phase extraction. Plasma containing 200 ng/ml of fluphenazine (F); Analytical mobile phase: 0.02M phosphate buffer, pH 5.0–methanol [1:4].

solvent stream. While the contents of the second pre-column were being separated on the analytical column the first pre-column was switched back to the solvent stream of pump A, the washing solvent, which removed traces of organic solvent and prepared the pre-column for the next injection.

2.4.5.3 *Multi-dimensional chromatography*

While on the subject of column switching, it is worth referring to the topic of multidimensional chromatography, a technique in which the sample is separated by switching between two or more columns possessing separate, but usually complementary, separation characteristics (e.g. gel permeation with reversed-phase

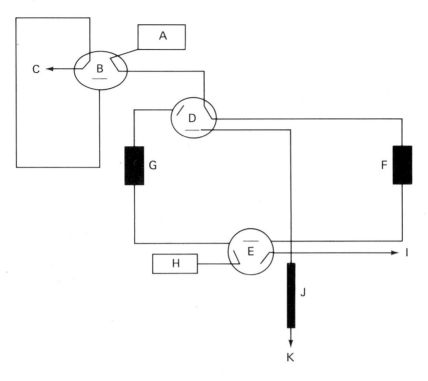

Fig. 2.25 — Alternating-column switching system for sample enrichment and backflush. (a):
pump A; (b): injection valve; (c) to waste; (d): switching valve 1; (e): switching valve 2; (f): pre-
column 1; (g): pre-column 2; (i) to waste; (j): analytical column: (k): to detector. Reproduced
with the permission of Dr W. Roth and Dr K. Thomas, GmbH, Germany.

chromatography). It is useful, like gradient elution, in situations where the 'general
elution problem' pertains, i.e. there is a requirement to separate a number of
components of widely differing capacity factors. It normally involves removing a
zone containing unresolved or partially resolved components from one column and
transferring it to another column which may be of a different type from the first, or be
of the same type but eluted with a different mobile phase. The zone is then further
separated on the second column, providing resolution which was not possible on the
first column. This is also the principle behind column switching when used for sample
clean-up; separation between the drug and biological components is achieved on the
short column, and separation of the remaining adsorbed materials is achieved on the
analytical column. Technically speaking, therefore, on-line solid-phase extraction is
a form of multidimensional chromatography, but some workers consider the technique to involve at least two modes of analytical separation and this can lead to some
confusion in terminology. The process where a portion of the eluent from one
column is transferred to another is known as 'cutting'. When the central portion of
chromatogram is isolated for further analysis, the technique is known as 'heart-
cutting', and was first described by Deans in 1968 [207].

As multidimensional chromatography involves the use of switching valves to channel zones and direct eluents between the columns, it may be automated by the incorporation of pneumatically actuated valves and a microprocessor control system. Automation improves reliability, sample throughput, analysis time, and minimizes sample loss, since analysis is performed in a closed-loop system. Phase exchanges from aqueous mobile phases to organic mobile phases, and *vice versa* in multidimensional liquid chromatography, may be realized by the incorporation of a purge-and-dry sequence. Furthermore, a new chemical dimension may be introduced by the inclusion of a chemical derivatization step between the first and second analytical columns.

Such systems may also be used for boxcar chromatography, an advanced column-switching technique which allows almost simultaneous injection of multiple samples for increased sample throughput [208]. The boxcar method overcomes the objection to liquid chromatography that it is limited in its throughput, because analysis rates as high as 50 samples per hour can be achieved with columns of conventional length [209], and maximal frequency of injections is only limited by the resolution capability of the column. Nazareth and co-workers [209] applied this arrangement to the analysis of phenobarbitone and primidone in serum. They found no memory effects from the previous sample, achieved 100% transfer between the second and third columns (i.e. the first and second analytical columns), and were able to analyse up to 40 samples per hour.

2.4.6 Supercritical fluid extraction (SFE)

Supercritical fluids have now a firmly established role in the field of chromatography, and their applications were discussed earlier. By virtue of their enhanced solvating power as compared with gases, supercritical fluids can also be used for the extraction of analytes from complex matrices, including the extraction of drugs from biological media. Although supercritical fluids offer solvating powers which are, at best, comparable to those of conventional solvents, they enjoy a number of distinct advantages over liquid solvents which have contributed to the continued interest in this unique extraction mode. For instance, because supercritical fluids have higher rates of mass transfer and lower viscosities than liquids, extractions are rapid; the polarity of supercritical fluids can be varied by changing the density, a variable which may be altered by changing temperature, or more commonly, the pressure in the system. Thus, the extraction may be optimized to suit the physico-chemical properties of the drug, and fractionation of a sample containing analytes of mixed polarities may be achieved by successive extraction with fluids of sequentially increasing polarities.

A significant advantage of SFE in trace analysis where pre-concentration is required is that removal or evaporation of the extracting solvent is more readily accomplished than with a liquid solvent. In addition, supercritical fluids, particularly carbon dioxide, are readily available in a pure form at reasonable cost.

Supercritical extractions may be carried out either off-line, or on-line usually coupled to gas or supercritical fluid chromatographic systems. The off-line approach allows the extraction and concentration of analytes for subsequent HPLC analysis.

In the off-line approach, such as that described by Hedrick and Taylor [210], compounds are extracted from aqueous solutions with supercritical CO_2 and subsequently deposited into a solvent or onto a solid adsorbent. The analytes may also be trapped in empty vessels, but since aerosols can form from small volumes, these systems are not well suited to trace analysis [211].

On-line SFE has received considerable attention in the past few years and has proven to be particularly useful when combined with SFC. SFE–SFC methods have been described for the determination of PAH in marine sediments [212], decongestants and analgesics in ointments [213], ouabain in alcohol [214], and steroids, vitamins and drugs in biological tissues [215–217]. On the other hand, SFE is also well-suited to GC analysis, and if a separation can satisfactorily be carried out by capillary GC, it would be advisable to adhere to this mode since it is technically simpler and provides greater resolution and faster analysis times [218]. Reviews have been published on the coupling of SFE to multidimensional SFC [219] and to both SFC and GC [220]. On-line SFE–HPLC is less widely used, possibly because of the technical complexity in the setting up of such a tandem system, but more likely because the types of compounds which lend themselves well to extraction by SFE are best suited to analysis by SFC or GC. The general technique of SFE has been discussed in a number of articles [218,221,222] and the coupling of SFE to a variety of chromatographic methods, including TLC, has also been reviewed [223].

Unfortunately, since the majority of drug compounds are relatively polar and are most conveniently determined by liquid chromatographic methods, the potential for SFE–GC or SFE–SFC as routine analytical methodologies in biopharmaceutical analysis would appear to be rather limited. For the same reasons, off-line SFE is also restricted in its application to drug extractions owing to the fact that the safest and most readily available supercritical fluid, CO_2, is usually too non-polar, even at high densities, to extract the majority of drug compounds efficiently. A way around this problem has been the inclusion of organic modifiers such as methanol into supercritical CO_2, but these tend to raise the critical temperature and to diminish the advantages of rapid solvent evaporation after extraction.

Methods using supercritical pentane, ethyl ether, methanol, and THF for the extraction of phenanthrene from coal tar pitch have been described [224], but so far little has appeared in relation to drug extractions using supercritical fluids other than CO_2. In relation to its solubilizing power, supercritical ammonia is a reasonable alternative to CO_2 as an extraction solvent, but in practice, its corrosiveness and the general hazards associated with its use render it unviable for routine processing of large numbers of samples. Other supercritical solvents such as pentane, although useful for some analytes, are unlikely to find wide application in biopharmaceutical analysis owing to the fact that their high critical temperatures could promote chemical changes or degradation of labile drugs. By comparison with solid-phase systems, the number of extraction media and, therefore, selectivities available in SFE are quite limited, and until a greater variety of useful supercritical solvents becomes available, it is unlikely that SFE will make serious inroads into the positions of either liquid or solid extractions as the principal methods of sample preparation in biopharmaceutical analysis.

2.5 PRACTICAL CONSIDERATIONS IN BIOPHARMACEUTICAL ANALYSIS

2.5.1 Storage of samples

Collection of an uncontaminated sample at the correct time in relation to the dose is vital in any projects involving drug analysis. If insufficient care is taken in collection and handling of biological specimens, data generated by using even the most sophisticated techniques may be invalidated. Considerable time may elapse between the collection of a sample and the analysis stage, and changes occurring during storage can be important. Common sources of error are the adsorption of the drug onto the walls of the container, and some drugs can undergo decomposition under the influence of temperature, light and certain pH conditions. Thus, the stability of the drug in a biological fluid should be determined at various temperatures so that the effect of storage on the ultimate analysis can be established. This is especially true if samples arrive irregularly and assays are carried out on a batch basis.

The usual procedure to ensure stability of samples is to store them at low temperatures, normally in a deep freeze, if the samples are to be held for an appreciable length of time. Fresh plasma or serum samples can usually be kept for 6 hr at room temperature, or for 1–2 days at 4°C. For longer term storage, samples should be held at −20°C. For instance, Jonkman *et al.* [225] have shown that theophylline is stable in blood for one day at 25°C, and stable for one month at −20°C. Storage at low temperatures serves to arrest, or at least minimize, enzyme activity which can lead to breakdown of the drug structure. An example of this is the decomposition of ester-type drugs by serum esterases [226], a problem which can also be overcome by the use of a fluoride-type anticoagulant which inhibits esterase activity. Metabisulphite is sometimes added as a preservative to prevent the oxidation (including photo-oxidation) of drugs during transport and storage, and is added to prevent the photodecomposition of LSD [227]. However, antioxidants can present their own set of problems, such as promoting the reversion of *N*-oxide metabolites to the parent compound, yielding erroneously high drug concentrations in subsequent analysis. Apart from drug transformations, changes to the biological material itself can occur as a result of putrefaction during storage, causing the appearance of endogenous (breakdown) products which can interfere with subsequent analysis. For instance, bacteria in decaying biological material have been shown to promote the reduction of the nitro group in nitrazepam to an amino group [166].

2.5.2 Sampling

2.5.2.1 *Analysis in blood, plasma and serum*

The choice must be made as to whether whole blood, plasma or serum is to be analysed. In drug analysis, the most commonly sampled body fluids are plasma and serum, because a good correlation between drug concentration and therapeutic effect is usually found [166]. The analyst could, however, be faced with no choice if the samples are haemolysed on arrival at the laboratory. It is important to avoid haemolysis if at all possible, since it prevents subsequent separation of plasma or serum. Care must be taken at the point of sampling to avoid excessive mechanical agitation, and if the sample is to be diluted, this should be done with isotonic saline.

On no account should the blood sample be frozen without treatment, because this would also cause haemolysis to occur.

Most drugs are concentrated in the plasma, yet there are a significant number, such as clorthalidone or cyclosporin, which have higher concentrations in erythrocytes. Assays for such drugs will show enormous differences depending on whether analysis is performed on whole blood or on plasma, and if plasma or serum is analysed, the expressions 'blood samples' or 'blood levels' should not be used to describe the analysis of their concentrations.

After collection of blood, a clot can be allowed to form, and the supernatant then collected after centrifugation is **serum**. Serum contains no fibrinogen (a protein involved in the clotting process) and coagulation is complete in about 30 min at room temperature. Alternatively, the blood can be collected in a tube containing an anticoagulant, and the supernatant which remains is **plasma**, which does contain fibrinogen, but since the anticoagulant effect is temporary, collected specimens should be centrifuged quickly to avoid eventual clotting. Plasma is more frequently used than serum in drug analysis, since the collected specimens can be centrifuged immediately, whereas the formation of serum is more time-consuming. Moreover, it is relatively easy to centrifuge blood which has been treated with anticoagulant, as the plasma separates quickly and the maximal volume can be recovered if required. In the separation of plasma, temperature can have a marked effect on the recovery. For example, plasma phenytoin levels have been shown to increase by 10% when a sample is equilibrated and centrifuged at 4°C rather than 24°C [228].

The type of sampling tube and the nature of the anticoagulant it contains can have a dramatic effect on results, particularly in relation to the presence of interfering peaks and measured drug levels. Bergqvist *et al.* found a significant decrease in the serum concentration of phenobarbitone, carbamazepine and phenytoin when samples were stored in gel-barrier sampling tubes [229].

As regards the type of anticoagulant used, Rapaka *et al.* [230] found that heparin caused the introduction of interference peaks in the chromatographic analysis of frusemide with fluorescence detection, whereas Vacutainer tubes containing EDTA as the potassium salt resulted in no interfering peaks with either fluorescence or ultraviolet absorption detection. In the same study, containers with sodium oxalate as anticoagulant were found to cause interference with ultraviolet, but not with fluorescence detection. In another study, heparin was shown to interfere with the radioenzymic and homogeneous immunoassay for aminoglycoside antibiotics such as gentamycin [231]. The influence of the type of anticoagulant on interfering plasma peaks is demonstrated in Fig. 2.26; these are drug-free plasma samples which were drawn at the same time and analysed by reversed-phase HPLC after on-line solid-phase extraction.

The type of anticoagulant used can also affect the apparent concentrations of recovered drugs. Hoskin *et al.* investigated the influence of collection tubes on the assay of morphine and two of its glucuronide metabolites [232]. They found the highest concentration of the drug and its two metabolites in serum collected in glass tubes, with similar results for plasma collected in heparinized plastic and glass tubes. However, concentrations of all three compounds were significantly lower when samples were collected in the glass tubes containing either citrate or oxalate as

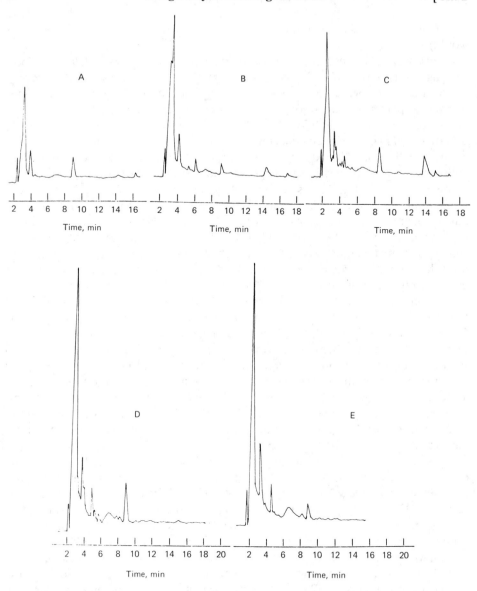

Fig. 2.26 — Effect of anticoagulant on the levels of plasma interferents recovered during on-line solid-phase extraction of 250 μl of drug-free plasma on a C-18 pre-column. (A): lithium heparin; (B): fluoride oxalate; (C): EDTA; (D): citrate, brand 1; (E): citrate, brand 2. Washing eluent: water; Analytical separation on Spherisorb 10-μm ODS1 (25 cm) wiith a mobile phase of 0.025*M* acetate buffer, pH 6.0–acetonitrile [1:1], M. T. Kelly, unpublished data.

anticoagulants. The effect of heparin and citrate on measured concentrations of various analytes in plasma has been reviewed by Smith *et al.* [233]. These authors found that measured concentrations for a number of common clinical analytes such as cholesterol and vitamins were lower in citrated plasma than in heparinized plasma, but that the difference in concentrations between serum and heparinized plasma was not statistically significant. They proposed that these observations were due to the increase in osmolality of citrated plasma over serum and heparinized plasma. The increased volume of plasma harvested with citrate was first observed by Foley *et al.* in 1968 [234].

The effect of plasticizers in the body and stoppers of blood-collection tubes in plasma levels of collected specimens is well known [235–238]. The presence of the plasticizer tris-(butoxyethyl)phosphate in stoppers has been shown to cause redistribution of drugs into erythrocytes and distortion of the apparent concentration in plasma [236]. An example of this phenomenon is where Stargel *et al.* [238] observed that binding of lignocaine to plasma proteins decreased from 56% to 28% when collected blood came into contact with the stopper of a Vacutainer tube. Other drugs which were affected in this way include quinidine, propranolol, alprenolol, some tricyclic antidepressants and some phenothiazines. Shang-Qiang and Evenson [239] also observed both increases and decreases in extracted concentrations of drugs, depending on the evacuation tube and the drug investigated. They also found that the coefficients of variation of 25 commonly measured drugs in plasma pools increased from 5% for samples collected in glass tubes to greater than 20% for samples collected in evacuated collection tubes.

Apart from influencing measured drug concentrations, collection tubes may also be a source of contaminants which can appear on the chromatogram, making determination of the compound of interest difficult or impossible. Shang-Qiang and Evenson measured by mass spectrometry a range of esters and other impurities in a number of commercial serum separator tubes which they identified as chemicals used in the preparation of the polyester gel [239]. Other workers found that contaminants from plastic collection tubes caused interference with the analysis of cyclosporin in whole blood and plasma [240], and these interferents were not detected when various kinds of glass tubes were used for sampling. Shang-Qiang and Evenson also carried out extractions followed by GC analysis of stoppers from various commercial evacuated collection tubes and found a number of peaks, two of which they identified as tris(2-butoxyethyl)phosphate and 2,2'-methylenebis(4-ethyl-6-*tert*-butyl)phenol, an antioxidant sometimes used in rubbers and plastics.

Many other types of contaminants can be encountered in biological samples, including antioxidants, pesticides, food additives, and vulcanizing agents. The problem of these interferents can only be minimized by careful manipulation of the sample. The need for close liaison between the analyst and the person taking the sample is most important, and a typical problem which can arise as a result of lack of communication between the analyst and medical personnel is when contamination of a blood sample by a local anaesthetic occurs while a catheter is being inserted by a nurse or doctor.

Care must also be taken in the thawing of samples, since localized concentration effects can occur at this stage. Frozen samples of plasma or serum should be brought to room temperature gradually; heating is not recommended since it can promote degradation of the drug, and after thawing, the sample should be subjected to vortex mixing for 10 sec to ensure homogeneity prior to analysis. Centrifugation of thawed plasma samples is advisable before direct injection onto either an HPLC column or a pre-concentration column, since fibrin may have formed which could clog the column or tubing systems.

2.5.2.2 Analysis in other biological fluids

Urine analysis is useful where a drug or rapidly formed metabolite is extensively excreted into it, and drug metabolites can be detected in urine for quite some time after they have become undetectable in blood. Analysis of drugs in urine is widely used in bioavailability studies and it is usually done on a single or 24-hr specimen. Because urine does not usually contain protein or lipids, it is regarded as a cleaner matrix to work with, and is occasionally introduced without pre-treatment into analytical systems. Alternatively, it is well-suited to liquid or solid extraction since most of the endogeneous constituents it contains are ionic in nature. The problem is that the drug usually appears in the urine as its more polar (for example glucuronide) metabolite and it may need to be hydrolysed before partitioning into a solid liquid organic phase.

Both pH and volume are important factors in urine drug analysis and must be recorded immediately on collection. Urinary pH can have a pronounced effect on the excretion of certain drugs, since unfavourable pH may promote reabsorption of the drug in the kidney rather than excretion into the glomerular filtrate, since basic drugs are excreted more readily into acid urine, while acidic drugs are excreted more readily into alkaline urine. Urinary pH may be affected by factors such as medication, e.g. with antacids, which if absorbed into the system can render the urine alkaline. The volume of urine may be affected by diet, with respect to, for example, sodium intake; urinary volume will be reduced by drugs which cause fluid retention, and urinary flow is increased by diuretic drugs. These factors should be taken into account when measuring the concentrations of drugs and metabolites.

If urine is allowed to stand at room temperature, bacterial action causes the conversion of urea into ammonium carbonate, and thence to ammonia, which effects an increase in pH resulting in the precipitation of inorganic phosphates. Urine can be preserved by freezing at $-20°C$ or by the addition of a preservative. Commonly used preservatives are toluene, boric acid or concentrated hydrochloric acid, though freezing is preferable to the latter approach since the preservative can interfere with subsequent drug analysis. In emergency testing, urine analysis is a useful tool for detecting tricyclic antidepressants, since their urine concentration is much higher than that in blood [241]. The determination of drugs of abuse in urine is used in the screening of patients on detoxification programmes; for example opiates are measured in urine to check patient compliance with the programme, and to screen for other drugs of abuse [171].

Less commonly analysed biological media are saliva and cerebrospinal fluid (CSF). Drug concentrations in saliva are sometimes assumed to represent free plasma levels, but this was found to be true only for a limited number of drugs, such as carbamazepine and phenytoin [152]; for other drugs, the correlations are less satisfactory or apparently non-existent [241]. It is not practical to analyse CSF routinely, and such analyses are normally only carried out when damage to the blood–brain barrier is suspected.

2.5.3   Drug metabolites

Metabolism normally proceed in two stages which have been classified as phase I and phase II metabolism [242]. Phase I involves simple reactions such as oxidation, reduction or hydrolysis to yield a moiety suitable for phase II metabolism. A typical phase II reaction is the conjugation of a polar function to the drug or its phase I metabolite. The most common phase II reaction is the conjugation of glucuronic acid to various functional groups, forming *O*-, *S*-, and *N*-glucuronides, depending on the moiety through which the reaction proceeds. Glucuronides differ markedly in their stability in acid and alkaline conditions, and the analyst must be aware that certain vigorous sample preparation procedures may cause reversion to the parent drug, thus yielding anomalously high concentrations in the analysis stage. Other important phase II reactions are acetylation and reactions with sulphuric acid to form sulphates.

As mentioned previously, the metabolites of a drug are generally more polar than the parent compound, and the co-analysis of a drug and its metabolites presents a considerable challenge to the biopharmaceutical analyst. In order to select a suitable analytical method and devise a practical extraction, it is usually necessary to be aware of the metabolic pathway of the drug, the chemical structure of the metabolites, as well as of their expected concentrations. Determination of the metabolic profile of the drug may be accomplished by the administration of a radiolabelled form of the compound, followed by two-dimensional TLC of a suitable extract. After development, the plate may be contacted with a radiation-sensitive photographic plate, or the various zones may be scraped off and measured by scintillation counting. The zones can then be extracted and the structural characterization of the fraction it contains can be carried out by nuclear magnetic resonance, infra-red and/or mass spectrometry. Final structural confirmation is obtained by synthesis of the identified compound and the synthetic metabolites may be then used in the development of a suitable analytical methodology. Because this process is laborious and time-consuming, analysts will often seek to predict metabolic profiles in advance, a goal which is becoming increasingly realistic with the support of modern computer programs and expert-system technology.

The presence of drug metabolites in a specimen can generate misleading results if the analytical method does not incorporate some kind of separation step to discriminate between them and the parent compound. This is particularly true if the metabolite differs significantly in its pharmacological activity from the parent drug. If the metabolite is inactive, it may be desirable to simply exclude it from the assay for the original drug, but if it is equiactive, or even more potent than the parent drug, it is generally desirable to measure its concentration separately. The advantage of chromatography over other methods is that it can accommodate the separation step without undue refinement of the clean-up stages. (The procedures used in the extraction of metabolites are discussed more fully in Section 2.4.4).

HPLC offers the possibility of measuring intact drug conjugates without derivatization. For instance, paracetamol can be detected in its conjugated form by direct injection of urine, with UV detection [243]. Few drugs, however, are present in high enough concentrations to permit this approach, and it is more usual to hydrolyse conjugates and release the parent drug for extraction. This is done chemically with hydrochloric acid or sodium hydroxide, or enzymatically with enzymes such as glucuronidase or sulphatase. Chemical hydrolysis decreases the yield of labile compounds which are sensitive to aggressive reagents, for instance certain benzodiazepines. Enzymatic methods, on the other hand, provide mild conditions which are less likely to cause degradation of the drug in question, though sulphatase can only hydrolyse aryl sulphates [166], and as glucuronidase can be inhibited by high salt concentrations, it is advisable to dilute the sample and to run control experiments to ensure that the enzyme is still active. Generally, a urine sample is divided into two fractions, one of which is subjected to hydrolysis. The latter gives the total (conjugated and unconjugated) concentration following extraction, whereas the unhydrolysed fraction yields the proportion of drug which is non-conjugated; the amount of conjugated drug is obtained by subtraction. This procedure has, for example, been used in the determination of oxmetidine and its sulphoxide metabolite [244].

2.5.4 Protein binding
Since many drugs are protein-bound to some extent, protein binding is a factor which must be taken into consideration when measurements are made in biological fluids. Therefore, it is common in pharmacokinetic and clinical studies to differentiate between free, protein-bound, and total drug levels. Albumin is the most significant binding protein for many drugs, particularly neutral and acidic drugs such as warfarin, phenytoin and valproic acid [245]. Basic drugs such as quinidine, propranolol, and the tricyclic antidsepressants bind not only to albumin, but also to other proteins such as alpha-acid glycoprotein and lipoproteins [246–248]. The extent of protein binding varies enormously one drug to another; for instance, warfarin and dicoumerol are 99% protein-bound, whereas caffeine is hardly bound at all [249]. For many years it was assumed that only the free fraction is active and available to interact with the drug receptor site. This assumption, which was made on the basis that drugs bound to proteins could not traverse capillary walls and cell membranes, has been challenged in a number of studies which demonstrated the passage of glycoprotein-bound propranolol, lignocaine, and some hormones across membranes [250], and the transport of lignocaine and propranolol into brain tissue [251].

Of great importance, however, is the way in which the equilibrium binding is affected by factors such as disease states and the co-administration of other drugs. This is because, for highly bound drugs, any change in the degree of binding will represent a large percentage change in the amount of unbound drug. The pathological factors which contribute greatly to disturbances in protein binding include liver and kidney diseases. Liver diseases associated with hypoalbuminemia generally feature higher unbound concentrations for drugs such as propranolol [252], benzodiazepines [253] and verapamil [254], and this is mainly because there are physically fewer albumin molecules to bind the circulating drug molecules. Other diseases, such as nephrotic syndrome, generate endogenous substances which may compete with the drugs for the available binding sites [255].

An important factor which can radically affect the degree of protein binding is the co-administration of other drugs. The classic example of this occurrence is when the non-steroidal anti-inflammatory drug (NSAID) phenylbutazone causes the displacement of warfarin from protein-binding sites to which it is extensively bound. This displacement significantly increases the amount of free warfarin circulating in the blood and the concomitant increase in anticoagulant effect can result in fatal internal haemmorrhage. Other NSAIDs also displace warfarin from binding sites, but when used in therapeutic doses, do not produce a clinically significant increase in pharmacological response [256].

Apart from possible potentiation in therapeutic response, another implication of a disturbance in the degree of protein binding is the rate at which a drug is cleared from the body. The unbound fraction of some drugs is cleared by passive mechanisms in the kidney and liver, and if the percentage unbound is increased, the drug may be removed at a faster rate than is desirable, necessitating more frequent dosing and possibly leading to exaggerated peaks and troughs in the concentration–time profile.

It is usually the total concentration of drug in plasma or serum which is measured rather than the free fraction. Unless specialized techniques which exclude bound drug, such as equilibrium dialysis or ultrafiltration, are employed, the measurement of free drug is difficult since many of the widely-used extraction techniques cause at least partial liberation of drugs from their binding sites. However, provided that protein binding of drugs is normal and constant, both within and between patients in a particular study, free drug concentration is a reasonably constant fraction of total drug. This relationship does not hold where protein binding is abnormal or affected by drug interaction as outlined above, or where there is individual variation in the extent of binding, for instance with phenytoin, where the free fraction would appear to offer superior correlation with therapeutic effect and toxicity than the total drug concentration [257,258]. The clinical and bioanalytical implications of changes in protein binding patterns are extremely complex and the clinical significance of choosing to measure bound or unbound drug fractions has been dealt with in detail by Kwong [259].

Protein precipitation techniques will frequently liberate drugs from their binding sites, as will liquid–liquid extraction. Some protein-bound drugs, such as propranolol, will bind strongly to reversed-phase packing material used in solid-phase extraction columns, and thus good recoveries will be obtained by using this method of drug removal from plasma [204]. In cases where the drug is very strongly protein bound, the relatively mild extraction techniques outlined above will not release the drug from binding sites, and methods such as acid and enzyme digestion have been employed to obtain a measure of the total drug concentration. Acid digestion can lead to hydrolysis of the drug, and the search for a more general procedure has led to the use of the proteolytic enzyme, *subtilisin Carlsberg*. Osselton *et al.* [260,261] employed this enzyme for the release of antibiotics and benzodiazepines: the procedure simply involved incubation with the enzyme, after which the sample was filtered over a plug of glass wool and analysed by chromatography. The enzyme was found to release the drug completely without affecting chemical structures. The enzyme *subtilisin Carlsberg* has been used to degrade liver tissues at high pH [262]; this has led to good recovery for drugs, though it was unsuitable for substances which degrade under alkaline conditions. However, later work on *subtilisin Carlsberg* revealed that reaction can be carried out at lower pH values with considerable improvement in drug recovery [263].

Werkhoven-Goewie *et al.* [264] demonstrated the application of enzyme hydrolysis with *subtilisin*, followed by HPLC analysis using automated on-line pre-concentration of secoverine. After enzymatic hydrolysis for 15 min at 55°C, the drug was liberated from its protein binding sites, and 1-ml volumes of the hydrolysate were injected onto short (2–30 mm) pre-columns containing cyano packing material. Up to 50 ml of treated plasma could be injected without appreciable build-up in pre-column back-pressure or loss of performance. As enzymatic hydrolysis produces a large number of small peptides and amino acids, the method may require an additional step in the sample clean-up protocol, or a selective (for instance, fluorescence) detection system.

2.5.5 Internal standardization

Optimal reliability of analysis is achieved though the use of internal standardization. The role of an internal standard is to correct for variation in instrument response, injection volume, or extraction or derivatization yield. It is important to add the internal standard at the earliest possible stage in order to achieve maximum compensation for procedural variations. When the standard is added immediately prior to introducing the sample into the analytical system (that is, after extraction and other manipulations), it is no longer deemed to be 'internal' and is often referred to as an 'external standard'. An internal standard must fulfil a number of physical and chemical criteria: it should be structurally similar to the drug, have a similar pK_a, exhibit the same partitioning behaviour, convert to derivatives to a similar extent and have similar volatility (for GC). In monophasic dilution (such as protein precipitation), the internal standard must only be soluble in the dilution solvent, and its selection then depends on analytical parameters. In liquid–liquid or liquid–solid extraction, the behaviour of the internal standard in the extracting liquid or solid is important. In chromatography, the internal standard should elute clear of any endogenous interferents, and close to, but clearly resolved from, the drug.

Most internal standards are compounds chemically similar to the drug to be assayed, and in general, the more its chemical structure resembles that of the drug, the better the control of variation achieved. It is essential also that the internal standard and the drug have similar responses in the detection scheme employed. For instance, in the highly sensitive electrochemical detection of imipramine at an oxidation potential of 100 mV, the use of nortriptyline as an internal standard is precluded because it is not oxidized at this potential [171]. It is important to note, however, that the use of an internal standard belonging to the same chemical class as the drug of interest can present more problems than is commonly assumed. Even when there are close structural similarities, such as among chemical homologues, pronounced differences in physicochemical behaviour may arise. In addition, while it is desirable that the internal standard belongs to the same chemical group as the drug being monitored, it is inadvisable to use a compound which is likely to be a metabolite of the drug *in vivo*. Structures to be avoided would include desmethyl analogues and a metabolite where an amine replaces a nitro group on the original compound. It is also important to avoid commonly prescribed drugs, or drugs likely to be co-administered with the compound of interest. Care must also be exercised in the selection of an internal standard where the patient's intake of medication is not closely monitored, especially in toxicology where little or no information can be collected about ingested drugs. The amount of internal standard introduced during the analysis must be selected so that the detector response-ratio falls within one range of the calibration curve. An upper limit of 2 is generally chosen as the maximum response ratio [44]. Ideally, the internal standard is added in equal volumes and concentrations to all samples and standards used in the preparation of a calibration curve. This would include a drug-free standard (the control), whereas the 'blank' is free of both standard and drug, but is spiked with the solvent in which the drugs are made up. The ratio of the detector response (peak height or area) for the drug and the internal standard is then used in the calibration and assay.

2.5.6   Calibration and evaluation

Samples of drug-free material (serum, urine) spiked with known drug concentrations should, where possible, be employed for calibration, though this is sometimes not possible in the case of tissue samples such as liver or muscle. Alternative external standardization based on extracting and analysing, or even simply measuring authentic drug standards, ignore the specific peculiarities of the biological matrix; for instance protein-binding would not be accounted for by using external standardization. However, even spiked plasma or serum standards are subject to criticism, since a drug added to blank matrix material might not be present in the same physicochemical state as the compound *in vivo* [44]. It follows that blank plasma used in the calibration should have been prepared with the same anticoagulant as that employed in taking patient samples: as outlined previously the type of anticoagulant can affect not only the appearance of interfering peaks, but also the amount of drug recovered from plasma.

In chromatography, peak-height or peak-area ratios (compound to internal standard) are usually plotted against concentration. The calibration graphs are calculated by linear regression analysis, but in order to avoid inaccuracies near the origin, weighted regression analysis should be performed. This follows because in simple linear regression of y upon x it is assumed that there is no variance in x and a constant variance in y [265].

Several criteria need to be evaluated in order to check the reliability and the overall performance of an assay. Most regulatory agencies, like the Food and Drugs Administration, have mandated Good Laboratory Practice regulations for clinical and non-clinical laboratory studies [266,267]. The parameters which are used to evaluate the overall performance of an assay include the drug stability, specificity, limit of detection, accuracy, linearity, and recovery.

Drugs stored under different conditions of heat, light, humidity and pH should be examined for possible decomposition. If the drug is to be stored for an appreciable length of time, it is necessary to establish its stability under prolonged storage conditions. The kinds of problems that can arise have been discussed previously, and the extent of any loss must be established before commencing the assay. If the amount of drug lost is not reproducible, it may be necessary to alter the analytical protocol.

It is necessary to determine whether endogenous compounds in the biological matrix will interfere with either the drug signal or that of the internal standard. This is readily achieved by taking drug-free plasma (serum, urine, etc) through the sample work-up and analytical processes. This procedure should be repeated with each batch during both calibration and assay of unknown samples, in case of changes in column performance or the development of artefacts which could have caused the appearance of peaks co-eluting with the drug or the internal standard. Drugs which are therapeutically combined with the analyte drug (for instance, antihypertensives with diuretics) should be checked for possible interference or cross-reactivity. Since caffeine is a common source of interference, it is usual for patients involved in a clinical trial to be given a caffeine-free diet. However, this is not always practicable, and if the samples to be analysed are likely to contain this compound, it may be necessary to exclude caffeine as a potential interferent.

A systematic search should be undertaken to locate the origin of any interference. It may, as mentioned above, arise from the diet, or it may be introduced from the equipment used in the assay. If it proves impossible to remove the source of an interference (for instance a vital co-administered drug), it may be possible to eliminate the unwanted signal by selective extraction, more selective detection schemes, or by revision of the analytical parameters.

The limit of detection (LOD) is defined as the lowest concentration of an analyte that the analytical system can reliably detect. In mathematical terms it may be expressed as 3σ above the gross blank signal [268], where σ is the standard deviation of the peak-to-peak noise. If there is a linear dependence of signal (S) on concentration (C), the following equation may be written:

$$d_{C_x}/C_x = d_{S_x}/S_x$$

where S_x is the signal corresponding to concentration C_x, and d_{S_x} is the signal variability corresponding to concentration variability d_{C_x}.

The limit of detection may vary from day to day, as the result of a change in detector response, for instance. However, once the LOD has been statistically defined, data obtained below this value are not valid and should not, therefore, be reported.

Assessment of accuracy is a fundamental problem because the true value can never be known with absolute certainty. The term 'accuracy' denotes the nearness of a measurement to its true value and is expressed in terms of error [269]. The accepted error is the difference between the observed and the true value. The relative error is expressed as a percentage of the true value, and is often used to express the accuracy of chromatographic assay. It is common to test the accuracy of the method in hand by comparing it with another reliable method. Chromatographic methods are frequently compared with immunoassay techniques, and the correlation between the two is taken as an index of the accuracy of the method which is being validated.

Precision or reproducibility of an assay is defined as the coefficient of variation (relative standard deviation) of the results at a certain drug concentration. In chromatographic methods of drug analysis, the reproducibility of a method should be determined on the basis of within-day (intra-assay), and between-day (inter-assay) results. In performing both inter- and intra-assays, usually four replicate analyses over the entire concentration range should be carried out. For intra-assay, the mean ratios of drug to internal standard are plotted as a function of concentration, to generate a regression curve. The individual ratios are then interpolated as unknowns on this curve to obtain new values of concentration. The mean and standard deviation are calculated, and precision is expressed in terms of the overall percentage coefficient of variation. For inter-assay, individual regression lines are generated for each set of data, and each set is then interpolated on its own regression line. The new values of concentration thus obtained are then used to determine inter-assay precision.

The linearity of the assay is defined by linear regression analysis of replicates of spiked biological standards in the expected concentration range of the drug to be assayed. Linearity is frequently quoted in terms of the correlation coefficient of the single regression line used for intra-assay calculations. If the signal-to-concentration function is not linear over the entire concentration range, as is sometimes the case where a wide range of concentrations is to be covered, it may be necessary to split the calibration curve into two portions to permit determination of unknown samples. With some techniques, non-linear calibration graphs are frequently obtained. Some workers propose special approaches which extend the linear range artificially [270], but much greater accuracy can be obtained by calculating the actual non-linear calibration graph by using weighted polynomial regression analysis [44].

The extraction recoveries of the drug and internal standard provide useful information and can be calculated either by running isotopically labelled compounds through the procedure, or by assaying drug-supplemented plasma (serum, urine) and comparing the peak-heights of the extracted standards with those of authentic (non-extracted) standards at the same concentrations. By generating a calibration curve for both extracted and non-extracted standards, the ratio of the two slopes can be obtained and used as a second measure of recovery.

Routine drug assays should be checked regularly for reliability using internal quality control schemes. This involves the preparation of plasma (serum, urine) pools to which defined amounts of drug are added. Whether it is advisable to inform the analyst of the concentrations in the quality control standards is a matter for some speculation. If the concentrations are unknown to the analyst, the process may be perceived to be a more rigorous and objective test of the system, but this policy may lead the analyst to believe himself to be under scrutiny rather than the analytical method, possibly undermining his self-confidence and ability to work efficiently. With each set of determinations on patient samples, a control sample is also analysed, and the result is interpreted in terms of established limits of variation. It is recommended that at least 10% of the total samples should consist of quality control standards interspersed amongst the patient samples and calibration standards. If the results of the quality control standards are not within 10% of their expected values, it is normal practice to repeat the batch [271].

Inter-laboratory surveys on a similar basis are currently organized by different institutions, and are often imposed by law. Control samples are distributed to the laboratories participating in such an external quality control programme and the results are statistically evaluated. In order for this type of system to be successful, it is necessary that equipment, materials and operating conditions are identical in each of the participating laboratories. In practice it can be difficult to realize this goal, since there is significant variation between batches of even the same brand and type of column, and differences in equipment performance can contribute to inter-laboratory variation in results.

REFERENCES

[1] J. Kricka, *Anal. Proc.*, 1983, **20**, 163.
[2] J. Hirtz, *Pharm. Drug Disp.*, 1986, **7**, 315.
[3] R. V. Smith, *TrAC*, 1984, **3**, 178.
[4] R. A. A. Maes and J. G. Leferink, *TrAC*, 1982, **1**, 228.
[5] J. Kochweser, in A. Richens and V. Marks, eds., *Therapeutic Drug Monitoring*, Churchill Livingstone, Edinburgh, 1981.
[6] J. K. Aronson, D. G. Grahame-Smith and F. M. Wigley, *Quart. J. Med.*, 1978, **186**, 111.
[7] S. H. Curry, *TrAC*, 1986, **7**, 315.
[8] V. Marks., *Ann. Clin. Biochem.*, 1979, **16**, 370.
[9] AA. D. Munro-Faure, J. Fox, A. S. E., Fowle, B. F. Johnson and S. Lader, *Postgrad. Med. J.*, 1974, **50**, Suppl. 6, 14.
[10] S. Kalman and D. R. Clark, *Drug Assay: The Strategy of Therapeutic Monitoring*, Masson, New York, 1979.
[11] P. Trinder, *Biochem. J.*, 1954, **57**, 301.
[12] N. S. Omer, I. G. Thwainey, N. H. Rasheed, S. Hussain and N. H. Borazan, *Iraqi Sci.*, 1989, **30**, 1.
[13] P. A. D. Edwardson, J. D. Nichols and K. Sugden, *J. Pharm. Biomed. Anal.*, 1989, **7**, 287.
[14] A. S. Dhake, N. N. Singh, A. Y. Nimbkar and H. Tipnis, *Indian Drugs*, 1989, **26**, 190.
[15] E. B. Muzhanovskii, A. F. Fartushnyi, A. I. Desov and A. M. Muzhanovskaya, *Farm. Zh. (Kiev)*, 1988, **5**, 47.
[16] R. Huang and X. Xu, *Yaoxue Xeubao*, 1989, **24**, 37.
[17] A. Amlathe and V. K. Gupta, *J. Indian Chem. Soc.*, 1989, **66**, 359.
[18] J. B. McConnell, H. Smith, M. Davis and R. Williams, *Br. J. Clin. Pharmacol.*, 1979, **8**, 506.
[19] L. Przyborowski, P. Zawisza and A. Wojtczak, *Farm Pol.*, 1988, **44**, 466.
[20] G. Misztal, L. Pryborowski and A. Smajkiewicz, *Chem. Anal. (Warsaw)*, 1988, **33**, 149.
[21] E. Dunckova, J. Cizmarik and V. Springer, *Farm. Obz.*, 1989, **43**, 55.
[22] A. F. Fell, *Anal. Proc.*, 1982, **19**, 398.
[23] I. M. Warner, J. B. Callis, E. Davidson, R. Gouterman and G. D. Christian, *Anal. Lett.*, 1975, **8**, 665.
[24] D. W. Johnson, J. B. Challis and G. D. Christian, *Anal. Chem.*, 1977, **49**, 747A.
[25] A. Savitsky and M. J. E. Golay, *Anal. Chem.*, 1964, **36**, 1627.
[26] J. Steiner, Y. Termonia and J. Deeltour, *Anal. Chem.*, 1972, **44**, 1906.
[27] D. E. Metzler, C. E. Harris, R. L. Reeves, W. H. Lawton and M., S. Maggio, *Anal. Chem.*, 1977, **49**, 864A.
[28] V. Knaub and V. A. Kartashov, *Farmatsiya*, 1989, **38**, 46.
[29] V. J. Hammond and W. C. Price, *J. Opt. Soc. Am.*, 1953, **43**, 924.
[30] A. T. Geise and G. S. French, *Appl. Spectrosc.*, 1955, **9**, 78.
[31] W. L. Butler and J. W. Hopkins, *Photochem. Photobiol.*, 1970, **12**, 451.
[32] G. Talsky, L. Mayring and H. Kreuzer, *Angew. Chem. Int. Ed. Engl.*, 1978, **17**, 785.
[33] A. F. Fell, *Proc. Anal. Div. Chem. Soc.*, 1978, **15**, 260.
[34] T. C. O'Haver, *Anal. Chem.*, 1979, **51**, 91A.
[35] A. F. Fell, *UV Spectrom. Group Bull.*, 1980, **8**, 5.
[36] D. Martinez and M. Paz Giminez, *J. Anal. Toxicol.*, 1981, **5**, 10.
[37] A. J. McBay, *Clin. Chem.*, 1987, **33**, 11(B), 33B.
[38] M. Lesne, *J. Pharm. Biomed. Anal.*, 1983, **1**, 415.
[39] A. F. Fell, D. R. Jarvie and M. J. Stewart, *Clin. Chem.*, 1981, **27**, 286.
[40] A. F. Fell, *Anal. Proc.*, 1982, **19**, 398.
[41] R. A. de Zeeuw, *J. Pharm. Biomed Anal.*, 1983, **1**, 435.
[42] S. H. Y. Wong, *J. Pharm. Biomed. Anal.*, 1989, **7**, 1011.
[43] R. S. Yalow and S. A. Berson, *Nature (London)*, 1959, **184**, 1648.
[44] A. P. DeLeenheer and H. J. C. F. Nelis, *Analyst*, 1981, **106**, 1025.
[45] P. Jatlow, *Human Pathology*, 1984, **15**, 404.
[46] B. K. Van Weeman and A. H. Schurs, *FEBS Lett.*, 1971, **15**, 232.

[47] A. Voller, D. Bidwell, G. Huldt and E. Engvall, *Bull. W.H.O.*, 1974, **51**, 209.

[48] K. E. Rubenstein, R. S. Schneider and E. F. Ullman, *Biochem. Biophys. Res. Commun.*, 1972, **47**, 846.

[49] E. F. Ullman, R. A. Yoshida, J. I. Blackmore, E. Maggio and R. Leute, *Biochem. Biophys. Acta*, 1979, **567**, 66.

[50] W. D. Odell and W. D. Daughada, *Principles of Competitive Protein Binding Assay*, J. P. Lippincott, New York, 1971.

[51] J. A. Miller, *Science News*, 1978, **14**, 444.

[52] M. Tswett, *Proc. Warsaw Soc. Nat. Sci.(Biol.)*, 1903, **14**, 6.

[53] A. J. P. Martin and R. L. M. Synge, *Biochem. J.*, 1941, **50**, 1358.

[54] J. T. James and A. J. P. Martin, *Biochem. J.*, 1952, **50**, 679.

[55] A. C. Giddings, *Dynamics of Chromatography, Part I: Principles and Theory*, Dekker, New York, 1965.

[56] C. Horvath, B. A. Preiss and S. R. Lipsky, *Anal Chem.*, 1967, **39**, 1422.

[57] J. F. K. Huber, *J. Chromatogr. Sci.*, 1969, **7**, 85.

[58] J. J. Kirkland, *J. Chromatogr. Sci.*, 1969, **7**, 7.

[59] R. M. Caprioli, J. G. Liehr and W. E. Siefert, in G. R. Waller and O. C. Dermer (eds), *Biochemical Applications of Mass Spectrometry*, First Supplementary Volume, Wiley, New York, 1980, p. 33.

[60] H. J. Kupferberg, *Clin. Chem. Acta*, 1970, **29**, 283.

[61] B. L. Karger, L. R. Snyder and C. S., Horvath, *An Introduction to Separation Science*, Wiley, New York, 1973.

[62] N. H. C. Cooke and I. Olsen, *J. Chromatogr. Sci.*, 1980, **18**, 512.

[63] C. F. Poole and S. A. Schuette, *Contemporary Practice of Chromatography*, Elsevier, Amsterdam, 1984.

[64] R. D. Rasmussen, W. H. Yokoyama, S. G. Blumenthal, D. E., Beergstrom and B. H. Reubner, *Anal. Biochem.*, 1980, **101**, 66.

[65] D. L. Massart and L. Buydens, *J. Pharm. Biomed Anal.*, 1988, **6**, 535.

[66] D. Dadgar and A. Power, *J. Chromatogr.*, 1987, **416**, 99.

[67] A. Power and D. Dadgar, *Anal. Proc.*, 1986, **23**, 416.

[68] D. Dadgar and A. Power, *J. Chromatogr.*, 1987, **421**, 216.

[69] E. Stahl, ed., *Thin Layer Chromatography*, Springer-Verlag, New York, 1969.

[70] J. Chamberlain, *Analysis of Drugs in Biological Fluids*, CRC Press, Florida, 1986.

[71] H. Knauf and E. Mutschler, *Eur. J. Clin. Pharmacol.*, 1984, **26**, 513.

[72] B. Brady, *Irish Chemical News*, 1990 (winter), 20.

[73] S. M. Han and N. Purdie, *Anal. Chem.*, 1984, **56**, 2822.

[74] B. J. Clarke and J. E. Mama, *J. Pharm. Biomed. Anal.*, 1989, **7**, 1883.

[75] D. R. Taylor, *Lab. Pract.*, 1986 (Jan.), 45.

[76] G. Gubitz, *J. Liq. Chromatogr.*, 1986, **9**, 425.

[77] V. M. Meyer, *Practical High Performance Liquid Chromatography*, Wiley, Chichester, 1988.

[78] D. J. Cram, G. Dotsevi and Y. Sogah, *J. Am. Chem. Soc.*, 1975, **97**, 1259.

[79] J. Zukowski and R. Nowakowski, *J. Liq. Chromatogr.*, 1989, **12**, 1545.

[80] W. H. Pirkle and J. M. Finn in J. D. Morrisson, ed., *Asymmetric Synthesis*, Vol. 1, pp. 87–224, Academic Press, New York, 1983.

[81] I. W. Wainer, T. D. Doyle, K. H. Donn and J. R. Powell, *J.Chromatogr.*, 1984, **306**, 405.

[82] E. Klesper, A. H. Corwin and D. A. Turner, *J. Org. Chem.*, 1962, **27**, 700.

[83] H. H. Hill, Jr., and M. M. Gallagher, *J. Microcolumn Sep.*, 1990, **2**, 114.

[84] H. H. Hill, Jr., and C. B. Shumate, *Chromatogr. Sci.*, 1989, **45**, 267.

[85] R. J. Wall, *J. Chromatogr. Anal.*, 1989, **3**, 16.

[86] D. J. Bornhop and J. G. Wangsgaaard, *J. Chromatogr. Sci.*, 1989, **27**, 293.

[87] S. Kueppers, B. Lorenschat, F. P. Schmitz and E. Klesper, *J. Chromatogr.*, 1989, **475**, 85.

[88] H. C. K. Chang and L. T. Taylor, *J. Chromatogr. Sci.*, 1990, **28**, 29.

[89] M. Lafosse, P. Rollin, C. Elfakir, L. Morin-Allory, M. Martens and M. Dreux, *J. Chromatogr.*, 1990, **505**, 191.

[90] M. A. Morrissey, W. F. Siems and H. H. Gill, jun., *J. Chromatogr.*, 1990, **505**, 215.

[91] C. R. Yonker, R. W. Gale and R. D. Smith, *J. Chromatogr.*, 1986, **371**, 83.

[92] S. M. Fields, K. E. Markides and M. L. Lee, *J. Chromatogr.*, 1987, **406**, 223.

[93] J. M. Levy and W. M. Ritchey, *J. Chromatogr. Sci.*, 1986, **24**, 242.

[94] W. Steuer, M. Schindler, G. Schill and F. Erni, *J. Chromatogr.*, 1988, **447**, 287.

[95] D. W. Later, B. E. Richter, D. E. Knowles and M. R. Anderden, *J. Chromatogr. Sci.*, 1988, **24**, 249.

[96] J. L. Janicot, M. Caude and R. Rosset, *J. Chromatogr.*, 1988, **437**, 351.
[97] R. M. Smith and M.. M. Sangai, *J. Pharm. Biomed Anal.*, 1988, **6**, 837.
[98] M. Ashrav-Khorassani, L. T. Taylor and P. Zimmermann, *Anal. Chem.*, 1990, **62**, 1117.
[99] L. J. Mulcahey and L. T. Taylor, *J. High Res. Chromatogr.*, 1990, **13**, 393.
[100] W. Steuer, J. Baumannn and F. Erni, *J. Chromatogr.*, 1990, **500**, 469.
[101] M. Petersen, *J. Chromatogr.*, 1990, **505**, 3.
[102] H. G. Janssen and C. A. Cramers, *J. Chromatogr.*, 1990, **505**, 19.
[103] M. Ashraf-Khorassani, M. G. Fessahale, L. T. Taylor, T. A. Berger and J. F. Deye, *J. High Res. Chromatogr., Chromatogr. Commun.*, 1988, **11**, 287.
[104] S. Hara, A. Dobashi, K. Kinoshita, T. Hondu, M. Saito and M. Senda, *J. Chromatogr.*, 1986, **371**, 153.
[105] P. Macaudiere, M. Caude, R. Rosset and A. Tambute, *J. Chromatogr.*, 1987, **405**, 135.
[106] P. Macaudiere, M. Caude, R. Rosset and A. Tambute, *J. Chromatogr. Sci.*, 1989, **27**, 383.
[107] R. D. Smith, J. L. Fulton, H. K. Jones, R. W. Gale and B. W. Wright, *J. Chromatogr. Sci.*, 1989, **27**, 309.
[108] T. L. Chester and J. D. Pinkston, *Anal. Chem.*, 1990, **62**, 394R.
[109] M. Ashraf-Khorassani, S. Shah and L. T. Taylor, *Anal. Chem.*, **62**, 1173.
[110] S. H. Y Wong, *Clin. Chem.*, 1989, **35**, 1293.
[111] A. Giorgetti, N. Pericles, H. M. Widmer, K. Anton and P. Daetwyler, *J. Chromatogr. Sci.*, 1989, **27**, 318.
[112] J. R. Wheeler and M. E. McNally, *Res. Dev.*, 1989, **32**, 134.
[113] S. F. Y. Li, *J. Chem. Technol. Biotechnol.*, 1989, **46**, 1.
[114] D. R. Luffer, W. Ecknig and M. Novotny, *J. Chromatogr.*, 1990, **505**, 45.
[115] R. M. Smith and M. M. Sanagi, *J. Chromatogr.*, 1990, **505**, 147.
[116] R. D. Smith, B. W. Wright and C. R. Yonker, *Anal. Chem.*, 1988, **60**, 1323A.
[117] W. M. A. Niessen, U. R. Tjaden and J. Van Der Greef, *J. Chromatogr.*, 1989, **492**, 167.
[118] H. Chi, Y. Wang, T. Chou and C. Jin, *Anal. Chim. Acta*, 1990, **235**, 273.
[119] M. R. Hackman, M. A. Brooks, J. A. F. De Silva and T. A. Ma, *Anal. Chem.*, 1973, **45**, 263.
[120] M. A. Brooks and M. R. Hackman, *Anal. Chem.*, 1975, **47**, 2059.
[121] J. R. Barreira Rodrigues, A. Costa Garcia and P. Tunon Blanco, *Analyst*, 1989, **114**, 939.
[122] J. M. Clifford in E. Reid, ed., *Assay of Drugs and other Trace Compounds in Biological Fluids*, North Holland, Amsterdam, 1976.
[123] H. Siegerman in A. J. Bard, ed., *Electroanalytical Chemistry*, Dekker, New York, 1979.
[124] M. A. Brooks, J. A. F. De Silva and L. M. D`Arconte, *Anal. Chem.*, 1973, **45**, 263.
[125] K. M. Kadish and V. R. Spiehler, *Anal.. Chem.*, 1975, **47**, 1714.
[126] J. Wang, *Electronalytical Techniques in Clinical Chemistry and Laboratory Medicine*, VCH, New York, 1988.
[127] M. Telting-Diaz, A. J. Miranda Ordiers, A. Costa Garcia, P. Tunon Blanco, D. Diamond and M. R. Smyth, *Analyst*, **115**, 1215.
[128] J. Wang, P. Tuzhi, M. S. Liin andd T. Tapia, *Talanta*, 1986, **33**, 707.
[129] J. Wang, M. S. Lin and V. Villa, *Anal. Lett.*, 1986, **19**, 2293.
[130] J. Wang, J. S. Mahmoud and P. A. M. Farias, *Analyst*, 1985, **110**, 855.
[131] R. Kalvoda, *Anal. Chim. Acta*, 1984, **162**, 197.
[132] A. Taira and D. E. Smith, *J. Assoc. Off. Anal. Chem.*, 1978, **61**, 941.
[133] J. Wang, P. A. M. Farias and J. S. Mahmoud, *Analyst*, 1986, **111**, 837.
[134] E. N. Chaney and R. P. Baldwin, *Anal. Chem.*, 1982, **54**, 2556.
[135] U. Forsman, *Anal. Chim. Acta*, 1983, **146**, 71.
[136] J. A. Squella, E. Barnafi, C. Perna and L. J. Ninez-Vergara, *Talanta*, 1989, **36**, 363.
[137] A. Miranda Ordiers, A. Costa-Garcia, J. M. Fernandez-Alvarez and P. Tunon Blanco, *Anal. Chim. Acta*, 1990, **233**, 383.
[138] D. W. Armstrong and S. J. Henry, *J. Liq. Chromatogr.*, 1980, **3**, 657.
[139] F. J. DeLuccia, M. Arunyanart and L. J. Cline-Love, *Anal. Chem.*, 1985, **36**, 1564.
[140] M. Arunyanart and L. J. Cline-Love, *J. Chromatogr.*, 1985, **342**, 293.
[141] F. J. DeLuccia, M. Arunyanart, P. Yarmchuck, R. Weinberger and L. J. Cline-Love, *LC-GC.*, 1985, **3**, 794.
[142] J. G. Dorsey, *Adv. Chromatogr.*, 1987, **27**, 167.
[143] L. P.. Stratton, J. B. Hynes, D. G. Priest, M. T. Doig, D. A. Barron and G. L. Asleson, *J. Chromatogr.*, 1986, **357**, 183.
[144] J. Haginaka, J. Wakai, H. Yasunda and T. Nagagawa, *Anal. Chem.*, 1987, **59**, 2732.
[145] G. R. Granneman and L. T. Senello, *J. Chromatogr.*, 1982, **229**, 149.
[146] A. Berthod, A. N. Asensio and J. J. Laserna, *J. Liq. Chromatogr.*, 1989, **2**, 2621.
[147] I. H. Hagestam and T. C. Pinkerton, *Anal. Chem.*, 1985, **57**, 1757.

[148] T. Nakagawa, A. Shibukawa, N. Shimono, T. Kawashima, H. Tanaka and J. Haginaka, *J. Chromatogr. Biomed. Appl.*, 1987, **64**, (*J. Chromatogr.*, **420**), 297.

[149] A. Shibukawa, T. Nakagawa, M. Migake, N. Nishimura and H. Tanaka, *Chem. Pharm. Bull.*, 1989, **37**, 1311.

[150] F. Li, C. K. Lim and T. J. Peters, *Biochem. J.*, 1986, **239**, 481.

[151] Z. K. Shibahi and R. D. Dyer, *J. Liq. Chromatogr.*, 1987, **36**, 2383.

[152] A. C. Mehta, *Talanta*, 1986, **33**, 67.

[153] R. W. Geise, *Clin. Chem.*, 1983, **36**, 1331.

[154] J. Blanchard *J. Chromatogr.*, 1981, **226**, 455.

[155] J. C. Mathies and M. A. Austin, *Clin. Chem.*, 1980, **26**, 1760.

[156] C. K. Lim and T. J. Peters, *J. Chromatogr.*, 1984, **316**, 397.

[157] A. M. Rustum, A. Rahman and N. E. Hoffman, *J. Chromatogr. Biomed. Appl.*, 1987, **65** (*J. Chromatogr.*, 421) 418.

[158] M. Somogyi, *J. Biol. Chem.*, 1945, **160**, 69.

[159] S. Lam and G. Malikin, *J. Liq. Chromatogr.*, 1989, **12**, 1851.

[160] S. J. Van der Wal and L. R. Snyder, *Clin. Chem.*, 1981, **27**, 1233.

[161] P. J. Meffin and J. O. Miners, in J. W. Bridges and L. F., Chausseaud, eds., *Progress in Drug Metabolism*, Wiley, New York, 1980.

[162] D. Jung, M. Meyersohn and D. Perrier, *Clin. Chem.*, 1981, **27**, 166.

[163] W. A. Joern, *Clin. Chem.*, 1981, **27**, 417.

[164] R. L. G. Norris, J. T. Ahokas and P. J. Ravenscroft, *J. Pharmacol. Meth.*, 1982, **7**, 7.

[165] H. Varley, A. H. Gowenlock and M. Bell, *Practical Clinical Biochemistry*, Heineman, London, 1976.

[166] R. Gill and A. C. Moffat, *Anal. Proc.*, 1982, **19**, 170.

[167] D. Chin and E. Fastlich, *Clin. Chem.*, 1974, **20**, 1382.

[168] D. M. Bailey and M. Kelner, *J. Anal. Toxicol.*, 1984, **8**, 26.

[169] S. M. Johnson, C. Chan, S. Cheng, J. L. Shimek, G. Nygard and S. K. Wahbba, *J. Pharm. Sci.*, 1982, **71**, 1027.

[170] R. Gill, A. A. T. Lopez and A. C. Moffat, *J. Chromatogr.*, 1981, **226**, 117.

[171] M. Van Damme, M. Leopold and A. K. Faouzi, *Clin. Toxicol.*, 1985–86, **23**, 589.

[172] L. D. Bower, M. L. Parsons, R. E. Clement, G. A. Eiceman and F. W. Karasek, *J. Chromatogr.*, 1981, **206**, 279.

[173] J. Ramsey, and D. B. Campbell, *J. Chromatogr.*, 1971, **63**, 303.

[174] M. K. Brunson and J. F. Nash, *Clin. Chem.*, 1975, **21**, 1956.

[175] G. Schill, in E. Reid, ed., *Assay of Drugs and Other Trace Compounds in Biological Fluids*, North Holland, Amsterdam, 1976.

[176] E. Tomlinson, *J. Pharm. Biomed. Anal.*, 1983, **1**, 11.

[177] M. G. Horning, P. Gregory, J. Nowlin, M. Stafford, K. Lertratanangkoon, C. Butler, W. G. Stillwell and R. M. Hill, *Clin. Chem.*, 1974, **20**, 282.

[178] K. E. Brooks and N. B. Smith, *Clin. Chem.*, 1989, **35**, 2100.

[179] D. Dadgar and M. Kelly, *Analyst*, 1988, **113**, 229.

[180] M. T. Kelly and J. F. Bloomfield, Manuscript in preparation.

[181] E. Reid and J. P. Leppard, eds., *Drug Metabolite Isolation and Determination*, Plenum Press, New York, 1983.

[182] G. G. Skellern, *Analyst*, 1981, **106**, 1071.

[183] L. E. Martin and E. Reid in J. W. Bridges and L. F., Chasseaud, eds., *Progress in Drug Metabolism*, Wiley, New York, 1981.

[184] L. R. Snyder, *Anal. Chim. Acta*, 1980, **114**, 3.

[185] J. W. Dolan, S. van der Wal, S. J. Bannister and L. R. Snyder, *Clin. Chem. Acta*, 1980, **114**, 3.

[186] J. C. Kraak, *TrAC.*, 1983, **2**, 183.

[187] R. J. Allan, H. T. Goodman and T. R. Watson, *J. Chromatogr.*, 1980, **183**, 311.

[188] T. C. Kwong, R. Martinez and J. M. Keller, *Clin. Chim. Acta*, 1982 **126**, 203.

[189] R. J. Ruane and I. D. Wilson, *J. Pharm. Biomed. Anal.*, 1987, **5**, 723.

[190] J. P. Anhalt and S. D. Brown, *Clin. Chem.*, 1978, **24**, 1940.

[191] S. K. Maitra, T. T. Yoshikawa and J. L. Hansen, *Clin. Chem.*, 1977, **23**, 2275.

[192] K. R. Lim, Y. J. Lee, H. S. Leee and A. Zlatkis, *J. Chromatogr.*, 1987, **400**, 285.

[193] D. Roberts, R. J. Ruane and I. D. Wilson, *J. Pharm. Biomed. Anal.*, 1989, **7**, 1077.

[194] B. C. Thompson, J. M. Kuzmark, D. W. Law and J. T. Winslow, *LG–GC*, **7**, 846.

[195] M. Bozguz, J. Bialka and J. Gierz, *Z. Rechtsmed.*, 1981, **87**, 287.

[196] N. Elahai, *J. Chromatogr.*, 1982, **20**, 483.

[197] Y. H. Caplan, R. C. Backer, M. Stajic and B. C. Thompson, *J. Forensic. Sci.*, 1979, **24**, 745.

[198] T. Ishida, K. Oguri and H. Yoshimura, *J. Pharmacobio.-Dyn.*, 1982, **5**, 521.
[199] R. A. Hux, H. Y. Mohammed and F. F. Cantwell, *Anal. Chem.*, 1982, **54**, 113.
[200] E. L. Johnson, D. L. Reynolds, D. S. Wright and L. A. Pachia, *J. Chromatogr. Sci.*, 1988, **26**, 372.
[201] E. L. Johnson, L. A. Pachia and D. L. Reynolds, *J. Pharm. Sci.*, 1988, **75**, 1003.
[202] R. D. McDowall, J. C. Pearce and G. S. Murkitt, *J. Pharm. Biomed. Anal.*, 1986, **4**, 3.
[203] R. D. McDowall, J. C. Pearce and G. S. Murkitt, *TrAC.*, 1989, **8**, 134.
[204] M. T. Kelly, M. R. Smyth and D. Dadgar, *Analyst.*, 1989, **114**, 1493.
[205] M. T. Kelly, M. R. Smyth and D. Dadgar, *J. Chromatogr.*, 1989, **473**, 53.
[206] W. Roth, K. Beschke, R. Jauch, A. Zimmer and F. W. Koss, *J. Chromatogr.*, 1981, **222**, 13.
[207] D. R. Deans, *Chromatographia*, 1981, **203**, 3.
[208] L. R. Snyder, J. W. Dolan and S. T. Van der Wal, *J. Chromatogr.*, 1981, **203**, 3.
[209] A. Nazareth, L. Jaramillo, B. L. Karger and R. W. Geise, *J. Chromatogr.*, 1984, **309**, 357.
[210] J. L. Hedrick and L. T. Taylor, *J. High Res. Chromatogr.*, 1990, **13**, 312.
[211] B. W. Wright, C. W. Wright, R. W. Gale and R. D. Smith, *Anal. Chem.*, 1987, **59**, 38.
[212] S. B. Hawthorne, D. J. Miller and J. J. Langenfeld, *J. Chromatogr. Sci.*, 1990, **28**, 2.
[213] J. W. King, *J. Chromatogr. Sci.*, 1990, **28**, 9.
[214] Q. L. Xie, K. E. Markides, M. L. Lee, *J. Chromatogr. Sci.*, 1989, **27**, 365.
[215] E. D. Ramsay, J. R. Perkins, D. E. Games and J. R. Startin, *J. Chromatogr.*, 1989, **464**, 353.
[216] M. Saito, Y. Yamauchi, K. Inomata and W. Kottkamp, *J. Chromatogr.*, 1989, **27**, 79.
[217] H. Engelhardt and A. Gross, *J. High Res. Chromatogr. Chromatogr. Commun.*, 1988, **11**, 38.
[218] S. B. Haawthorne, *Anal. Chem.*, 1990, **622**, 633A.
[219] J. M. Levy, R. A. Cavalier, T. N. Bosch, A. F. Rynaski and W. E. Huhak, *J. Chromatogr.*, 1989, **27**, 341.
[220] M. R. Andersen, J. T. Swanson, N. L. Porter and B. E. Richter, *J. Chromatogr. Sci.*, 1989, **27**, 371.
[221] J. W. King, *J Chromatogr Sci.*, 1989, **27**, 355.
[222] D. R. Luffer, W. Ecknig and M. Novotny, *J. Chromatogr.*, 1990, **505**, 79.
[223] R. W. Vanoort, J. P. Shervet, H. Lingeman, G. J. DeJong and U. A. Th. Brinkman, *J. Chromatogr.*, 1990, **505**, 45.
[224] D. P. Ndiomu and C. F. Simpson, *Anal. Proc.*, 1989, **26**, 393.
[225] J. H. G. Jonkman, J. P. Franke, R. Schoenmaker and R. A., De Zeeuw, *Int. J. Pharm.*, 1982, **10**, 177.
[226] R. V. Smith and H. J. Escalona-Castillo, *Microchem. J.*, 1977, **22**, 305.
[227] H. Hellberg, *Acta. Chem. Scand.*, 1957, **11**, 219.
[228] O. Borga, I. Petters and R. Dalqvist, *Acta. Pharm. Suecia*, 1978, **15**, 549.
[229] Y. Bergqvist, S. Eckerbom and L. Funding, *Clin. Chem.*, 1984, **30**, 465.
[230] R. S. Rapaka, J. Roth, T. J. Goehl and V. K. Prasad, *Clin. Chem.*, 1981, **27**, 1470.
[231] D. J. Krogstad, G. G. Granich, P. R. Murray, M. A. Pfaller and R. Valdes, *Clin. Chem.*, 1982, **28**, 1517.
[232] P. J. Hoskin, O. Alsayed-Omar, G. W. Hanks, A. Johnson and P. Turner, *Ann. Clin. Biochem.*, 1989, **26**, 182.
[233] J. C. Smith, Jr., S. Lewis, J. Holbrook, K. Seidel and A. Rose, *Clin. Chem.*, 1987, **33**, 815.
[234] B. Foley, S. A. Johnson, B. Hackley, J. C. Smith Jr. and J. A. Halstead, *Proc. Soc. Exp. Biol. Med.*, 1968, **128**, 265.
[235] O. Borga, K. M. Piafsky and O. G. Nilsen, *Clin. Pharmacol. Ther.*, 1977, **22**, 539.
[236] V. P. Shah, G. Knapp, J. P. Skelly and B. E. Cabana, *Clin. Chem.*, 1982, **28**, 2327.
[237] R. Janknegt, M. Lohmann, P. M. Hooymans and F. W. Merkus, *Pharm. Weekbl. Sci. Ed.*, 1983, **5**, 287.
[238] W. W. Stargel, C. R. Roe, P. A. Routledge and D. G. Shand, *Clin. Chem.*, 1979, **25**, 617.
[239] J. Shiang-Qiang and M. A. Evenson, *Clin. Chem.*, 1983, **29**, 456.
[240] S. Maguire, F. Kyne and D. UaConaill, *Clin. Chem.*, 1987, **33**, 1493.
[241] R. A. de Zeeuw in E. Reid, ed., *Trace Organic Sample Handling*, Horwood, Chichester, 1981.
[242] R. T. Williams, 'Deoxication Mechanisms and the Design of Drugs' in *Symposium on Biological Approaches to Cancer Chemotherapy*, Academic Press, London, 1960.
[243] J. H. Knox and J. Jurand, *J. Chromatogr.*, 1978, **149**, 297.
[244] G. S. Murkitt, R. M. Lee and R. D. McDowall, *Anal. Proc.*, **21**, 246.
[245] M. Wandell and J. Wilcox-Thole, in M. Mungall, ed., *Applied Clinical Pharmacokinetics*, Raven Press, New York, 1983.
[246] K. M. Piafsky, *Clin. Pharmacokinet.*, 1980, **5**, 346.
[247] M. H. Bickel, *J. Pharm. Pharmacol.*, 1975, **27**, 733.
[248] J. J. Valner, *J. Pharm. Sci.*, 1977, **66**, 447.
[249] M. Rowland and T. N. Tozer in *Clinical Pharmacokinetics: Concepts and Applications*, Lea and

Febiger, Philadelphia, 1980.

[250] W. M. Pardridge, *Endocr. Rev.*, 1981, **2**, 103.

[251] W. M. Pardridge, R. Sakiyama and G. Fierer, *J. Clin. Invest.*, 1983, **71**, 900.

[252] A. J. Wood, D. M. Kornhauser, G. R. Williamson, D. G. Shand and R. A. Branch, *Clin. Pharmacokinet.*, 1978, **3**, 478.

[253] R. K. Roberts, G. R. Wilkinson, R. A. Branch and S. Schenker, *Gastroenterology.*, 1978, **36**, 479.

[254] K. M. Giancomini, N. Masoud, F. M. Wong, *J. Card. Pharmacol.*, 1984, **6**, 924.

[255] M. M. Preidenberg and D. E. Drayer, *Clin. Pharmacokinet.*, 1984, **9** (Suppl. 1), 18.

[256] P. M. Aggeler, R. A. O'Reilly, K. Leong and P. E. Kowitz, *N. Engl. J. Med.*, 1967, **32**, 496.

[257] H. E. Booker and B. Darcey, *Epilepsia*, 1973, **14**, 177.

[258] C. J. Kilpatrick, S. Wanwimolruk and L. M. H. Wing, *Br. J. Clin. Pharmacol.*, 1984, **17**, 539.

[259] T. C. Kwong, *Clin. Chem. Acta*, 1985, **151**, 193.

[260] M. D. Osselton, M. D. Hammond and P. J. Twichett, *J. Pharm. Pharmacol.*, 1977, **29**, 460.

[261] M. D. Osselton, I. C. Shaw and H. M. Stevens, *Analyst*, 1978, **103**, 1160.

[262] M. D. Osselton, *J. Forensic. Sci. Soc.*, 1977, **17**, 189.

[263] M. D. Hammond and A. C. Moffat, *J. Forensic Sci. Soc.*, 1982, **22**, 293.

[264] C. E. Werkhoven-Goewie, C. De Ruiter, U. A. Th. Brinkman, R. W. Frei, C. J. De Jong, C. J. Little and O. Stahl, *J. Chromatogr.*, 1983 **255**, 79.

[265] D. A. Schoelleer, *Biomed. Mass Spectrom.*, 1976, **3**, 265.

[266] *Good Laboratory Practices Regulations for Non-Clinical Laboratory Studies*, Federal Register, (43 FR 59986) 43, 1978, No. 247, Part II.

[267] *Tentative Guidelines for Clinical Laboratory Procedure Manuals*, U.S. National Committee for Clinical Laboratory Standards, Vol. 1 (No. 13), 369–424.

[268] Guidelines for Data Acquisition and Data Quality Evaluation in Environmental Chemistry, *Anal. Chem.*, 1980, **52**, 2242.

[269] D. A. Skoog and D. M. West, *Fundamentals of Analytical Chemistry*, Holt, Reinhart and Winton, New York, 1975.

[270] B. J. Millard, ed., *Quantitative Mass Spectrometry*, Heyden, London, 1978.

[271] D. Dadgar and M. R. Smyth, *TrAC.*, 1986, **5**, 115.

3

Analysis in the brewing industry

Ian McMurrough
Guinness Brewing Worldwide Research Centre, St. James Gate, Dublin 8, Ireland

3.1 INTRODUCTION

Brewing is the oldest of all applications of biotechnology and was practised, albeit in a rudimentary way, in Ancient Egypt. The processes then used had in common with more modern counterparts only the exploitation of an active yeast, to convert fermentable carbohydrates into ethanol. Brewing, moreover, was one of the first manufacturing activities to benefit from the Industrial Revolution and was transformed from a long-standing 'cottage industry' into a large-scale enterprise that demanded intensive capital input and specialized brewing plant. The activities of yeasts have fascinated many notable scientists such as Buchner and Pasteur, and even today, considerable research effort is expended in continuing investigations into the raw materials of brewing, the processes employed and the quality of the products. Given such a long history it might seem to those outside the industry that there can be little scope for further scientific or technological development, but this is not so. Brewing is a mature industry, with a global capacity to produce about 1000 million hectolitres of beer annually. The demand is shrinking slowly, however, causing intensive competition between brewing companies to maintain or expand their market share and profitability. To remain competitive in the present climate, breweries must be operated at the highest levels of efficiency, so effective process controls must be exercised constantly. Added to this, consumer expectations for greater product quality and diversity set ever-increasing challenges to the brewer, while stricter legislative standards and specifications impose further demands.

Effective control over the sequences of chemical, biochemical and microbiological events through which the raw materials of brewing are transformed into the final product requires the making of many measurements. For instance, one of the earliest known industrial applications of the humble thermometer was in the brewing context. Temperature can now be measured in a modern brewery by sophisticated, though rugged, electronic instrumentation capable of feeding output signals to computer-programmed heating and cooling systems. Several other possibilities exist for similar in-line measurement and control systems, depending on the availability of the appropriate detectors [1,2]. By using instrument outputs transmitted by feedback loops to controllers, continuous automatic control over many factors in beer production is achievable. Whereas some physical parameters such as pressure, mass, volume, flow-rate, etc., are amenable to such control, in-line instrumentation for the monitoring and control of dissolved gases (e.g. oxygen, carbon dioxide) and specific gravity are not yet sufficiently robust to be beyond improvement. However, most of the measurements, even in the most technically refined of breweries, are made by off-line systems, for which an appropriately equipped laboratory is required.

Many methods of analysis have been devised and standardized for use throughout the brewing world, under the aegis of organizations such as the European Brewery Convention, the Institute of Brewing and the American Society of Brewing Chemists. These assay methods have been validated for routine process or quality control and are usually expected to provide, by relatively simple laboratory tests, most of the basic information required for day-to-day operations. It would be out of place in this chapter to elaborate further on these tests (gravimetric, photometric, volumetric, etc.), so suffice it to say that they serve their purpose.

For many research or investigative needs, however, the shortcomings of these methods are often all too evident. As contemporary analytical techniques became more and more powerful, the chemical complexities of beer and the raw materials have been revealed. Indeed, beer now seems as chemically complex as any of the other matrices encountered by analysts in industry or medicine. Beer is a mixture of alcohols, esters, aldehydes, acids, carbohydrates, proteins, vitamins, polyphenols and a plethora of other substances, all of which make a contribution, however subtle, to the product quality. These substances are derived either directly from the raw materials or by chemical and biochemical evolution during the brewing process. Total control over the composition of the final product requires in-depth knowledge of the fates of specific chemical entities and of their interplay with many other substances present in close association. Providing this knowledge can be a severe test for the analyst, who must choose the most effective of the tools available and apply them within realistic time-scales. It is the aim of this chapter to describe some of the contemporary instrumental methods used for research and control purposes in the brewing industry, chosen to exemplify the different approaches that can be adopted in fulfilling different criteria.

3.2 THE BREWING PROCESS

A comprehensive review of the many different variations of the brewing process, whereby standard raw materials can be converted into a range of different beer products (e.g. lager, ale and stout) would be out of place in this chapter. Detailed texts [3–7] have been written recently on this topic, which more than adequately make good the deficiencies of the following scant overview.

Traditionally, the brewing process has consisted of a linear sequence of processing events.

(1) Malting.
(2) Mashing.
(3) Boiling.
(4) Fermentation.
(5) Conditioning.
(6) Stabilizing.

3.2.1 Malting

Malted barley is the key ingredient of the brewing process, from which is obtained the fermentable sugars and all the other requirements for the growth of yeast. Grains of barley (*Hordeum vulgare*) contain about 60–65% of their dry weight as starch, the main carbohydrate reserve material, but unmodified starch is not fermentable by yeasts. The first stage in obtaining a fermentable extract is the modification of the barley grain by its conversion into malt. Dried barley grains are steeped in water for about two days, during which time their water content increases by about four times to 40–45%. The process of germination is then initiated under conditions of controlled aeration and temperature, and the grains undergo many internal changes preparatory to the emergence of new shoots and roots. Among the changes that are crucial to the preparation of high quality malt is the development of carbohydrases and proteases. These enzymes partially degrade the internal structure of the grain, thereby rendering the grain more suitable for subsequent extraction. This germinative process must be terminated prematurely after about five days, when the grains are considered to be optimally modified. Accordingly, the grains are dried rapidly by a flow of hot air in a purpose-built kiln. At this stage it is important not to over-heat since this could cause losses in the required enzymic activities.

3.2.2 Mashing

In this process, the malted grains are extracted with hot water to yield a nutritionally rich liquid termed the 'sweet wort'. The malted grains are first finely ground in a mill, then mixed with water at a pre-determined temperature, to form a thick slurry or mash. This mash is then held in a vessel for about one hour to allow the malt enzymes to act on their corresponding substrates. At this stage different beer products require different temperature regimes (50–65°C) to promote either the proteolytic or amylolytic activities. The end-products of enzymic action (fermentable carbohydrates and amino acids) are then washed out from the mash bed by spraying or sparging with hot water (78°C). In some breweries the mashing and sparging take place in the same vessel (mash tun) while other breweries employ a separate vessel for mashing (mash conversion vessel) and for sparging (lauter tun, kieve). Whatever system is used, wort separation takes place by filtration through narrowly-slotted metal plates in the base of the vessel. The liquid run-off (sweet worts) from the filter units is collected, while large volumes of water are passed through the grain bed to ensure as complete an extraction as is possible.

3.2.3 Boiling

The sweet worts collected after mashing must be sterilized before any contaminating micro-organisms grasp the opportunity to multiply. Sweet worts are boiled in specialized vessels called kettles or coppers for 1–2 hr, and several events important to the final beer quality take place, apart from sterilization of the wort. By tradition, female flowers of the hop plant (*Humulus lupulus*) are added to the kettle before boiling commences, to bitter the wort. Hops are sometimes added also as a second charge shortly before boiling is stopped, to enhance aroma. Proteins extracted from the mash, including enzymes with residual activities, are denatured and precipitated with polyphenols as 'break' or 'trub'. Water amounting to as much as 15% of the initial wort volume is lost as vapour, along with some undesirable volatiles, so that the wort becomes more concentrated and refined. Finally, the solids precipitated during boiling are allowed to settle from suspension, along with any hop debris, so that a clarified and sterile hopped wort can be withdrawn from the kettle.

3.2.4   Fermentation

Hopped worts must be cooled to a temperature (8–18°C) suited to the growth of the yeast strain in use. Typically, brewing yeasts are strains of *Saccharomyces cerevisiae* or *Saccharomyces uvarum* (formerly *S. carlsbergensis*). The yeast is pitched in controlled amounts into oxygenated hopped wort held in a fermenter. Oxygen dissolved in the wort is rapidly consumed and the yeast enters into a period of rapid growth during which cell numbers will increase twenty to thirty-fold. The energy required for this growth is obtained by glycolysis of the fermentable sugars, which are the main carbon source for synthetic pathways. In this way, the contents of fermentable sugars and amino acids progressively diminish while the end-products of metabolism, including ethanol and carbon dioxide, accumulate. Eventually the wort becomes so depleted of nutritional content that further growth of the yeast cannot be sustained. It is very important for the brewer to recognize the approach of this decline in yeast growth, before the yeast dies and begins to autolyse. Simple measurements of the wort gravity are, however, sufficiently reliable indicators of the progress of a fermentation to allow the brewer to terminate the process at the appropriate point, by physically separating the yeast from the fermented wort (i.e. by settlement, centrifugation, filtration). Even after the removal of yeast the fermented wort is not ready for sale as beer, since certain by-products of yeast metabolism will make it unpalatable and the physical stability of the product might require extension.

3.2.5   Conditioning

Processes vary between different types of beer (i.e. stout, ale, lager), but usually involve a controlled secondary fermentation and storage in a vat. For this purpose a small amount of fresh yeast and unfermented wort are added to a much greater volume of the fermented wort that is held at a specified temperature (−2–18°C). During the very restricted yeast growth that ensues, certain unwanted flavour-active substances are broken down and removed by the yeast from the wort. When this process is conducted at very low temperature (lagering) the rate is correspondingly slow but the colloidal stability of the product is improved by the precipitation of protein–polyphenol complexes. To accelerate the settlement of solids some brewers add isinglass finings, a pure form of collagen, which co-precipitates with insoluble matter.

3.2.6 Stabilization

The beer drawn from the storage vat invariably requires further clarification, either by centrifugation or filtration. Lager beers especially are expected to be free from haze, and the term 'colloidal instability' relates to their tendency to reform haze subsequent to filtration and bottling. To extend the period (shelf-life) between bottling and the generation of unacceptable haze it is customary to combine filtration with treatments with powdered silica gel and polyvinylpolypyrrolidone (PVPP). These two substances adsorb high molecular weight proteins and polyphenols respectively, which are the principal components of beer haze.

Ultimately, the product is processed to the point at which it qualifies for the appellation 'bright beer'. As such, the product must conform with many specifications which define its strength, colour, flavour, carbon dioxide content, etc. Moreover, some brewers may wish to stabilize their products further by the addition of anti-oxidants, foam enhancers or preservatives, but there are many regulations that control, and sometimes ban, their use. Thereafter, all that remains for the brewer is to ensure hygienic operations in the final racking (keg draught) or packaging (bottles, cans).

3.3 SCOPE FOR INSTRUMENTAL ANALYSIS METHODS IN THE BREWING INDUSTRY

In addition to the empirical tests and analyses that are used routinely, there are a number of applications that require specialized instrumentation. These applications arise in the context of quality assessment of the raw materials, including barley, hops (and hop products) and water. Instrumental methods are useful also for following the transformation of hop resins into the bitter principles formed during wort boiling, or tracing the fates of individual sugars during fermentation. In the area of product specifications it is recognized that the consumer judges the quality of beer on four principal criteria.

(1) Foam or head.
(2) Clarity.
(3) Colour.
(4) Flavour.

A wide range of chemical substances are involved in determining acceptance on these grounds, and all require specialized techniques for their assay. Head quality depends on positive contributions from substances such as glycoproteins and hop components being greatly in excess of any negative influences due to lipids. The clarity of a beer can be compromised by the interactions of proteins and polyphenols. Traces of oxygen and certain metals (Fe, Cu) can accelerate the formation of such protein–polyphenol hazes. Beer flavour is determined by both volatile and non-volatile substances, and is so complex that some components still remain to be identified. Among the simple volatile substances are alcohols, esters, aldehydes, dimethyl sulphide, hop oils, di-acetyl, etc. The non-volatile group includes hop-derived bitter principles, polyphenols that are believed to contribute bitterness and astringency, nucleotides that are expected to have a general enhancing effect, and carbohydrates that bestow body to the product.

3.4 CRITERIA FOR INSTRUMENTAL ANALYSES

Brewing is a business, so any instrumental method contending for a role in the brewing chemist's repertoire will be judged according to how well it serves an identified need, and at what cost. A typical hierarchy of criteria for the endorsement of an analysis procedure within a commercially orientated laboratory is as follows.

(1) Rapidity, reliability and versatility.
(2) Minimal sample preparation.
(3) Minimal manual input.
(4) Unattended operations capability.
(5) Minimal use of expensive consumables.
(6) Low capital investment.
(7) Computer-interface capability.

The importance placed on each criterion will clearly vary according to different work-place circumstances, It would indeed be ideal if a single instrument could serve all purposes while complying with all the aforementioned requirements. Unfortunately the beer matrix is so complex that no such possibility exists. There are, however, two broad categories of analytical technique, namely spectrometry and chromatography, that contend for first place in the popularity ranking.

3.5 APPLICATIONS FOR SPECTROMETRIC METHODS

The Beer–Lambert Law, which linearly relates the concentration of a substance in solution to an absorbance of a beam of monochromatic light, is one of the most powerful and widely used relationships used by analytical chemists. (The 'Beer' in the title refers of course to one of the two eponymous law-makers rather than to the main subject of this chapter.) One of the most familiar applications of this principle is in the widespread use of ultraviolet/visible(UV/VIS)-spectrophotometric assay methods. Several methods of this type are used in brewery laboratories but few provide selective information. For instance, several colorimetric methods are available for measuring Total Carbohydrates, Total Proteins, Total Flavanols, Total Reducing Sugars, etc. [8–10]. Many of the methods which are more selective, such as those for specific cations and anions, are being replaced by more convenient systems that demand little sample preparation or other manual manipulations. Of the several other spectrometric methods it is without doubt that Atomic Absorption Spectrometry (AAS) and Near Infra-Red Spectrometry (NIRS) are used most extensively in the brewing world.

3.5.1 Atomic-absorption spectrometry
The principle of this method is that the concentration of an element is measured by the absorption of radiation with a characteristic frequency by free atoms of the

element. The strength of the method is that atoms absorb only a very narrow range of wavelengths as compared to molecular species. The radiation source is a hollow-cathode lamp, with the cathode made of the element to be determined, which emits a specific sharp resonance line. Atomization of the element can be achieved by introducing a fine spray of the test solution though a nebulizer into an air/acetylene or nitrous oxide/acetylene flame. When very high sensitivity is required, an alternative atomization system with an electrically heated graphite furnace is used. Whatever sample dissociation device is used, a monochromator is needed to isolate the resonance line in the transmitted light, which is then detected by a photomultiplier. Modern instruments use computerized data capture and data management, and are equipped for automatic sample-feed. Calibration of the system is simple because the detector is highly selective, and for the analysis of some metals the beer or wort sample can be injected directly after degassing. Sample preparation rarely requires anything more complicated than centrifugation of turbid samples, or dilution of the sample to the dynamic range of the calibration.

Atomic-absorption analyses are used extensively for the measurement of several elements in the raw materials, processing aids and in the final product. These elements can be considered in two categories:

Category 1. Alkali metals (Ca, Mg, Na, K)
Category 2. Trace elements (Cu, Fe, Pb, Zn, Cd, Co, Mn, Ni, Cr, Sn, Al, As).

Category 1 elements are usually present in beers and worts in relatively high concentrations but can vary widely depending on the product. The typical ranges of variation are:

$Na = 20\text{--}40$ mg/l
$K = 200\text{--}800$ mg/l
$Ca = 20\text{--}100$ mg/l
$Mg = 50\text{--}100$ mg/l

Calcium is the most important of these metals from the brewing standpoint, and worts may have to be supplemented by additions of gypsum to the mash if they would otherwise be calcium-deficient. Calcium is important for regulating the pH of the mash, for stabilizing the enzyme α-amylase, and for assisting the precipitation of certain proteins during wort-boiling. Beers prepared from worts that are calcium-deficient may also suffer from a tendency for white precipitates of calcium oxalate crystals to form slowly on storage. Both calcium and magnesium catalyse the transformation of hop resins to the bitter principles formed during wort boiling.

Category 2 elements are usually present at <0.1 mg/l and are often not detectable at all by flame AAS. Certain elements (e.g. Pb, Cd, Co, As, Al) are monitored to ensure that legally proscribed limits are not exceeded. Concentrations of elements such as iron and copper must be controlled since both can catalyse the oxidation of beer components, with deleterious effects on haze, shelf-life or flavour. Some trace elements, such as zinc, are required by yeast for vigorous fermentation, and may have to be added to the wort; whereas others such as copper can be toxic to yeast, even in low concentration. Zinc is known to be involved in several aspects of yeast biochemistry, particularly in the synthesis of acetate esters and higher alcohols.

A wide variety of samples are routinely examined by AAS analysis. Apart from the obvious application in determining the compositions of worts and beers, samples of filter powders (e.g. kieselguhr) and other processing aids also require attention. Filter aids can potentially contribute unwanted metals, so the extent to which beer can leach minerals from the powders must be assessed. Similarly, the metal packages in which beer is supplied to the trade are another possible source of metal 'pick-up'. Kegs and cans are thoroughly lacquered at manufacture to ensure that there is no beer-to-metal contact, but tests must be performed to check on the perfection of lacquer coverage or any possible accidental damage to the film. In this context, the brewery chemist is involved particularly in measuring possible product contamination with iron or aluminium.

3.5.2 Near infra-red spectrometric analysis

Near infra-red reflectance spectrometry is a technique which was revived in the Beltsville Agricultural Research Center, Maryland, U.S.A., during efforts made (almost twenty-five years ago) to improve the measurement of the moisture content of grain. With the knowledge that light was absorbed strongly at 1935 nm and very weakly at 1680 nm by water, by measuring the relative light absorption at these two wavelengths a direct measurement of the water content of grain was sought [11]. The method chosen was to scan the diffuse light reflected in the NIR region (800–2500 nm) from the surface of a finely-ground grain sample. As the method was developed it was recognized that light absorption by other grain constituents, such as protein, carbohydrate and oil, interfered with the relationship between grain moisture and absorbance. To counter the effects of these interferences it was necessary to take measurements at different wavelengths (2180 nm, 2100 nm and 2310 nm respectively) that corresponded to the absorbance characteristics of the interfering components. Two additional reference wavelengths (1680 nm and 2230 nm) were needed also. It was then realized that not only had the measurement of water content become more accurate, but that the measurement of all three interferents was possible. From this finding has followed the development of several instruments for exploiting near infra-red spectrometry, using either a reflectance, transmission or transflectance mode of radiation collection. Fortuitously, developments in computer technology paralleled advances in optical components so that the capabilities for intensive data-collection and precise data-processing were made available within a sufficiently close time-frame to facilitate such progress.

In its essentials, NIR spectroscopy approaches an ideal measurement system, displaying numerous advantages:

(1) Samples can be solid, slurry or liquid
(2) Samples are not destroyed by measurement
(3) Sample preparation is minimal
(4) No consumable chemicals involved
(5) Little operator training involved
(6) Measurement time is short (<1 min)
(7) Precision is good
(8) Adaptable to in-line measurements.

Perhaps the only disadvantage is that calibration requires primary or reference methods (usually wet-chemical) to analyse the constituents of a 'teaching set' of samples. Clearly, the accuracy of the NIR spectrometric measurements derived subsequently can be no better than that of the reference method used for calibration. Nevertheless, the advantages far outweigh the disadvantages, and NIRS is proving to be a technique of sizeable worth in the brewing industry, with a wide range of applications.

3.5.2.1 Instrumentation

A high-energy incident ray striking the surface of a medium can produce regular (specular) reflections (Fig. 3.1) while part of the refracted ray is modified by selective absorptions and transmitted to the medium boundary [12]. Depending on the nature of the boundary, the transmitted ray may be refracted yet again on its emergence into the adjoining medium, or may be returned by back-scattering or reflection to the first surface, where it will emerge as a diffuse secondary reflection. Indeed, all three possibilities may occur simultaneously. It is, however, only the transmitted and secondary reflected rays that contain useful absorption information, owing to their interaction with the medium.

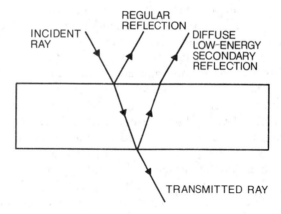

Fig. 3.1 — Simultaneous transmission, refraction, reflection, dispersion, absorption and re-emission of light interacting with an optically dense solid.

Various commercial instruments are designed for collecting exiting radiation in either the diffuse reflectance mode (usually for the analysis of solids) or transmission mode (for liquids). A third mode, diffuse transflection, is a combination of transmission and reflection and involves placing a reflector behind the sample so that the transmitted energy is returned to the same side of the sample as the incident beam.

The essential features of a suitable instrument include:

(1) spectral scanning or point measurements
(2) low-level light detection
(3) data-storage and data-handling capacity.

A simple representation of these basic requirements for NIR reflectance measurements is shown in Fig. 3.2. A high-energy light source is needed to obtain penetration of the sample, and instruments differ in the way this beam is produced [13,14].

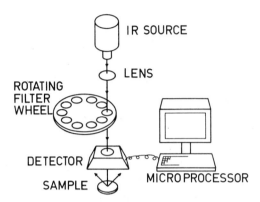

IR SOURCE

LENS

ROTATING
FILTER
WHEEL

DETECTOR

SAMPLE

MICRO PROCESSOR

Fig. 3.2 — Diagram of fixed filter Near Infra-Red Reflectance measurement instrument.

(a)   Fixed filter instruments embody a rotating filter wheel containing up to twenty different filters. Light at specific wavelengths is passed sequentially through each filter to the sample. Obviously, the number of measurable data points in the absorption spectrum is limited by the number of filters employed.

(b)   A broader spectral cover than can be obtained with fixed filter systems can be accomplished by using three or four tilting filters mounted on a rapidly rotating 'paddle wheel' arrangement [13, 14]. Typically, this type of system can scan 300–700 specific wavelengths in 0.1 sec.

(c)   Unlimited spectral coverage can be obtained by passing the light source through a special variable filter and then reflecting it off a large oscillating holographic grating. Scans over the whole near infra-red range are possible in 0.1 sec.

Extremely sensitive detectors are necessary to provide accurate measurements of the NIR absorptions. For this purpose, lead sulphide photoelectric detectors are well suited, combining high sensitivity in the required spectral range, low noise and fast response. Their effectiveness in this application is enhanced by their installation within an integrating sphere. Many instruments automatically convert the measured transmittance or reflectance to absorbance (i.e. $\log 1/T$ or $\log 1/R$) prior to reporting and storing the values.

3.5.2.2   Generation of NIR spectra

The infrared (2500–50000 nm) absorption spectrum of a compound is a very characteristic property caused by the absorption of the frequencies that correspond to internal bond vibrations. Atoms joined by covalent bonds are capable of absorbing electromagnetic radiation energy so that the bonds vibrate with greater amplitude. Energy is transferred to vibrating atoms when the frequency of the oscillating dipole of an individual vibration is the same as the oscillating electric dipole of the irradiating light. At a certain frequency (fundamental) a transition from the ground state to the first vibrational excited state will occur, and this usually corresponds to wavelengths in the mid-IR range (2500–15 000 nm). Apart from the strong absorption that takes place in the infrared region, additional absorption also occurs at frequencies of roughly twice and three times the fundamental frequency. These absorptions are due to a small number of molecules being excited to a second and third level, and are known as second and third overtones. NIR absorption bands are read, therefore, at wavelengths of about one-half and one-third of the wavelength at which maximum absorption occurs [12]. Owing to the small mass of the hydrogen atom and its large dipole moment when combined with certain other atoms, the bulk of the absorption in the NIR range is caused by C–H, N–H and O–H overtones. The NIR spectrum of chemically complex entities, such as barley grains, can consequently contain quantitative information on the contents of carbohydrates and proteins, but it is presented as a composite of overlapping spectra of the principal components (Fig. 3.3). To derive quantitative results from the raw spectral data, a transformation by complex algorithms is required.

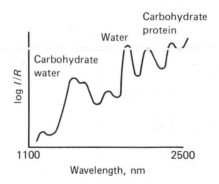

Fig. 3.3 — Idealized NIR reflectance spectrum of ground barley showing the positions of the major absorption bands.

3.5.2.3   Interpretation of NIR spectra — (multivariate chemometrics)

The application of multivariate chemometrics to the interpretation of NIR absorbance data is a typical example of selectivity enhancement. Whereas the NIR absorbance spectrum of barley grains, for example, can be obtained very quickly, it has little relevance in itself until it is related to the chemical properties of interest. In contrast with univariate calibration, several measurements must be taken to obtain a relationship with chemical composition. Multivariate calibration involves mathematical modelling of the relationship between the concentration of the analyte and all the relevant spectral measurements. By using measurements taken at several wavelengths on a 'teaching set' of samples, a calibration equation is developed, which displays the systematic effects on the spectrum caused by slight variations in different aspects of composition. Stepwise multiregression procedures are employed to estimate the effects of changes in composition and compensate accordingly in the computation of the composition of similar samples submitted for analysis [12, 15].

The various steps involved in achieving an effective calibration are as follows:

(1) Select representative samples (e.g. barley grains)
(2) Develop reproducible sample preparation procedure (e.g. fine grinding)
(3) Obtain NIR measurements and record optical data at all required wavelengths
(4) Analyse selected samples by reference method for constituents of interest (e.g. protein, starch, water)
(5) Determine calibration wavelengths and coefficients
(6) Validate calibration with separate sample set.

To obtain a robust calibration, the teaching set should be drawn from the same population as the test samples and be sufficiently large (100–200 samples) to encompass the whole range of expected compositions. Notably, care must be taken to grind barley grains uniformly to a specified particle size (e.g. 0.5 mm) and henceforth to protect from moisture changes prior to analysis. After the NIR measurements, certain refinements may be applied as mathematical data pretreatments, including averaging, linearization and scatter corrections. Ultimately, it is on the precision and accuracy of the reference methods used to measure water content, carbohydrate content, and protein content (as Kjeldahl Nitrogen) that the reliability of the NIR predictions rests.

Correlation of the 'wet-chemical' data with the NIR measurements is complicated because the NIR spectrum of barley (Fig. 3.3) is the product of overlapping absorbances from the three main absorbing constituents, i.e. water, protein and starch, each of which has its own unique optical signature or profile [13]. An appropriate mathematical model is required, therefore, to derive quantitative analytical results from the absorbance data. The most common approach is to use data collected at a fixed number of specific wavelengths rather than from the entire NIR spectrum, which is why instruments with fixed filters are popular. Multiple linear regressions are conducted to determine calibration coefficients relevant to the assay, using mathematical derivatives of the absorbance values as the independent variables, and the known concentrations of constituents in the 'learning set' of samples as the dependent variables. For instance, in the calibration for water content in barley the relative absorption of water could be expressed as:

$$\alpha_w = R_1/R_2,$$

where R_1 is the energy reflected at the absorption point and R_2 is the energy reflected at the reference point. The presence in barley of other major constituents, such as protein and carbohydrate, which can absorb at the wavelengths chosen for the measurement of water must also be accounted for. Relative absorptions for carbohydrates (α_c) and proteins (α_p) are calculated from those parts of the spectrum most sensitive to their presence so that the water content can be expressed as:

$$\text{Percentage of water} = K_0 + K_1 a_w + K_2 a_c + K_3 a_p$$

where the K-values are unique proportionality constants for barley, a_w is the primary correlation term, and both a_c and a_p are corrective terms. Similar multilinear equations for predicting percentages of carbohydrate or protein make up a general set for the major barley components. The K-values are determined from the 'learning set' of samples, and then the calibration is verified with more samples before the instrument is ready for use on 'unknowns'. Once in use, the calibration must be tracked regularly for drift over time, by the analysis of check samples. Other methods that can be used for transforming the basic optical data are more complex, but are expected to provide higher accuracy and precision. The higher order algorithms, such as normalized second derivatives, can reject interferences and thereby decrease the number of regression terms required [13].

3.5.2.4 Applications

In the foregoing description of the technique, attention was focused on the determination of water, protein and carbohydrate in ground barley, by reflectance spectroscopy. This is one of the more important off-line applications in the brewing industry, where it is required to know the moisture contents and protein contents of many different barley samples at the intake points. A trustworthy NIR result can now be obtained in less than one minute, whereas formerly the separate determinations of moisture and protein (as N) took about one hour and five hours respectively.

For liquid samples such as beer and wort, other possibilities exist for NIR transmission and transflectance modes of measurement, leading to in line monitoring of constitutents and rapid corrections for out-of-specification excursions through closed-loop control. The alcohol content and original gravity of beer are crucial quality features which must be controlled during blending operations prior to kegging and bottling. While NIR instrumentation appears acceptable for monitoring [16], views differ on the ruggedness of NIR analysers when used for control in a brewing environment [2,17]. Other potential applications currently under evaluation include measurement of total carbohydrates, fermentable carbohydrates, total soluble nitrogen and free amino-nitrogen in worts.

3.6   APPLICATIONS OF CHROMATOGRAPHIC METHODS

In traditional 'wet-chemical' analysis of complex matrices, efforts are made to separate and quantify a substance by eliminating interferences so that univariate measurements can be made relative to a readily available reference standard. Quite often the procedures are laborious and time-consuming but the reward is in the yield of relevant and selective information. Modern chromatography has been a boon to analysts since it provides the means for speedily separating analytes in very complex matrices so that each can be measured selectively by a suitably sensitive detector. Chromatography has become a very important analytical tool in the brewing industry, and is not only replacing some of the older methods but has also created possibilities for the measurement of components hitherto beyond the compass of traditional analytical capabilities. In this section, the scope for chromatography in the brewing context, for the quantitative determination of volatile and non-volatile constituents, will be reviewed. Examples have been chosen to illustrate the versatility of modern chromatography, through the use of different columns, different mobile phases and different detection systems.

3.6.1 High-performance liquid chromatography (HPLC)

Clearly, HPLC is a method with obvious advantages, such as the ability to separate components in complex mixtures rapidly and then to selectively and sensitively quantify their presence. There are, not surprisingly, limits to the capabilities of HPLC systems to handle very complex matrices like beer and wort, in which the concentrations of major components may be several orders of magnitude greater than the minor components. Faced with such a situation, the usual response of the analyst is to resort to some form of sample pretreatment or 'clean-up', before attempting a chromatographic separation. Unfortunately, some pretreatments may not only be time-consuming and laborious but they may also be a possible source of analytical error. Opportunities for automating sample preparation should, therefore, be availed of wherever possible, not only to save time but also to decrease errors. Even with a fully automated HPLC system, however, samples must inevitably be dealt with individually as they progress through the sample queue. This circumstance creates pressure on the analyst to minimize on-column analysis time, which is usually in conflict with the objective of component resolution. Opportunities sometimes present themselves for reconciling these two mutually opposed objectives by technological or methodological acuity. Certainly, the analysis of the raw materials of brewing, coupled with the requirements for process control and product control, offer the analyst ample scope for utilizing the full repertoire of chromatographic expertise.

3.6.1.1 Fermentable carbohydrates

The assay of fermentable carbohydrates in worts is relatively easy, because they are collectively by far the most concentrated of the solutes present. Accordingly, a simple isocratic HPLC system (Fig. 3.4) and a refractive index (RI) detector suffices for this assay. The RI detector is an example of a bulk property detector, a category which measures differences in some physical property of the solute in the mobile phase, in comparison with the mobile phase alone. Other examples include dielectric constant and conductivity detectors; all suffer from poor sensitivity and limited dynamic range. In the present application, however, there is no requirement for high sensitivity or selectivity. The RI detector measures the refraction of a light beam caused by optically-active solutes. All molecules have the ability to bend light, but the range of RI-index values for most organic compounds is small, so the detection method is insensitive though universal. Since the mobile phase will contain molecules with the light-bending property, RI detectors are constructed to measure on the differential principle, whereby the degree of refraction caused by mobile phase in a reference cell is compared with that in the sample cell. Unfortunately, because the differential detection principle is necessary, gradient elution is not possible, which is another disadvantage of RI-detectors. Three types of refactive index detectors are used for HPLC; the deflection-refractometer, the Fresnel-refractometer and the

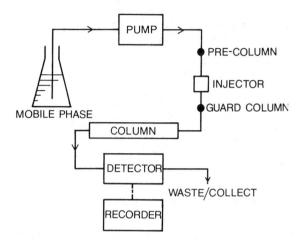

Fig. 3.4 — Diagramatic layout of HPLC system for isocratic elution.

interferometer detector [18,19]. All share the general disadvantages of RI detectors in that they are totally non-selective and cannot be used in conjunction with gradient elution. On the other hand, modern RI detectors have overcome earlier criticisms of insensitivity and the limit of detection has been progressively lowered from about 50 μg of sugar to about 20 ng.

As an alternative to RI detection, a mass detector is arguably more useful. The mass detector [20] (which must not be confused with a mass spectrometer) will produce a response with all non-volatile thermally stable analytes in an eluent stream. This detector works by first nebulizing the entire effluent from the chromatographic column, in a heated air stream. All non-volatile components of the eluate, such as carbohydrates, form a fine cloud of particles in the detector cell and reflect light to a photomultiplier. The volatile solvents in the eluent are evaporated and are not detected. Theoretically, the detector response is independent of chemical structure so this can be considered as a universal detector, for which separate calibrations relating to each separated component should not be necessary. We found, however, that the response factors for monosaccharides, disaccharides and trisaccharides are different for a reason that as yet is not understood. The mass detector has one great advantage over the RI detector in that it is compatible with gradient elution. Moreover, the mass detector displays greater stability than the RI detector, but is slightly less sensitive, with a lower detection limit of about 1 μg.

Of the two major types of HPLC column used for carbohydrate chromatography, the system most widely used contains a speciality propylamino-bonded silica stationary phase [21]. While these columns perform well when new, our experience is that they deteriorate rapidly. A more economical alternative that achieves almost the same results depends on modification of a standard silica column with a polyamine [22] such as tetraethylenepentamine. Modification is readily achieved by including the polyamine (0.01–0.1%) in the mobile phase. Both aminopropyl-bonded and amine-modified columns are usually eluted with mixtures of acetonitrile and water, the composition being chosen according to the resolution required. The proportion of acetonitrile in isocratic mobile phases is varied from 85% through 80% and 70% to 65% for the separation of pentoses, monosaccharides, disaccharides and small oligomers respectively. Since wort contains a mixture of all these sugar classes, the mobile phase used must achieve an acceptable compromise between resolution and analysis time. A mobile phase containing 75% of acetonitrile achieves satisfactory resolution of hexoses from monomers to tetramers (Fig 3.5), but pentoses are poorly

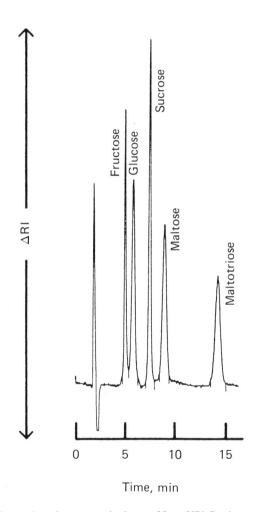

Fig. 3.5 — Separation of sugar standards on a 25-cm HPLC column of aminopropyl silica by elution with 75% (v/v) acetonitrile/water at 1 ml/min, detected refractometrically.

separated and higher oligomers are not eluted from the column in less than 30 min (Table 3.1). Even so, wort contains only very small amounts of pentoses, and of the

Table 3.1 — Effect of mobile-phase composition on retention of sugars
eluted from aminopropyl-silica column

Sugar*	Retention time, min	
	Acetonitrile-water 80–20	Acetronitrile-water 75–25
Ribose	4.2	3.2
Xylose	5.2	3.5
Arabinose	5.8	3.9
Fructose	6.8	4.3
Glucose	8.9	5.0
Sucrose	15.8	6.0
Maltose	21.2	8.3
Iso-Maltose	26.8	11.1
Maltotriose	—	13.7
Panose	—	15.8
Maltotetraose	—	24.8

* Sugars printed in italics are principal fermentable carbohydrates
of wort.
— Not measured.

remaining wort carbohydrates only fructose, glucose, sucrose, maltose and maltotriose are fermentable by the normal strains of brewers yeast. The 75% acetonitrile:
water mobile phase, therefore, appears optimal for the determination of the most
important carbohydrates in wort (Fig. 3.6). The higher dextrins require gradient
elution and mass detection for their analysis (Fig. 3.7).

Despite the successes of amino-columns in separating a wide range of sugars [21],
there is growing interest in the use of ion-exchange chromatography. Fermentable
sugars and their polymers have values of pK_a of around 12, so they can be separated
as ionic species with high pH eluents. When electrochemical detectors are used, the
lower limit of detection for sugars is at the ppb level.

Preparation of wort and beer samples for chromatography on amino-columns by
normal phase partition is uncomplicated. It is advisable to adjust the liquid composition of the sample to resemble that of the mobile phase, by adding two to three parts
of acetonitrile to one part of wort. This decrease in water content of the sample
minimizes local variations in composition at the point of injection. Moreover, high-
molecular-weight dextrins, proteins and other macromolecules in the sample are
precipitated before injection, so they can be removed by centrifugation or filtration.
If this procedure is not followed, it is likely that macromolecules will be precipitated
on contact with the mobile phase and thereafter cause blockage of either the guard
column or analytical column. Precipitation is a useful general method for pretreating
worts and beers for HPLC.

Results obtained by HPLC for the carbohydrate composition of a stout wort with
an original gravity of 1044 degrees are shown in Table 3.2. Original gravity is an
expression used by brewers to define the strength of the wort and this relates to a
specific gravity of 1.044 g/ml. The original gravity of a wort is one of the key
measurements made in all breweries and it depends largely on the content of soluble

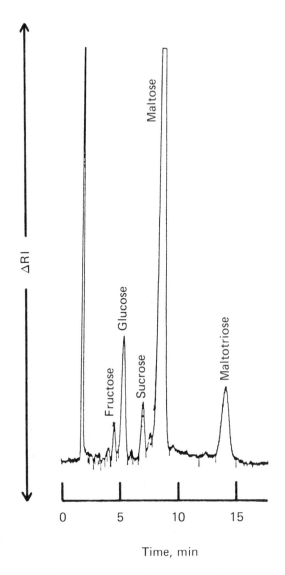

Fig. 3.6 — Separation of fermentable sugars in 25-μl sample of diluted wort (1 volume of wort to 2 volumes of acetonitrile). Chromatographic conditions as for Fig. 3.5.

carbohydrates. If all the carbohydrate in wort was sucrose, an original gravity of 1044 degrees would correspond to a carbohydrate content of 10.9% (w/w). This is the basis of a very useful way of comparing wort and beer strength, since brewers can refer to standard tables that relate original gravity to notional sugar content over a wide range. As shown in Table 3.2, only about 70% of the total carbohydrate content of the wort was measured by HPLC as fermentable sugars. The non-fermentable sugars were determined after fermentation by colorimetry, and they consist of the

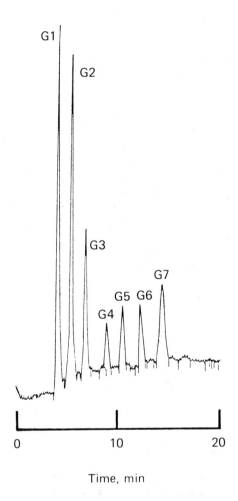

Fig. 3.7 — Separation of a standard mixture of glucose oligomers on a 25-cm HPLC column of aminopropyl silica, by elution with a gradient of increasing water content (20%–50% in 30 min) in acetonitrile, monitored by a mass detector. Peaks identified by number of glucose units in oligomer.

dextrinous end-products of amylolysis. Whereas the dextrins are not normally converted into alcohol, they do contribute to the body of the beer. For the preparation of some speciality beers the dextrins are degraded further to glucose, through the addition of amyloglucosidase during wort preparation.

HPLC is the most convenient method for following the fates of individual carbohydrates during the course of wort fermentation. Detailed information of this type is not needed for routine control, but is of importance when changes in procedure are contemplated. Such possible changes could be in the use of new raw materials or added enzymes, or in the application of an altered mashing regime. In

Table 3.2 — Carbohydrate composition of typical brewer's wort
(original gravity = 1044 deg)

Sugar		Content (g/100 ml)
Fructose		0.24
Glucose		0.94
Sucrose		0.19
Maltose		4.78
Maltotriose		1.58
Total fermentable (by HPLC)		7.73
Non-fermentable (by end-fermentation and colorimetry)		2.90
	SUM	10.63

this respect, it is important to control not only the total amounts of fermentable carbohydrates formed, but also the relative amounts of the different sugars. As shown in Fig. 3.8, the monosaccharides are metabolized first by the yeast, then during the phase of active growth, the maltose and maltotriose contents decrease rapidly. In contrast, the content of maltotetrose, the smallest of the non-fermentable dextrins, remains unchanged throughout the course of fermentation.

3.6.1.2 *Hop resins*

A mixture of substances termed 'hop resins' are formed in small sacs called lupulin glands, at the base of the bracts in the hop cone. The principal components of the resin are termed α-acids and β-acids (Fig. 3.9); each is a family of isoprenyl-substituted phloroglucinol analogues. While the β-acids are of little brewing value, the bitterness of beer is derived almost entirely from the α-acids. The α-acids are not bitter themselves, but they can isomerize to form intensely bitter iso-α-acids during wort boiling (Fig. 3.10). The transformation is inefficient, however, with a yield not much better than 30%. Methods for increasing this yield have occupied hop chemists for several decades and the result has been the introduction of several new hop products. Notable amongst these are the pre-isomerized hop extracts. These are prepared by extracting the resins from hops using a suitable solvent (hexane, methylene chloride, liquid carbon dioxide, methanol, ethanol, etc.). Then the isomerization is catalysed with calcium ions at high pH. The result is a refined liquid product containing a high concentration of iso-α-acids with little or no contamination from α-acids or β-acids. While it is not possible to dispense entirely with the addition of hops (or hop extract) to the kettle, it is an advantage to have a large proportion of the total beer bitterness derived from a pre-isomerized extract addition, made during the final stages of beer processing.

The simultaneous HPLC analysis for iso-α-acids, α-acids, and β-acids in hop products usually requires gradient elution, because of the big differences in the polarities of these analytes. Solutions of hop resins in methanol are suitable for

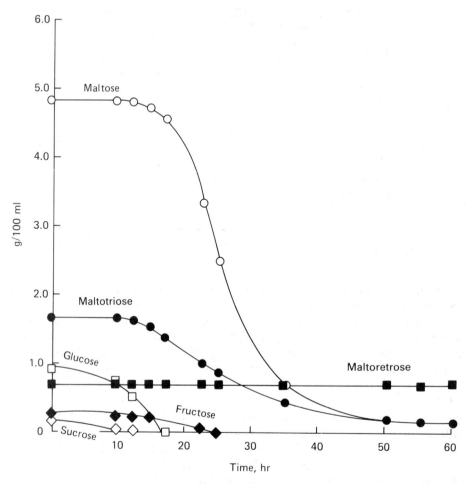

Fig. 3.8 — Contents of sugars measured by HPLC in samples taken during the small-scale
fermentation of stout wort.

chromatography, after chilling and filtration to remove any unwanted waxes that
might be extracted from hops. A wide range of bonded octadecyl-silica reverse-
phase columns can be used for elution with mobile phases containing water,
methanol and orthophosphoric acid [23]. As shown in Fig. 3.11, the iso-α-acids and
hop resins can be resolved readily into their analogues. Monitoring of the eluate
stream at 280 nm detects most hop components, but at 330 nm the detection of α-
acids and β-acids is more selective. Resolution of the iso-α-acids can be increased
greatly by substituting acetonitrile for methanol in the mobile phase. In Fig. 3.12 the
resolution of iso-α-acid analogues into *cis-* and *trans*-isomers is shown. Iso-α-acids
are slightly dissociated in methanol/water mobile phases unless the pH is lower than
that obtained by addition of modest amounts (1–2%) of phosphoric acid. In contrast,
there appears to be very little dissociation in acetonitrile/water mobile phases, which

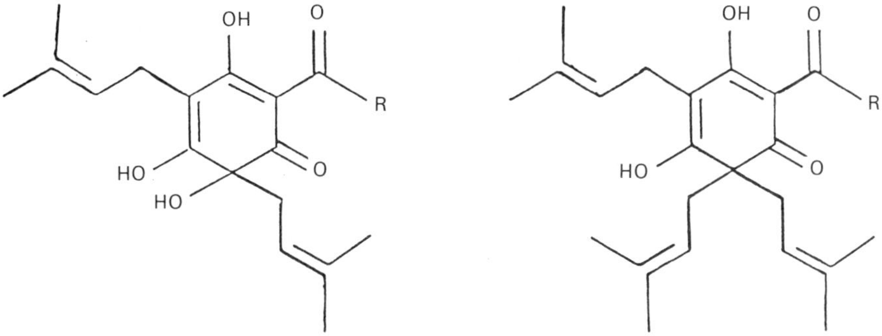

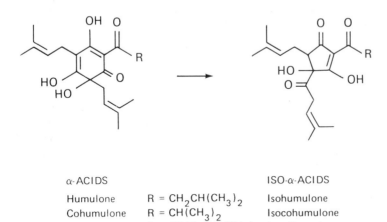

α-ACIDS

		β-ACIDS
Humulone	R = CH$_2$CH(CH$_3$)$_2$	Lupulone
Cohumulone	R = CH(CH$_3$)$_2$	Colupulone
Adhumulone	R = CH(C$_2$H$_5$)(CH$_3$)	Adlupulone

Fig. 3.9 — Structures of α-acid and β-acid resins from hops.

α-ACIDS

		ISO-α-ACIDS
Humulone	R = CH$_2$CH(CH$_3$)$_2$	Isohumulone
Cohumulone	R = CH(CH$_3$)$_2$	Isocohumulone
Adhumulone	R = CH(C$_2$H$_5$)(CH$_3$)	Isoadhumulone

Fig. 3.10 — Isomerization of α-acids to iso-α-acids.

is perhaps the reason for the superior chromatography. Chromatographic separations such as those shown in Figs 3.11 and 3.12, or variants thereof, are useful in the brewery for evaluating the quality of hop and hop products and are also used by the manufacturers to control the isomerization process.

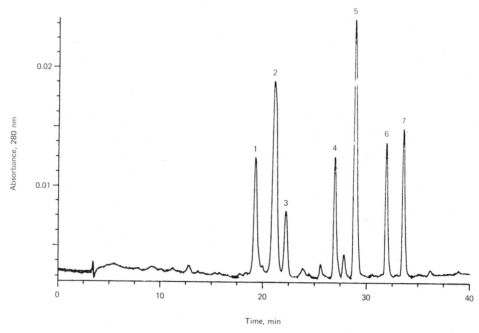

Fig. 3.11 — Separation of iso-α-acids, α-acids and β-acids by elution from a 25-cm HPLC column of C18-silica with a gradient of increasing methanol content (70%–95% in 30 min) in water containing 1% of orthophosphoric acid. Peaks identified; 1 = isocohumulones, 2 = isohumulones, 3 = isoadhumulones, 4 = cohumulone, 5 = humulone/adhumulone, 6 = colupulone, 7 = lupulone/adlupulone.

3.6.1.3 Phenolics

Among the many substances that are termed 'secondary plant products' [24], the phenolics constitute one of the largest and most chemically diverse groups. Their significance to the brewing industry is that all flowering plants contain phenolics, and both barley and hops are typical in this respect. The phenolics of barley and hops can be considered in three broad categories (Fig. 3.13).

(*i*) Monocyclic phenolics are the precursors of 15-carbon structures and are hydroxy- or methoxy-derivatives of either benzoic acid or cinnamic acid. Usually, only small amounts of the free acids are found in plants, but various esters of the acids are plentiful. Free acids are found in beer in small amounts where they have a very slight effect on flavour. The importance of this sub-group is that the acids can be decarboxylated, either by thermal fragmentation or by microbial activity, to form highly flavoured phenols with undesirable medicinal taints.

(*ii*) Flavonol glycosides are pigments that are abundant in leaves and petals so they are, as expected, present in the flowers of hops. Chemical complexity within this sub-group arises in part from the possibility for three different aglycone structures differing in the degree of hydroxylation on the phenolic B-ring (Fig. 3.13). Added to this are many more possibilities for variation in the attached glycosides (mono-, di-,

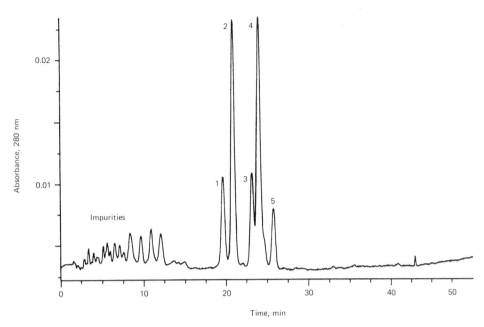

Fig. 3.12 — Separation of iso-α-acid analogues in pre-isomerized extract by elution from a 25-cm HPLC column of C18-silica with a gradient of increasing acetonitrile content (50%–95% in 45 min) in water containing 1% of orthophosphoric acid. Peaks identified; 1 = *trans*-isocohumulone, 2 = *cis*-isocohumulone, 3 = *trans*-isohumulone, 4 = *cis*-isohumulone/*trans*-isoadhumulone, 5 = *cis*-isoadhumulone.

tri-). Their presence in small amounts in beer appears not to have any significant effect on quality.

(*iii*)   Flavanol oligomers or proanthocyanidins are the simple members of the condensed tannin family. The series of compounds present in barley grains derives from combinations of the monomers, (+)-gallocatechin and (+)-catechin whereas the oligomers found in hop flowers are combinations of (−)-epicatechin and (+)-catechin. Individual dimers and trimers that have been isolated from barley, hops, beer and wort and structurally identified [25] have all been either procyanidins or prodelphinidins. These designations refer to the identity of the dominant flavylium ion formed on acid hydrolysis of the inter-flavin links. Apart from having differences in stereochemistry, the monomers differ in the number of hydroxyl groups carried on the phenolic B-ring (Fig. 3.13). These vicinal hydroxyl groups are the source of interest in these compounds in the brewing industry, since it has been proposed that their ready oxidation to *o*-quinones leads to further polymerization, and ultimately to haze formation in beer [4].

Reverse-phase HPLC on octadecyl-silica is an ideal method for separating phenolics, owing to their having both hydrophilic and hydrophobic regions in their structures. Accordingly, phenolics can be separated on the basis of slight differences in polarities arising from patterns of functional group substitution (hydroxyl, methoxyl, glycosyl) and by differences in molecular size (monomers, dimers and

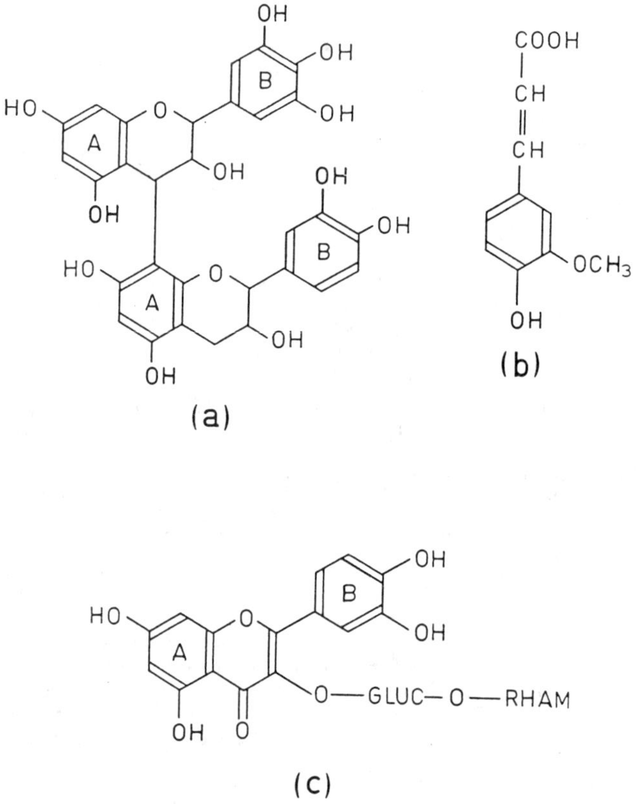

Fig. 3.13 — Structures of some phenolic substances found in beer: (a) is a dimeric flavanol, consisting of units of (+)-gallocatechin (upper half) and (+)-catechin (lower half). This substance is a proanthocyanidin from barley and is designated prodelphinidin B3. (b) is a cinnamic acid derivative designated ferulic acid. (c) is the major flavonol glycoside from hops. The structure shown is the 3,6-O-α-L-rhamnosyl-D-glucoside of the flavonol quercetin and is designated rutin.

trimers) or stereochemistry. Detection of separated compounds can be by absorption measurement at 280 nm, which is an acceptable general purpose wavelength, but selectivity can be increased by measuring at different wavelengths, or by using fluorescence or electrochemical detection. Phenolic substances are usually present only in low concentrations [25], in either raw materials or products, so a sample pretreatment is usually in order.

(a) Phenolic acids
Acidified beer or wort samples are extracted first by shaking with iso-octane to remove non-polar substances, and secondly by shaking with ethyl acetate to recover the phenolic acids [26]. The concentrated ethyl acetate extract can be used for chromatography on C_{18}-silica eluted with a gradient of increasing methanol content in 2.5% aqueous acetic acid (Fig. 3.14). The total contents of phenolic acids in beers

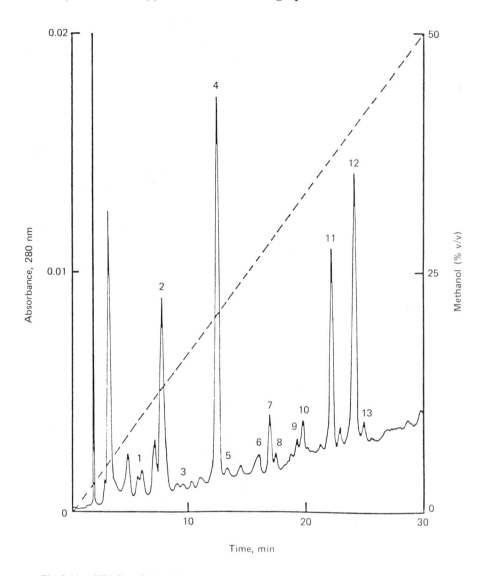

Fig. 3.14 — HPLC profile (solid line) of phenolic acids and simple flavanols extracted from beer with ethyl acetate, and separated by elution from a 25-cm C18-silica column eluted with a gradient of increasing methanol content in water containing 2.5% of acetic acid. The broken line indicates the percentage of methanol in the mobile phase. Identified peaks; 1 = gallic acid, 2 = prodelphinidin B3, 3 = protocatechuic acid, 4 = procyanidin B3, 5 = 4-hydroxybenzoic acid, 6 = (+)-catechin, 7 = vanillic acid, 8 = caffeic acid, 9 = syringic acid, 10 = (−)-epicatechin, 11 = p-coumaric acid, 12 = ferulic acid, 13 = sinapic acid.

and worts is usually less than 10 mg/l in Irish beers, though levels of up to 40 mg/l have been measured in some German speciality beers [26]. The sensitivity and selectivity of the assay can be increased by employing electrochemical detection

[27,28]. Phenolic acids in barley grains or hops are extracted from finely ground samples with acetone/water (3/1) which is then saturated with sodium chloride to promote phase separation. Phenolic acids are recovered from the upper organic phase, which is concentrated by evaporation [26].

(b)   Flavonols
Flavonols are usually present in hops in conjugation with various carbohydrates, so free aglycones are not often present. Acetone/water (3/1) is an excellent extractant for all phenolics and such an extract of hops can be used directly for chromatographic assay of flavonols, even though many different types of phenolics may be extracted simultaneously. Selective detection at 365 nm permits the assay of flavonols in crude extracts, and this is made possible by the presence of the B-ring cinnamyl system [29]. Separation of the flavonol glycosides is accomplished on C_{18}-silica by elution with an increasing concentration of tetrahydrofuran containing 2.5% acetic acid (Fig. 3.15).

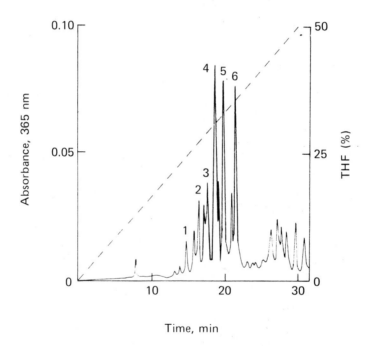

Fig. 3.15 — Separation of flavonol glycosides from hops by elution from a 30 cm HPLC C18-silica column with a gradient of increasing tetrahydrofuran (THF) content in water containing 2.5% acetic acid. The broken line indicates percentage of THF in the mobile phase. Identified peaks; 1 = quercetin triglycoside, 2 = quercetin neohesperidoside, 3 = quercetin rutinoside, 4 = quercetin glucoside, 5 = kaempferol rutinoside, 6 = kaempferol glucoside. (Reproduced with permission, from *J. Chromatog.*, 1981, **218**, 683–693.)

Tetrahydrofuran is the preferred organic modifier in this application since it effects a better separation than methanol [30]. Isocratic elution with acidified 50% methanol:-water is acceptable for the separation of the aglycones, myricetin, quercetin and kaempferol [31], where capacity factor differences depend solely on the number of

B-ring hydroxyls (see Fig. 3.13). The aglycones are released quantitatively from the glycosides by refluxing the extracts in methanolic HCl [23].

Beer and worts contain low concentrations (1–2 mg/l) of flavonol glycosides, so sample concentration is a prerequisite. Proteins and dextrans are first precipitated by addition of three volumes of acetone, then after centrifugation, the clarified solution is evaporated to dryness so that the residue can be dissolved in a small volume of methanol [31].

(c) Flavanols

Extraction of finely-ground barley grains with acetone/water (3/1) dissolves oligomeric flavanols and tannins. Such crude extracts can be separated by gradient elution from C_{18}-columns into discrete peaks of oligomeric flavanols, but the higher molecular weight tannins usually either migrate as a poorly resolved smudge or remain immobile [32]. In Fig. 3.16 the separation by elution with a gradient of

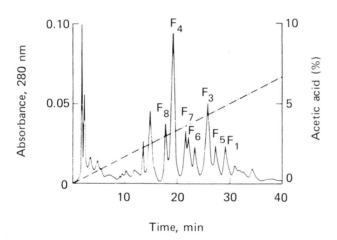

Fig. 3.16 — Separation of flavanol oligomers from barley by elution from a 30-cm HPLC column of C18-silica with a gradient of increasing acetic acid in water (broken line). Identified peaks:

monomers:	F_1 = (+)-catechin,
dimers:	F_3 = procyanidin B3,
	F_4 = prodelphinidin B3,
trimers:	F_8 = prodelphinidin trimer,
	F_7 = prodelphinidin trimer,
	F_6 = prodelphinidin trimer,
	F_5 = procyanidin C2.

(Reproduced by permission of the SCI, from *J. Sci. Fd. Agric.*, 1983, **26**, 62.)

increasing acetic acid is shown. With a gradient of acetic acid, the interference from tannins, which obscure the baseline, is less marked than when methanol is used as the mobile phase modifier. A tannin-free solution of oligomeric flavanols can be prepared by conventional chromatography on Sephadex LH20 eluted with methanol, but this is tedious [32]. Flavanol oligomers are extracted into the wort from both barley malt and hops. During the brewing process their contents decrease but there still remain measurable quantities of monomers, dimers and trimers in the beer (Fig. 3.17). Since these substances are the precursors of haze, some beers require special

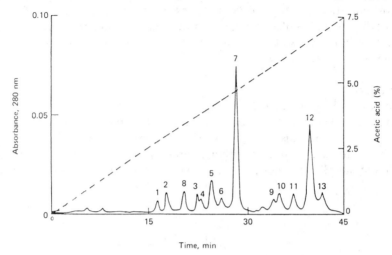

Fig. 3.17 — Separation of flavanols recovered from beer by sorption onto Sephadex LH20, followed by desorption with acetone/water (3/1). Samples of the concentrated extract were eluted from 30 cm HPLC column C18-silica with a gradient of increasing acetic acid in water (broken line). Identified peaks; 1, 3, 4, 6 = trimeric flavanols from barley, 2 = prodelphinidin B3, 5 = procyanidin B3, 7 = (+)-catechin, 8, 9, 10, 11, 13 = hop proanthocyanidins, 12 = (−)-epicatechin.

stabilization procedures to decrease the contents of these polyphenols to below a safe threshold. HPLC methods [33] provide much more information on the success or failure of such operations than can be obtained by empirical colorimetric assays. A beer sample can be passed through a short column of Sephadex LH20 to sorb the polyphenols, since their medium displays a strong sorptive capacity for many aromatic nuclei. After washing the column with water to remove contaminating proteins, carbohydrates, etc., the polyphenols are recovered quantitatively by elution with acetone/water (3/1). Concentration of these acetone washings yields a solution suitable for HPLC. Detection of flavanoid polyphenols is done usually by absorption measurements at 280 nm, though electrochemical detection is another possibility [27,28].

(d)   Phenols

Phenols, if present at all in beers, are usually in very low concentrations (ppb), but even then may have a significant and usually deleterious effect. To detect such low concentrations, a fluorescence detector is required, with excitation wavelength 268 nm and emission detection wavelength 298 nm. Traces of phenol, gaiacol, *m*-cresol and *p*-cresol amounting to a total content of about 2 ppb have been measured in beer by using gradient elution, as used for phenolic acids. For the detection of phenolic acids by fluorescence, the excitation wavelength recommended is 330 nm and the emission wavelength is 435 nm [34].

(e)   Anthocyanidins

These compounds form the basic structure from which many plant pigments are derived, but they are not found in a free state in barley grains or in hops. They are formed, however, on acidic hydrolysis of the proanthocyanidins, which are named according to the identity of the anthocyanidins so formed [35]. The most common anthocyanidins (see Fig. 3.18) are pelargonidin(4'-OH), cyanidin(3',4'-di-OH) and delphinidin(3',4',5'-tri-OH) and these absorb light strongly in the region 465–550 nm, so appear orange-red to magenta to the human eye.

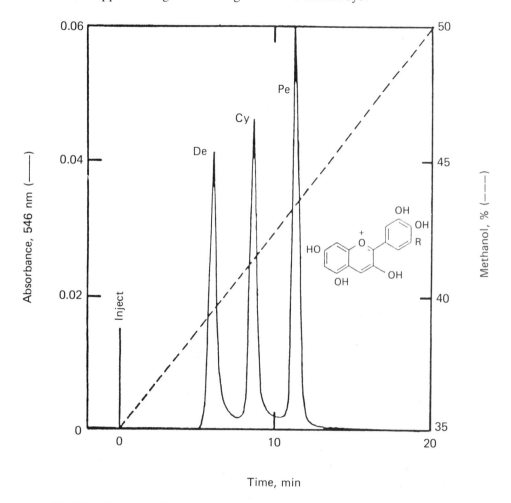

Fig. 3.18 — Separation of anthocyanidins by elution from a 30-cm HPLC column of C18-silica with gradient of increasing methanol concentration (– – –) in water containing 2.5% of acetic acid. Identified peaks; De = delphinidin, Cy = cyanidin, Pe = pelargonidin. Inset; structure of cyanidin (R=H), structure of delphinidin (R=OH).

Thin-layer chromatography is a useful technique for separating and identifying anthocyanidins in hydrolysates, but HPLC provides quantitative information, which is useful when more than one anthocyanidin hydrolysis product is formed. Separation of the anthocyanidins is readily achieved on C_{18}-silica eluted with a short steep methanol gradient. Detection can be made sufficiently selective by monitoring at 546 nm, which is a compromise wavelength at which all three anthocyanidins absorb (Fig. 3.18). Detection can be made even more selective by using multichannel monitoring at the absorption maxima of the various cationic species (pelargonidin, 520 nm, cyanidin, 553 nm, delphinidin, 546 nm).

3.6.1.4 Inorganic anions

Modern ion chromatography has greatly facilitated the determination of many inorganic constituents of brewing water, worts and beers, which were formerly measured by time-consuming procedures. Most of the anions of interest can be determined now in a single chromatographic run that requires minimal sample preparation — usually nothing more than dilution and clarification. The essential components of one popular type of ion chromatograph are shown in Fig. 3.19 and

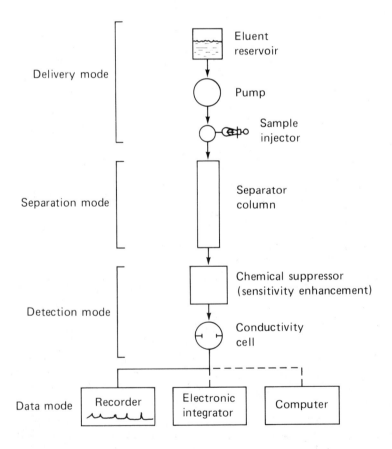

Fig. 3.19 — Configuration of a typical ion chromatograph.

both the theoretical and practical aspects of its many applications have been reviewed comprehensively [36]. Such a system can be used for different separating techniques including ion-exchange chromatography, ion-exclusion chromatography and ion-pair chromatography. For the separation of the major anions in beer and wort, the separation mechanism is anion exchange, for which special stationary phases have been developed. A stationary phase that is stable over a wide range of pH is required, so beads (particle diameter, 10–25 µm) with a core of surface-sulphonated polystyrene/divinylbenzene resin are used. Held around the bead core by electrostatic and van der Waals forces are many, much smaller (0.1-µm diameter) particles of aminated latex, on which are exposed the quaternary ammonium groups responsible for the ion-exchange function. For the separation of inorganic anions in beer a carbonate/bicarbonate buffer is used as the eluent, and the different ionic species are separated according to their relative rates of migration through the stationary phase (Fig. 3.20). Before the eluate arrives at the detector it passes

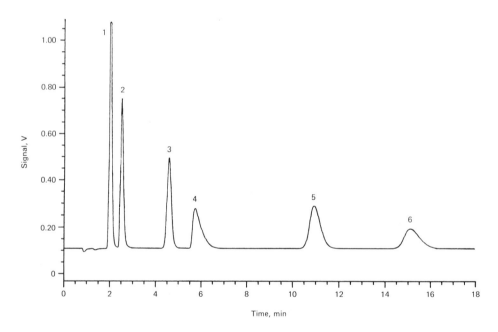

Fig. 3.20 — Separation of anion standards by elution from a 25-cm HPLC ion-exchange column of Dionex HPIC-AS4 with 1.6mM Na$_2$CO$_3$/2.1mM NaHCO$_3$. Detection was by conductivity, after post-column ion suppression with 25mM H$_2$SO$_4$. Identified peaks; 1 = chloride, 2 = nitrite, 3 = phosphate, 4 = nitrate, 5 = sulphate, 6 = oxalate.

through a device which suppresses the high background conductivity of the components of the mobile phase. Accordingly, the chemical suppressor greatly increases the sensitivity of detection for many sample ions. This suppression of background conductivity is achieved by causing a flow of eluate along one surface of a cation-exchanger configured as an ultrathin membrane. The other surface of the membrane is exposed to a counterflow of a regenerant solution, for which purpose a very dilute solution of sulphuric acid serves well in the present application. During passage of the eluate through the suppressor, the sodium ions of the buffer are exchanged for hydrogen ions, thereby converting the most abundant molecular species into weakly dissociated, and therefore weakly conducting, carbonic acid. At the same time the sample ions are converted into the corresponding acids, so their conductance responses are increased.

In many breweries, ion chromatography is used routinely for the measurement of chloride, nitrate, phosphate and sulphate [37,38]. The ratio of sulphate to chloride affects beer flavour, and both ions can affect the brewing process in several other ways [3,4]. Moreover, in recent years concern for environmental levels of nitrates has grown considerably. Possibilities exist for nitrates present in foodstuffs and beverages to be converted into nitrites and thence to nitrosamines, which are known to be carcinogens. The levels of nitrates in brewing liquor should, therefore, be checked regularly, as should the contents in barley malt and hops. When nitrate levels in wort are held low and brewery hygiene is high, the possible dangers arising from nitrosamine production are removed. Since the nitrate levels usually found in beer are less than European Community guidelines for potable water (25 mg/l), the intake of nitrates from beer-drinking seems insignificant. Contents of oxalate in beers can be measured from the same chromatogram as for the inorganic anions (Fig. 3.20); several other organic acids are measureable by anion-exclusion chromatography.

3.6.1.5 Organic acids
The contents of several organic acids in beers are readily measured by using the ion chromatograph (Fig. 3.19) and the appropriately selected separation column and chemical suppressor. A suitable column contains a totally sulphonated cation-exchange resin, and with this is used very dilute ($0.001M$) hydrochloric acid as eluent. The mechanism of the ion-exclusion process is explained [36] by the extensive hydration of the sulphonic acid groups, sufficient to surround the stationary phase with a shell of ordered water molecules bounded by a hypothetical negatively charged membrane. This hypothetical structure is termed the 'Donnan membrane' and is perceived as being permeable only to electrically neutral compounds. In the acidic milieu of the mobile phase, the molecules of the weaker acids are not dissociated and can therefore penetrate the membrane for further interaction with the resin surface, whereas charged species are excluded. Separation of organic acids depends not only on the relative avidities of adsorption but also on steric factors. As for ion-exchange chromatography, conductimetric detection is considerably enhanced for ion-exclusion chromatography by use of the appropriate ion-suppressor system. For this purpose, the regenerant is 5 mM tetrabutylammonium hydroxide (TBH), used as a counterflow in a micromembrane suppressor. The membrane is a sulphonated polyethylene derivative, and has a higher permeability for quaternary ammonium bases than for hydronium ions. Hydronium ions in the eluate pass through the membrane and react with hydroxide ions in the regenerant to form water. Simultaneously, the tetrabutylammonium cations in the regenerant pass through the membrane into the eluate where they interact with organic anions and chloride ions to form salts of the acids. The net effect is to decrease the conductivity

of the eluate by replacing the hydronium ion with a less conductive species, and to increase the conductivities of the weak acid analytes by promoting their ionization.

An example of the separation of organic acid standards relevant to beer analysis is shown in Fig. 3.21. With such a system, the effects of numerous beer processing

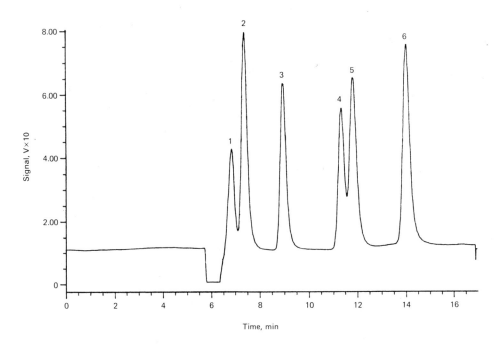

Fig. 3.21 — Separation of organic acid standards by elution from a 25-cm HPLC column of Dionex HPICE-AS1 with 1mM HCl. Detection was by conductivity, after post-column ion suppression with 5mM tetrabutylammonium hydroxide. Identified peaks; 1 = pyruvic, 2 = citric, 3 = malic, 4 = lactic, 5 = succinic, 6 = acetic acids.

parameters on acid production have been investigated [39]. In general, it was found that the concentrations of most acids in beer increased rapidly during active growth of yeast, each to 200–300 mg/l. Acetic acid was the exception in that concentrations varied cyclically from about 50 mg/l to 150 mg/l. In another study [40], the results obtained by chromatography were found to agree well with those obtained by more time-consuming enzymic assays.

3.6.1.6 *Iso-α-acids in beer*
Beer bitterness is one of the primary quality features and, accordingly, demands strict control at several stages in the brewing process. Photometric methods [8–10] used for the routine assessment of bitterness in the finished product are favoured mainly for their convenience, though it is recognized that these non-specific measurements can sometimes produce misleading results. The major contributors to beer bitterness, the iso-α-acids, are extracted from acidified beer, along with several

other beer components derived from both hops and barley, by shaking with iso-octane (2,24-trimethylpentane). The absorbance of the iso-octane extract at 280 nm is taken as a measure of the beer bitterness. Instruments are now available for automating all these operations so that the total analysis time can be as short as 3 min per sample. The major problem with this method is that some of the measured absorbance is due to non-bittering substances and this amount can vary. Clearly, a specific measurement of iso-α-acids by HPLC is a more meaningful measure of bittering substances; the main disadvantage is that it is much slower than the routine alternative.

Measurement of iso-α-acids by HPLC requires some attention to sample clean-up, but this is not necessarily a laborious process when modern automated equipment is used. Fortunately, beer does not usually contain sufficient α-acids and β-acids to impede the analysis, and the main potential interferences are either high-molecular-weight substances that might impair column efficiency or UV-absorbing species that elute close to the peaks of the iso-α-acids.

There are three main methods for off-line beer sample pretreatment:

(a)   Precipitation
This is by far the simplest of all pretreatments and involves merely the addition of three volumes of methanol to one volume of beer [41]. This procedure accomplishes nothing more than the precipitation of high-molecular-weight substances, such as proteins and polysaccharides, that would otherwise be precipitated on the chromatographic column on contract with the mobile phase. After they have been precipitated, these substances can be removed by centrifugation or filtration. None of these procedural operations is readily amenable to simple automation and, moreover, the sample becomes diluted by the process. Furthermore, the treated sample still contains many UV-absorbing species, so long elution times are required to separate the iso-α-acids from such interferences.

(b)   Liquid–liquid extraction
Extraction of one volume of acidified beer with two volumes of iso-octane produces an organic phase containing iso-α-acids and relatively few interferences [41]. Unfortunately, many beers produce intractable emulsions when subjected to this pretreatment, which compromises the efficiency of extraction, and inclusion of an internal standard does not always overcome the potential loss in accuracy [41]. Dilution of the analytes also results from this method. Whereas some of the objections to the procedure can be countered by using smaller phase ratios, or by partial evaporation of the iso-octane phase, there still remains the criticism that it is not good practice to inject solvents that are not totally miscible with the mobile phase. Even this disadvantage can be overcome by total evaporation of the iso-octane, followed by dissolution of the residue in mobile phase; but this adds further to the complexity of the procedure and, inevitably, to the risk of oxidizing the labile analytes.

(c)   Solid-phase extraction
The rationale is to use a small amount of a solid material in the form of a packed column or cartridge, to sorb the analytes of interest from the sample while non-

sorbed interferences are washed to waste. A wide variety of materials with divergent functionalities are available, the most popular of these being supplied in the form of small cartridges through which the sample or wash solvent are made to percolate. Rarely is it found that the solid phase is totally specific for the analytes and several other species may be sorbed too. Possibilities for the selective removal of such interferences by washing the solid phase prior to the recovery of the analytes is the overwhelming advantage of this method over all others. For the present application, a cartridge containing C_{18}-silica is usable for off-line sample pretreatment [41]. After the cartridge has been conditioned by washing with methanol and dilute phosphoric acid, a small volume (5 ml) of beer is applied and washed through with more dilute acid. Interferences are then washed from the column with acidified methanol/water mixtures, chosen so that the methanol content is substantially lower than that of the mobile phase. This is based on the premise that only those wash solvents that contain insufficient methanol for the analytes to distribute in their favour will effect the selective removal of unwanted substances. Exactly the same principles apply on transposing the method for on-line duty. The solid phase method is readily adaptable to this mode of operations and, what is more, automation of the system is easy [42].

In Fig. 3.22 the system used for manual on-line sample preparation is shown schematically. The C_{18}-silica solid phase packed in the small precolumn (Col. 1) is the same as that used for packing the main analytical column (Col. 2). The lines between columns and eluants are controlled by a two-position six-port valve, and this determines the flow of liquids through two interconnectable loops. When configured in the 'purge mode', wash solvent flows from Pump 1 through the precolumn to waste, while mobile phase from Pump 2 by-passes the precolumn and is routed directly to the analytical column. Sample pretreatment commences after the precolumn has been washed well with 1% phosphoric acid (Wash 1). A sample (10–100 µl) of beer is then injected onto the precolumn, whereupon the iso-α-acids become bound, along with some more polar interferences. The precolumn is then purged sequentially by the passage of acidic 20% methanol (Wash 2) and acidic 50% methanol (Wash 3) to remove UV-absorbing interferences which are routed to waste. The precolumn is then switched in-line with the analytical column so that chromatographic mobile phase, containing 70% methanol, is directed onto the precolumn. When contacted by the low polarity mobile phase, the iso-α-acids on the precolumn are desorbed and carried forward in the flow to the analytical column, where they are separated (Fig. 3.23). Automation is an obvious option because the operations involved consist merely of solvent selection through a low-pressure switching, valve and flow redirection through a high-pressure switching valve. At least two equipment manufacturers supply programmable automated valve stations suitable for this purpose. Remarkably, when such a device was coupled with an automated sample feed and applied to the analysis of iso-α-acids in beer, the precision and accuracy of the results obtained were significantly better than was obainable by manual column-switching [42]. Different wash protocols were devised to achieve different objectives, such as analyte enrichment or contaminant removal, and the system was sufficiently reliable to be operated unattended, so that up to ninety samples were processed every 24 hours over several days of uninterrupted running.

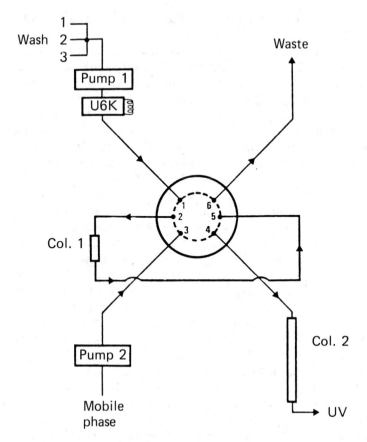

Fig. 3.22 — Configuration of equipment for manual sequential step elution, showing a two-position six-port valve. In the purge mode, the lines are connected: 1–2, 3–4, and 5–6. In the analysis mode the lines are connected: 1–6, 3–2, 5–4. Col. 1 = precolumn, Col. 2 = analytical column, U6K = manual injector. Washes 1, 2 and 3 are purging solvents.

3.6.1.7 *Iso-α-acids and α-acids in wort*

The main reason for this analysis is to provide information on the kinetics of the transformation of α-acids to iso-α-acids during wort-boiling (see Fig. 3.10). The extent to which this transformation occurs is of significant economic implication to the brewing industry since any unconverted α-acid is wasted. The isomerization of α-acids can be monitored by taking samples from wort kettles at intervals during boiling. Samples must be cooled quickly and stabilized against further chemical change and possible microbial infection. This can be easily accomplished by treating each sample with three volumes of cold methanol. Not only does this procedure stabilize the sample, but it also provides a degree of sample clean-up by precipitating carbohydrates and proteins. After centrifugation of the sample, the clarified supernatant can be used for HPLC without further treatment. In most instances the sample will contain iso-α-acids, α-acids and β-acids, and while it is essential to measure the first two groups of substances as accurately as possible, there is usually

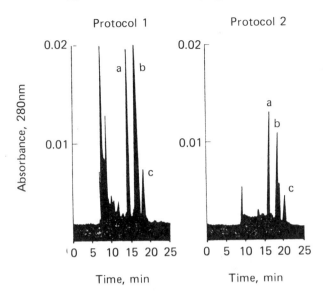

Fig. 3.23 — Separation of iso-α-acids by elution from a 15-cm HPLC column of C18-silica with mobile phase containing methanol:water:orthophosporic acid (75:24:1) at 1 ml/min, after treatment of beer sample by two different automated sequential step clean-up protocols. Protocol 2 was similar to Protocol 1 but included the washing of the precolumn with 0.2 ml of mobile phase. Identified peaks; a = isocohumulones, b = isohumulones, c = isoadhumulones.

no interest in the amounts of β-acids present. The presence of β-acids complicates chromatography because these substances emerge late and thereby prolong runtime. It is possible, of course, to use gradient elution to speed the emergence of the later peaks without loss of resolution, but isocratic elution is strongly favoured for rapid sample throughput and high reproducibility. As an alternative to gradient elution, however, column-switching with isocratic elution is now possible, and can be conducted easily with modern automated valve stations.

A configuration of equipment found suitable for the rapid analysis of wort samples for iso-α-acids and α-acids, with automated multiple column-switching [42] is shown schematically in Fig. 3.24. The essential features are the precolumn (Col. 1, 1-cm), the first analytical column (Col. A, 15-cm) and the second analytical column (Col. C, 10-cm), all of which contained the same C_{18}-silica stationary phase. Only one mobile phase was used, and this consisted of 75% methanol containing 0.01M tetrabutylammonium hydroxide (TBH) adjusted to pH 5.0. This pH was chosen [42] to ensure that β-acids were strongly retained on the stationary phase, as was displayed during conventional isocratic elution of a wort sample (see Fig. 3.25). In the example shown of isocratic elution, the iso-α-acids were not well separated from fast-running interferences, and neither were they well resolved from one another. The run time was determined by the retention time of the last β-acid to elute, and the resolution of colupulone, lupulone and adlupulone was far greater than would have been necessary, even if the β-acids were of interest. The separation conditions were suited only to the α-acids, which were resolved into cohumulone, humulone and

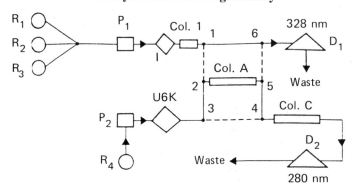

Fig. 3.24 — Schematic diagram of equipment configured for automated multiple column switching, shown in analysis mode. R_1, R_2, R_3, R_4 = solvent reservoirs, P_1 = purge pump, P_2 = analytical pump, I = autoinjector, U6K = manual injector, D = detectors, Col. 1 = precolumn, Col. A = 15 cm. C18-silica column, Col. C = 10-cm C18-silica column.

Timed events programme;

0 min, inject,	(1–2), (3–4), (5–6)
2.4 min,	(1–6), (3–2), (5–4)
7.0 min,	(1–2), (3–4), (5–6)

adhumulone in an acceptable run-time. When multiple column-switching was used, the general separation was improved considerably by varying the length of column (and hence the number of theoretical plates) to which each group of analytes was exposed during the run. Accordingly, only those fractions of eluate that contained compounds of interest were diverted from the outlet of one column for more complete resolution on another column. The method was a development of the general tactic of 'heart-cutting' and 'box-car chromatography' by column switching [43].

The column-switching method decreased analysis time by two-thirds. On autoin-jection of a stabilized wort sample, the β-acids were strongly bound to the precolumn (Col. 1) while both α-acids, iso-α-acids and interferences progressed to the first analytical column (Col. A). The fast-running interferences were allowed to elute from Column A so that they were carried to detector 1, then at a preset time the eluate flow was redirected to the Column C for about five minutes. During this period the iso-α-acids were eluted from Column A onto the Column C, while the β-acids were eluted from the precolumn and carried to detector 1. When all the iso-α-acids had been transferred from Column A to Column C, the programmed controller switched the eluate of Column A away from Column C and back towards detector 1. Eventually, the α-acids were eluted from the first column as three resolved peaks which were detected selectively at 328 nm (Fig. 3.25). The iso-α-acids were moni-tored at 280 nm by detector 2 and were seen as two well resolved peaks of isocohumulone and isohumulone/isoadhumulone (Fig. 3.25) that were not over-lapped by interferences. Overall, the analysis time was 15 min and the results were presented as two chromatographs per sample. Notably, peaks in the iso-α-acid chromatogram were integrated with very high precision, because of the quality of the separation. The early part of the α-acid chromatogram was crowded with interfering substances and β-acids, but since none of these were of interest they were ignored.

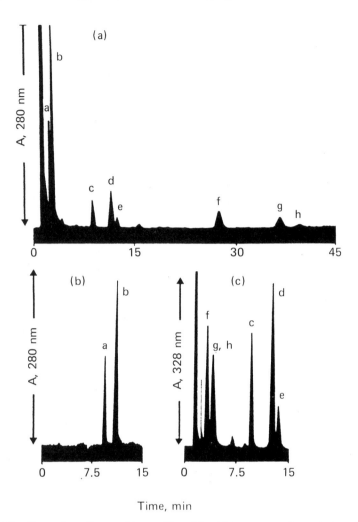

Fig. 3.25 — Separation of iso-α-acids, α-acids and β-acids in wort sample on elution with 75% methanol, containing $0.01M$ tetrabutylammonium hydroxide adjusted to pH 5.0; (A) by conventional HPLC on 15-cm C18-silica column, (B) and (C) by automated multiple column-switching. Identified peaks; a = isocohumulones, b = isohumulones/isoadhumulones, c = cohumulone, d = humulone, e = adhumulone, f = colupulone, g = lupulone, h = adlupulone.

This system has been used for unattended analysis of many thousands of samples and has proved to be extremely reliable [42]. Typical applications include studies of the effects of boiling at elevated temperature and pressure on reaction kinetics [44] and the use of new hop products such as the various different types of extracts that are now available.

3.6.1.8   *Fast protein liquid chromatography (FPLC)*
Instrumentation for rapid analaytical separation of proteins by chromatography has been developed by the Pharmacia Fine Chemicals Company and has been applied

widely over the past decade. The system shares many of the features of HPLC except that the operating pressures are only 100–200 p.s.i. and the columns are packed in glass rather than stainless steel. Different column packings are available for separations to be effected by molecular sieving, cation and anion exchange or by reverse-phase chromatography. So far, the potential of this versatile system has yet to be fully realized in the brewing context. The predominant proteins of beer are relatively small-molecular-weight polypeptides, consisting of two groups with average molecular weights of about 40 000 and 10 000. The relative importance of each group with respect to haze formation and foam characteristics remains open to question, despite many efforts to relate quality features to specific protein fractions. FPLC offers the promise of high resolution fractionation of beer polypeptides according to differences in those properties thought to be significant, i.e. molecular weight, charge, and hydrophobicity.

3.6.2   Gas chromatography (GC)
Gas chromatography and tasting panels form the mainstay of flavour research and routine beer flavour quality control. Most flavour-active components in beer are relatively small-molecular-weight volatile components that are readily measureable by GC and a wealth of information is now available on their impacts on flavour [4,6]. Flavour is a very complex sensation [4,45] consisting mainly of taste and odour (olfaction), but is also affected by other characters such as smoothness, coolness, dryness and pungency (mechano-, thermo- and nociception). Added to this is the sensation of texture which relates to beer in the character of 'palatefulness' or 'body'. The importance of individual components on beer flavour can be estimated by determining their taste thresholds. These are the lowest concentrations of each component at which it can be recognized by taste. Determination of threshold values is complicated by possible variations between the responses of different tasters and by the effects of different carrier solutions (i.e. lager, ale, stout). Moreover, it is known that the phenomenon of synergism exerts considerable influence in overall flavour impact, whereby the net effect is greater than the sum of the contributions from each individual constituent. Notwithstanding all the possible problems associated with sensory evaluation, the judgements of trained flavour panels are respected when applied to either the maintenance of existing product quality, or the development of new products. Precise analytical data is needed, however, to define uniquely any excursions from normality and to provide information on how variations in process or procedure might affect product quality.

More than 400 components of beer have been identified and most of these are sufficiently volatile to be measurable by GC without derivatization. Since beer wort is subjected to lengthy boiling during its production, most of the volatile substances derived directly from malt or hops are lost by evaporation. It is not surprising, therefore, that most of the volatiles of beer arise during fermentation as by-products of yeast metabolism [4,46]. Whereas a few components are present in high concentration (e.g. ethanol), many others are present in barely detectable amounts. The problem for the analyst is to measure the important flavour contributors present in an aqueous medium that contains a multitude of possible interferences. For most applications the method used is static headspace analysis, whereby beer constituents are measured indirectly by sampling the vapour phase in equilibrium with the beer sample held in a closed vessel [18]. A small volume of beer (10–50 ml) is saturated with either ammonium sulphate or sodium chloride and then sealed with a rubber septum in a vessel sized to provide a headspace-to-liquid ratio of 6.5. After holding the vessel for 30 min in a water-bath at 35°C, a sample (0.1–2.0 ml) of the headspace gas is withdrawn into a syringe by a needle passed through the septum, and the sample is then injected directly onto an appropriate column. Headspace analysis can now be conducted by reliable automated samplers so that unattended operations are possible.

Included in the volatile spectrum of beer components that can be analysed by head-space gas chromatography are the following:

(a) Alcohols and esters
(b) Carbonyls
(c) Sulphur compounds
(d) Organic acids
(e) Amines
(f) Phenols
(g) Pyrazines
(h) Hop essential oils.

In the following sections, examples have been chosen from this list to typify the use of particular detection systems and because they are of major significance in the control of beer flavour.

3.6.2.1 Alcohols and esters
Many alcohols other than ethanol are found in beer. These higher alcohols, or fusel alcohols, are formed by yeast during protein synthesis by decarboxylation and reduction of keto-acids (Fig. 3.26). The keto-acids are formed from amino acids by

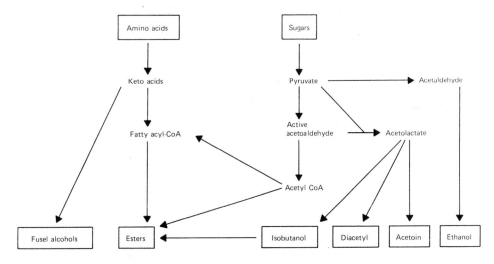

Fig. 3.26 — Biosynthetic routes to the formation of beer flavour volatiles by yeast.

deamination or transamination, so the quantitative and qualitative composition of both the nitrogen source and the carbon source for the yeast will ultimately influence beer flavour [46]. High fermentation temperatures tend to promote the formation of higher alcohols, and the amounts formed depend also on the strain of yeast used. The concentration of higher alcohols formed may be increased by stirring a fermenting wort and by aeration of the wort prior to pitching with the yeast inoculum. Even the amount of yeast used for pitching can have a slight effect, with larger pitching rates tending to produce lower concentrations of higher alcohols. Ester formation is also influenced by a variety of factors and most of the evidence links their synthesis with the lipid metabolism of the yeast [47]. Esters are particularly important components of beer because of their strong fruity flavours, so their measurement and control is of keen interest. Ester formation is influenced by fermentation parameters similar to those responsible for increased formation of higher alcohols, except that increased oxygenation depresses the formation of esters.

A routine method for measuring the most important lower boiling point volatiles in beer has been validated recently in collaborative trails involving representatives from most of the major breweries in the British Isles [48]. Capillary columns containing Carbowax 400, operated at 50°C, provide adequate resolution of the major volatiles of interest (Fig. 3.27). The concentrations of these components in beer varies typically from 0.1 to about 50 ppm, and flame ionization detection (FID) is the usual method of monitoring. The FID is close to being the universal detector, so nearly all volaties produce a significant response. The advantages of this detector are its wide applicability, high sensitivity, fast response range and simplicity in construction [18]. Surprisingly, the mechanism of ion production is still poorly understood. The FID response is the highest for hydrocarbons and much smaller responses are obtained with compounds containing sulphur, nitrogen and halogens. The FID detector is, therefore, not suited for all applications in beer flavour analysis and other, more selective, detectors have their place.

3.6.2.2 Vicinal diketones
Diacetyl and 2,3-pentanedione are formed during almost all brewery fermentations but few beers on sale to the public will contain more than traces of these *vic*-diketones. Diacetyl is unwanted in beer because of its powerful buttery flavour which is detectable at about 0.05 ppm. The low concentrations of diketones in beer is testimony to the effectiveness of the conditioning processes employed by brewers to remove them. The precursors of the diketones, 2-acetolactate and 2-acetohydroxy-butyrate, are formed during active growth of yeast [46] and leak out of the cells into beer. Oxidative decarboxylation of the acetohydroxy acids occurs spontaneously, giving the corresponding diketones (Fig. 3.26). At the end of the fermentation, 'green' beer can contain not only enough diacetyl to make it unpalatable, but also surplus concentrations of the precursors. During storage of the beer the precursors are allowed to break down to diketones, and the diketones in turn are removed by prolonged contact with yeast. Yeast has the capacity to reduce diacetyl to acetoin and further to 2,3-butanediol.

For the sensitive measurement of diketones, an electron capture detector (ECD) is used [18,49]. The ECD measures a reduction in the standing current when some

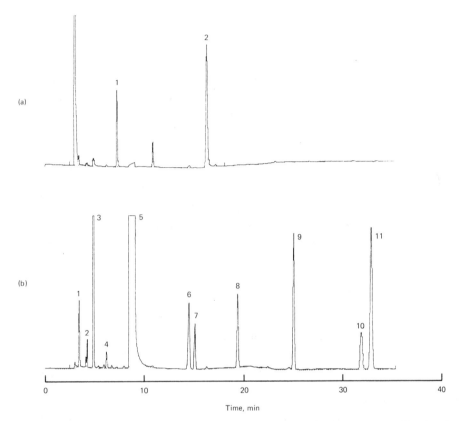

Fig. 3.27 — Gas chromatographic separation of headspace volatiles from beer on a 50-m fused silica capillary column (0.32-mm i.d.) coated with Carbowax 400, temperature programme 30–55°C. Split sample detection by — (a) ECD, peaks numbered; 1 = diacetyl, 2 = 2,3-hexanedione (int. std.): (b) FID, peaks numbered; 1 = acetaldehyde, 2 = acetone, 3 = ethyl acetate, 4 = ethyl proprionate, 5 = ethanol, 6 = isoamyl acetate, 7 = propanol, 8 = isobutanol, 9 = butanol (int. std.), 10 = 2-methyl butanol, 11 = 3-methylbutanol.

types of molecules, such as compounds containing conjugated carbonyl groups, disulphide groups or halogen atoms, pass between its electrodes. This form of structure-specific detector is second only to the FID in popularity because of its high sensitivity to certain beer components and its very poor sensitivity to many other substances. The ECD permits the detection of small amounts of diketones (Fig. 3.27) in the presence of much greater amounts of potential interferences. One of the brewer's main aims in the conditioning process is to ensure that contents of diacetyl are decreased below 0.05 ppm.

3.6.2.3   Dimethyl sulphide (DMS)
The sulphury smell associated with most lager beers is due mainly to dimethyl sulphide, but this volatile substance is not often present in significant amounts either in ale or stout. The current consensus concerning the origin of DMS in lager focuses on the presence in malt of *S*-methylmethionine (SMM), which is formed during the

germination of barley. S-Methylmethionine readily breaks down to DMS at elevated temperatures, so the amount of SMM in malt is dependent on the severity of the kilning process. The more lightly kilned malts used for lager brewing are more likely to contain higher residual levels of SMM than malts used for ale or lager brewing [46]. The mashing, wort-boiling and fermentation conditions used for ale and stout brewing may also favour the thermal decomposition of SMM more so than is encountered in lager brewing. For all these reasons, therefore, a lager beer typically contains 20–60 ppb of DMS, whereas this compound may be undetectable in ales and stouts. Measurement of this indispensable feature of lager quality requires specific and sensitive detection, for which a flame photometric detector is the first choice [50].

The flame photometric detector [18] utilizes a hydrogen diffusion flame to decompose sulphur-containing compounds (and phosphorus-containing compounds) to fragments which are then excited to a higher electronic state. When these excited molecules subsequently return to the ground state, they emit characteristic band spectra and it is this emission which is monitored with a photomultiplier, through a 392 nm bandpass filter. Such a detector was used to obtain the traces shown in Fig. 3.28, which demonstrate the differences in DMS contents in lager and stouts.

3.6.3 Quantitative thin-layer chromatography of lipids

Conventional thin-layer chromatography (TLC) is a technique that has contributed greatly over several decades to the analysis of lipids, but until the development of modern high-performance systems, quantitative analysis was tedious and subject to sizeable error. Before the advent of modern instrumentation, quantification often demanded the painstaking excision and recovery of separated components from developed plates. Since then, several ingenious devices have been invented to simplify and improve quantitative recovery. One such apparatus which developed into a commercial reality is the IATROSCAN Automated Thin Layer Scanning System [51]. In this system, the components of interest are separated on unique thin layer columns called 'Chromarods', before their detection and determination with a flame ionization detector (Fig. 3.29). A Chromarod consists of a thin layer of either silica or alumina fused to the surface of a 10 cm quartz core. Samples (1-μl) are applied to one end of the rods prior to their development with a solvent or mixture, chosen from the usual range used for conventional TLC. Up to ten Chromarods can be developed simultaneously in the special rod holder, in about 20 min. Each rod in the holder is then scanned automatically in sequence, by passage through a hydrogen-flame burner. As each Chromarod passes the flame zone (in 3 min) the separated organic compounds generate ions and thereby produce a change in current between the negative pole of the flame burner and the positive pole of the collector electrode. The amplified current is fed as an analogue signal to produce the chromatogram while the integral of the analogue output is used to calculate peak areas. Passage through the burner reactivates the Chromarods, which can be re-used up to 100 times.

One extra advantage of the system is that it offers options for partial scanning followed by redevelopment. For instance, neutral lipids, glycolipids and phospholipids from barley were separated by sequential development with three solvents of

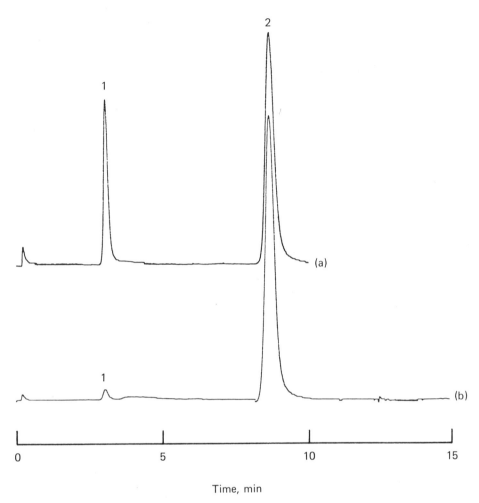

Fig. 3.28 — Gas chromatographic separation of dimethyl sulphide (peak 1) and ethyl methyl sulphide (peak 2, int. std.) on a 1.4-m Teflon column (0.3-mm i.d.) packed with 40/60 Carbopack B HT-100, temperature programme 60–110°C. Detection by FPD, trace (a) for lager beer, trace (b) for stout beer.

increasing polarity [51]. Neutral lipids were mobile when Chromarods were developed first with hexane:ether (100:12) whereas all other lipids remained at the origin. The neutral lipids were quantified by feeding the Chromarods through the hydrogen burner, from the front end up to a point about 1 cm short of the origin (Fig. 3.30). The rods were then developed with chloroform:acetone:acetic acid:water (10:90:2:3) so that the glycolipids were spread throughout the column length, while phospholipids remained at the origin. Repetition of the partial scanning procedure quantified the glycolipids and reactivated the rods. Finally, the phospholipids were separated by development with chloroform:methanol:water (84:35:3).

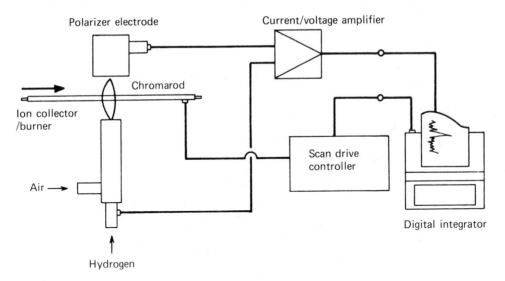

Fig. 3.29 — Schematic diagram of IATROSCAN Mk-5 TLC/FID analyser. (Reproduced by permission of World Distributors, Newman-Howells Associates Ltd., Llanwrtyd Wells, Powys, Wales, U.K.)

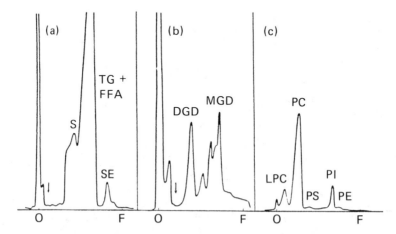

Fig. 3.30 — Chromatographic profiles of barley lipids obtained by sequential development of Iatroscan Chromarods with solvent mixtures of increasing polarities. Partial scanning between developments was to points indicted by arrows. (A) Neutral lipids, (B) Glycolipids, (C) Phospholipids. Identified peaks; S = sterols, TG = triglycerides, FFA = free fatty acids, SE = sterol esters, DGD = digalactosyl diglyceride, MGD = monogalactosyl diglyceride, LPC = lysophosphatidyl choline, PC = phosphatidyl choline, PS = phosphatidyl serine, PI = phosphatidyl inositol, PE = phosphatidyl ethanolamine. (Reproduced from [51], by permission of Oxford University Press.)

Such a system was used to good effect to study the fate of barley and malt lipids during the brewing process [51] and the consequences on beer quality. The lipid content of wort was shown to be influenced by the method of mashing, and autolysis of yeast at high temperature was shown to cause release of free fatty acids from phospholipids into beer. The presence of lipids in wort can have a significant influence on fermentation (Section 3.6.2) and lipids in beer can depress foam quality.

3.7 THE SERVO CHEM AUTOMATIC BEER ANALYZER (SCABA)

The strength of a beer is usually judged by its alcohol content, so several methods of measurement have beeen devised [4], based on the separation of alcohol from the beer by either distillation, microdiffusion, dialysis or gas chromatography. When distillation is used, the alcohol recovered can be determined by measurements of either the specific gravity or refractive index of the distillate. Clearly, measurement by such a method is time-consuming and the need for a more rapid though equally reliable system motivated the development in Sweden of an automated instrument by Servo Chem in collaboration with a brewing company [52]. A flow diagram of the SCABA instrument, which is capable of analysing twenty beer samples per hour for alcohol, original gravity, colour and pH, is shown in Fig. 3.31. A sample (50-ml) of degassed beer is pumped into a glass coil surrounded by circulating water from a precision thermostat ($20\pm0.01°C$) and from there to a bubble separator. The sample stream is then split into two, one part being directed to the densitometer while the other enters the evaporator cell. In the evaporator the alcohol in the sample rapidly vaporizes in the upward-flowing air stream, forming an equilibrium alcohol/air mixture at the top of the cell. The vapour mixture flows then into the alcohol detector, which consists of Wheatstone bridge circuit in which two of the filaments are coated with a catalyst. At the catalyst-coated filaments, the ethanol is oxidized to acetic acid and the heat generated causes a change in the resistance of the filaments. Accordingly, the change in the current in the bridge is proportional to the amount of alcohol oxidized. The system is not therefore, selective for ethanol, since higher alcohols that are readily vaporized will be measured also.

A density measurement is made on the sample stream directed to the U-shaped oscillator. This U-tube vibrates with a natural frequency which responds to changes in the density of the fluid flowing through it. From the densitometer the sample flows to a colorimeter and then to the pH meter. A microcomputer transforms the raw data into usable results, such as the alcohol content and original gravity of the beer [52]. The results agree well with those obtained by distillation or refractometry [52] and the precision of the method has been validated by collaborative testing [53].

3.8 FLOW-INJECTION ANALYSIS

The official method used at present for the measurement of beer bitterness involves the extraction of iso-α-acids from an acidified beer sample with a measured amount of iso-octane ([8–10] and Section 3.6.1.6). An automated method which uses a unique micro-phase separator and flow-injection analysis has been described recently [54]. The commercially available apparatus operates by injecting a fixed

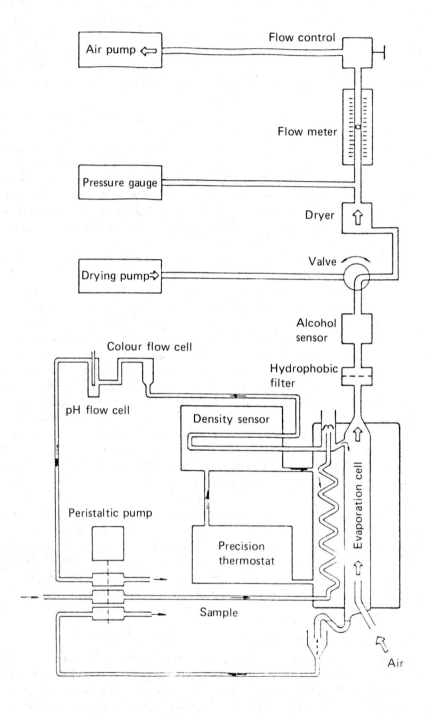

Fig. 3.31 — Schematic diagram of Servo Chem Automatic Beer Analyzer (SCABA).

volume of degassed beer into an unsegmented carrier stream containing $0.3M$ hydrochloric acid, and is then mixed on-stream with iso-octane, in a 2-m Teflon dispersion coil. The stream then passes as an emulsion to the phase separator, which embodies a tubular microporous membrane. Separation of the phases takes place by the permeation of iso-octane across the membrane, for collection in the extractant stream, while the impermeable aqueous stream is channelled to waste. The separated organic phase then progresses to a UV-detector for monitoring at 275 nm.

After calibration with beers of known bitterness contents, the system is capable of analysing a sample every 6.5 min with a relative standard deviation of 2%. When compared with the more laborious manual method, a correlation coefficient of 0.994 was obtained over a wide range of bitterness values [54]. One feature of the system which is conducive to greater precision than is usually obtained with the manual method is that the automated method does not depend on total extraction of analytes.

3.9 OVERVIEW AND PROJECTIONS

In the foregoing account, reference has been made to the different ways in which flexible instrumentation can be adapted to the special needs of the brewing chemist, and only one example has been given of an instrument designed specifically for beer analysis (i.e. SCABA, Section 3.7). Indeed, brewing chemists have always been quick to sieze whatever opportunities have been presented through advances in analytical/separation techniques. Some methodologies, such as immunochemistry and electrophoresis, have made only fleeting impact to date, but will no doubt be revisited many times in the future, as occasion demands. In contrast, the value of techniques such as AAS, NIRS, GC and HPLC is already firmly established, and we can expect many of the official methods of analysis to depend on such instrumentation in the future. Modern instruments have been especially valuable as research tools, in areas such as polyphenol analysis, for the determination of bittering substances, and in carbohydrate assays. Such has been their impact that we can expect many more new developments in the near future, even to the point at which advanced instrumental methods feature prominently in the standard manuals of routine analysis.

REFERENCES

[1] W. Knol, M. Minekus, S. A. G. F. Angelino and J. Bol, *Monatsschrift fur Brauwissenschaft*, 1988, **41**, 281.
[2] F. H. White, *The Brewer*, June, 1988, 234.
[3] D. E. Briggs, J. S. Hough, R. Stevens and T. W. Young, *Malting and Brewing Science, Vol. 1, Malt and Sweet Wort*, 2nd Ed., Chapman and Hall, London, 1981.
[4] J. S. Hough, D. E. Briggs, R. Stevens and T. W. Young, *Malting and Brewing Science, Vol. 2, Hopped Wort and Beer*, 2nd Ed., Chapman and Hall, London, 1982.
[5] J. R. A. Pollock, ed., *Brewing Science, Vol. 1*, Academic Press, London, 1979.
[6] J. R. A. Pollock, ed., *Brewing Science, Vol. 2*, Academic Press, London, 1981.
[7] J. R. A. Pollock, ed., *Brewing Science, Vol. 3*, Academic Press, London, 1987.
[8] *Analytica-EBC*, Fourth Edition, Brauerei- und Getranke-Rundschau, Zurich, 1987.
[9] *Recommended Methods of Analysis*, Institute of Brewing, 1977.
[10] *Methods of Analysis of the American Society of Brewing Chemists, Seventh Revised Edition*, American Society of Brewing Chemists, St. Paul, MN, U.S.A., 1976.

[11] K. H. Norris, in *Analytical Applications of Spectroscopy*, C. S. Creaser and A. M. C. Davies, eds., Royal Society of Chemistry, 1988, p. 3.
[12] I. Murray, in *Analytical Applications of Spectroscopy*, C. S. Creaser and A. M. C. Davies, eds., Royal Society of Chemistry, 1988, p. 9.
[13] R. D. Rosenthal, *An Introduction to Near Infrared Quantitative Analysis*, Neotec Instruments, Inc., 1978.
[14] W. F. McClure and R. E. Williamson, in *Analytical Applications of Spectroscopy*, C. S. Creaser and A. M. C. Davies, eds., Royal Society of Chemistry, 1988, p. 109.
[15] E. Stark, in *Analytical Applications of Spectroscopy*, C. S. Creaser and A. M. C. Davies, eds., Royal Society of Chemistry, 1988, p. 21.
[16] Anon., *Food Production*, May 1990, 10.
[17] Anon., *Beverage World International*, April 1989, 34.
[18] C. F. Poole and A. Schuette, *Contemporary Practice of Chromatography*, Elsevier Science Publishers B.V., Amsterdam, 1984., Chap. 5, p. 353.
[19] S. Ahuja, *Selectivity and Detectability Optimizations in HPLC*, Wiley, New York, 1989, Chap. 12, p. 505.
[20] T. H. Mourey and L. E. Oppenheimer, *Anal. Chem.* 1984, **56**, 2427.
[21] P. E. Shaw, *Handbook of Sugar Separations in Foods by HPLC*, CRC Press, Inc., Boca Raton, Florida, 1988.
[22] K. Aitzetmuller, *J. Chromatogr.*, 1978, **156**, 354.
[23] I. McMurrough and J. R. Byrne, in *Food Analysis by HPLC*, L. Nollet, ed., Dekker, New York, 1991, Chap. 11.
[24] J. B. Harborne, in *Secondary Plant Products*, E. A. Bell and B. V. Charlwood, eds., Springer-Verlag, Berlin, 1980, Chap. 6.
[25] B. Jende-Strid, in *Beer Analysis*, H. F. Linskens and J. F. Jackson, eds., Springer-Verlag, Berlin, 1988, p. 109.
[26] I. McMurrough, G. P. Roche and K. G. Cleary, *J. Inst. Brew.*, 1984, **90**, 181.
[27] P. J. Hayes, M. R. Smyth and I. McMurrough, *Analyst*, 1987, **112**, 1197.
[28] P. J. Hayes, M. R. Smyth and I. McMurrough, *Analyst*, 1987, **112**, 1205.
[29] I. McMurrough, *J. Chromatogr.*, 1981, **218**, 683.
[30] I. McMurrough and G. P. Hennigan, *J. Chromatogr.*, 1983, **258**, 103.
[31] I. McMurrough, G. P. Hennigan and M. J. Loughrey, *J. Agric. Fd. Chem.*, 1982, **30**, 1102.
[32] I. McMurrough, M. J. Loughrey and G. P. Hennigan, *J. Sci. Fd. Agric.*, 1983, **34**, 62.
[33] I. McMurrough, G. P. Hennigan and M. J. Loughrey, *J. Inst. Brew.*, 1983, **89**, 15.
[34] F. Garcia-Sanchez, C. Carnero and A. Heredia, *J. Agric. Fd. Chem.*, 1988, **36**, 80.
[35] C. F. Timberlake and P. Bridle, in *The Flavonoids*, J. B. Harborne, T. J. Mabry and H. Mabry, eds., Chapman and Hall, London, 1975, Chap. 5.
[36] J. Weiss, *Handbook of Ion Chromatography*, Dionex Corp., Sunnyvale, U.S.A., 1986.
[37] S. J. Priest, *Ferment*, 1989, **2**, 41.
[38] J. C. Jancar, M. D. Constant and W. C. Herwig, *J. Am. Soc. Brew. Chem.*, 1984, **42**, 90.
[39] D. C. O'Donnell, J. S. McIntosh, J. S. P. Fernando, J. M. Sue and B. K. Blenkinship, *Proc. Conv. Inst. Brew.*, (*Austr. and N.Z. Sect.*), Hobart, 1986, p. 221.
[40] L. E. Barber, *J. Am. Soc. Brew. Chem.*, 1990, **48**, 44.
[41] I. McMurrough, M. V. Lynch, F. Murray, M. Kearney and F. Nitzsche, *J. Am. Soc. Brew. Chem.*, 1987, **45**, 6.
[42] I. McMurrough, J. Byrne, E. Collins, M. R. Smyth, J. Cooney and P. James, *J. Am. Soc. Brew. Chem.*, 1988, **46**, 51.
[43] L. R. Snyder, J. W. Dolan and Sj. Van der Wal, *J. Chromatogr.*, 1981, **203**, 3.
[44] I. McMurrough, K. Cleary and F. Murray, *J. Am. Soc. Brew. Chem.*, 1986, **44**, 102.
[45] K.-H. Plattig, in *Sensory Analysis of Foods*, 2nd Ed., J. R. Piggot, ed., Elsevier, Barking, England, 1988, p. 1.
[46] S. Engan, in *Brewing Science, Vol. 2*, J. R. A. Pollock, ed., Academic Press, London, 1981, p. 93.
[47] H. A. B. Peddie, *J. Inst. Brew.*, 1990, **96**, 327.
[48] C. D. Baker, *J. Inst. Brew.*, 1989, **95**, 267.
[49] G. A. F. Harrison, W. J. Byrne and E. Collins., *Proc. Eur. Brew. Conv., 10th Congr.*, Stockholm, 1965, p. 352.
[50] G. A. F. Harrison, and C. M. Coyne, *J. Chromatogr.*, 1969, **41**, 453.
[51] H. Byrne, M. Loughrey and R. Letters, *Proc. Eur. Brew. Conv., 19th Congr.*, London, 1983. p. 659.
[52] H. Korduner and R. Westlius, *Proc. Eur. Brew. Conv., 18th Congr.*, Copenhagen, 1981, p. 615.
[53] ASBC Subcommittee report, *J. Am. Soc. Brew. Chem.*, 1988, **46**, 131.
[54] K. J. Switala and K. G. Schick, *J. Am. Soc. Brew. Chem.*, 1990, **48, 18**.

4

The analytical laboratory in the speciality sealants/adhesives industry

Raymond G. Leonard
Analytical Laboratory, Research and Development Department, Loctite (Ireland) Ltd., Whitestown Industrial Estate, Tallaght, Dublin 24.

4.1 INTRODUCTION

The analytical laboratory in a small-to-medium sized speciality chemicals industry finds itself in a rather unusual position. Unlike its counterparts in larger industries, it is much closer to the key profit/cost activity centres such as manufacturing and sales. Of necessity, therefore, the work of the laboratory has a bias towards the 'applied' rather than the research aspect in its work. The smaller laboratory needs to be more adept at juggling its resources to cope with a variety of demands on its capabilities. Prioritization of tasks becomes particularly important in order that the production and sales/marketing functions should have access to analytical support on an urgent basis to assist in the resolution of critical problems in their respective areas. Furthermore, in the case of the analytical laboratory operating in a smaller concern, there are a minimum of 'buffer' or 'filter' functions between the laboratory and the requesting function. The situation places the onus on the analytical department to 'interpret' or place its results in context when reporting them. This is important in order that corrective action may be taken at the earliest opportunity to remedy the problem. It also means that in order to be as effective as possible, the analytical laboratory needs to be involved in the whole process, from the identification of the problem, through the discussion of possible causes, definition of sampling strategy, choice of analytical techniques(s) and on to the resolution of the problem.

With respect to its relationship with functions such as manufacturing and marketing, the analytical laboratory is therefore perceived very much as a 'problem solver'. In this regard, many novel and interesting problems often come the way of the analytical laboratory, and this will be illustrated in the case of our laboratory through the discussion on case histories later in this chapter.

The span of techniques offered by the analytical laboratory is directly related to the depth and diversity of the technological base from which the company operates. In our case, the technological basis is relatively wide, and hence the analytical laboratory has had to develop its capabilities over a relatively wide range of modern analytical techniques.

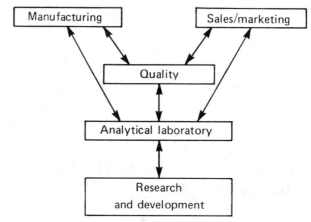

Scheme 4.1 — Analytical laboratory interactions.

Owing to its size limitations, however, the smaller analytical laboratory may not always possess the capabilities to solve every problem within its existing resources, but rather it may in certain circumstances act as a conduit for directing samples to appropriate external laboratories. In this case, the analytical laboratory would provide essential pre-analysis details regarding sample preparation, potential inter-ferences, likely concentration of analyte(s), etc.

A potential problem that sometimes arises in smaller organizations is the confusion that arises between the respective roles of the Analytical. and Quality Control (QC) functions. In our organisation they are quite distinct entities, with their own individual reporting structures operating to very different dictates. There is, however, a considerable degree of interaction between the analytical and QC laboratories, with the analytical laboratory often being involved in:

(i) development of chemical test procedures (for raw material, intermediates and finished products) for use by the QC laboratory;

(ii) operator training — provided to QC for all new test procedures;

(iii) providing assistance/advice with regard to selection/purchase of new analyti-cal instrumentation by QC;

(iv) confirmation of QC results — as requested by them;

(v) provision of more detailed analyses of samples as deemed appropriate.

4.2   COMPOSITION AND CHEMICAL BEHAVIOUR OF ANAEROBIC SEALANTS AND ADHESIVES

A summary of the range of products manufactured by Loctite is given in Table 4.1. For the purpose of this discussion, however, the examples that we cite in this chapter are drawn primarily from the areas of anaerobic sealants and acrylic adhesives.

Anaerobic sealants [1,2] are essentially liquids which remain stable (i.e. in the original liquid form) in the presence of oxygen. When confined between two closely fitting metal parts (e.g. nut and bolt), thereby excluding oxygen, the monomeric constituents in the liquid rapidly polymerize. This polymerization is rendered

Table 4.1 — Summary of major products manufactured by Loctite

1. Anaerobic products (methacrylate ester-based) — sealing, retaining, locking and gasketing applications
2. Instant adhesives (alkyl cyanoacrylate ester-based)
3. Acrylic (toughened) adhesives.
4. Ultra-violet curing adhesives and sealants (acrylate/methacrylate ester-based)
5. Epoxy adhesives (specialized)
6. Silicone adhesives/sealants
7. Rust treatment/prevention products

especially favourable if the metal is able to catalyse the redox-based "cure" process. These sealants consist of a complex mixture of methacrylate (or less commonly, acrylate) esters, free radical initiators, accelerators, free radical inhibitors, metal-chelating agents, plasticizers, thickeners, inert fillers, pigments and dyestuffs. These constituents are very carefully adjusted to give the correct balance between speed of cure and stability.

The polymerizable methyacrylate monomer is typically the dominant component in the composition. The monomer may be present as a single component or as mixtures. Some examples of commonly used methyacrylate esters are given below.

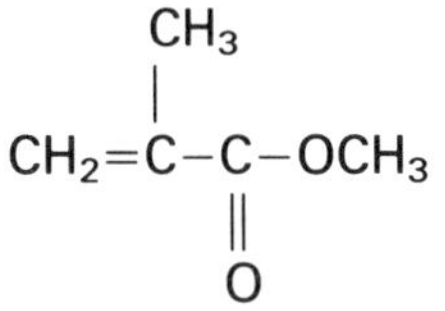

$$CH_2=\underset{\underset{O}{\parallel}}{\overset{\overset{CH_3}{|}}{C}}-C-OCH_3$$

Methyl methacrylate (I)

$$CH_2=\underset{\underset{O}{\parallel}}{\overset{\overset{CH_3}{|}}{C}}-C-O(CH_2-CH_2-O)_n-\underset{\underset{O}{\parallel}}{\overset{\overset{CH_3}{|}}{C}}-C=CH_2$$

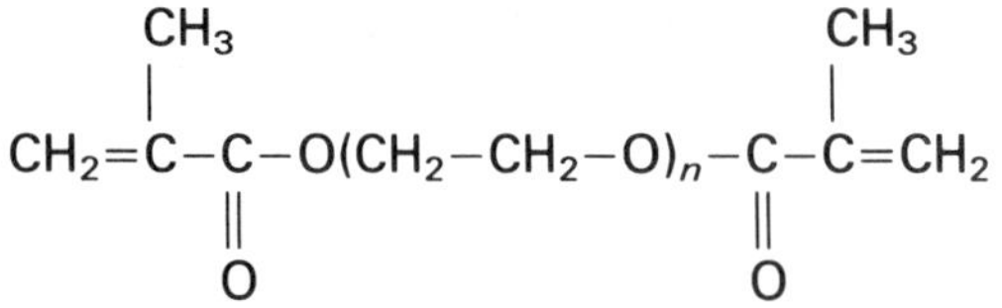

$n=3,4,5$ typically

Polyethylene glycol dimethylacrylate (II)

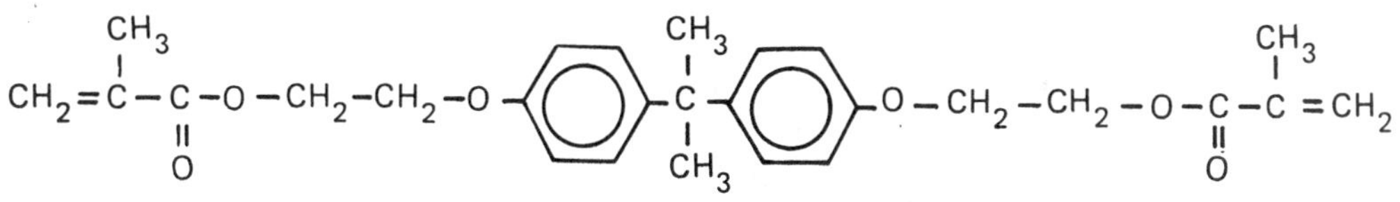

$$CH_2=\underset{\underset{O}{\parallel}}{\overset{\overset{CH_3}{|}}{C}}-C-O-CH_2-CH_2-O-\phi-\underset{\underset{CH_3}{|}}{\overset{\overset{CH_3}{|}}{C}}-\phi-O-CH_2-CH_2-O-\underset{\underset{O}{\parallel}}{\overset{\overset{CH_3}{|}}{C}}-C=CH_2$$

Ethoxylated bisphenol A dimethyacrylate (III)

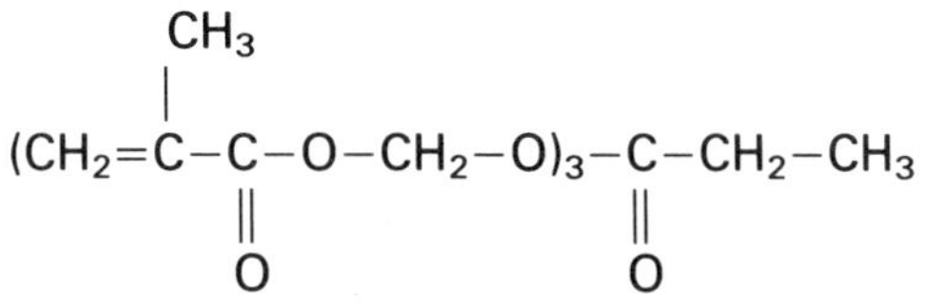

$$(CH_2=\underset{\underset{O}{\parallel}}{\overset{\overset{CH_3}{|}}{C}}-C-O-CH_2-O)_3-\underset{\underset{O}{\parallel}}{C}-CH_2-CH_3$$

Trimethylolpropane trimethacrylate (IV)

The cure systems typically incorporate an organic peroxide, an organic reducing agent and an organic acid. The peroxides used tend to be alkyl hydroperoxides such as:

Cumene hydroperoxide (V)   *tert*-Butyl hydroperoxide (VI)

Peresters and diacylperoxides are also used occasionally. Typical examples would be the following compounds:

tert-Butyl perbenzoate (VII)   Benzoyl peroxide (VIII)

The organic reducing agents are usually aromatic amines, such as:

N,N-Dimethyl-*p*-toluidine (IX)   Tetrahydroquinoline (X)

A variety of organic acids including *p*-toluenesulphonic acid and maleic acid have been used as cure accelerators. The other essential components in any anaerobic or acrylic formulation are the stabilizing components which are required to prevent any premature polymerization, either during manufacture of subsequent storage of the formulation prior to use. The free radical inhibitors may include the following compounds.

Hydroquinone (XI)   *p*-Methoxyphenol (XII)   Pyrogallol (XIII)

In certain instances, metal sequestering agents are incorporated to preclude any premature polymerization inititated by trace levels of contaminating transition metal ions — in particular copper(II), iron(II), nickel(II) and cobalt(II).

Viscosity modifiers, such as plasticizers and thickeners, inorganic fillers, silica-based thixotropic agents, pigments, and dyestuffs are also used in order to attain the desired physical form and appearance.

A rather simplifed reaction scheme of the redox-based cure chemistry of anaerobic sealants is shown below [3].

$$ROOH + M^{n+} + H^+ \rightarrow RO^{\bullet} + M^{(n+1)+} + H_2O$$
$$ROOH + M^{(n+1)+} \rightarrow ROO^{\bullet} + M^{n+} + H^+$$
$$2ROO^{\bullet} \rightarrow 2RO^{\bullet} + O_2$$
$$RO^{\bullet} + H_2C{=}CH_2 \rightarrow ROCH_2CH_2^{\bullet} \text{ etc.}$$

The above reaction scheme highlights the key catalytic role played by the active metal surface. In essence, the metal surface functions as an intrinsic component of the cure mechanism. The corollary of this is, however, the necessity to exclude active metal contamination from the sealant manufacturing process through rigorous control of both raw materials and plant equipment/processing conditions.

The analytical laboratory required to support this technology must of necessity develop considerable capabilities, competance and expertise in the area of both organic and trace metal analysis. In the latter case, particular expertise is required with regard to those middle row transition metals which have recognized redox catalytic properties.

4.3 TYPICAL APPROACHES TO THE ANALYSIS OF ANAEROBIC SEALANTS AND ADHESIVES

4.3.1 Monomers

The presence and/or identity of the polymerizable component(s) in a formulation is routinely established by a combination of nuclear magnetic resonance [4] (NMR) and infrared (IR) spectroscopy [5] together with gas (GC) and high-performance liquid chromatographic (HPLC) techniques. In the case of anaerobic sealants, which cure through a room temperature redox–initiated mechanism, the monomeric components are typically mono-, di- or tri-functional methacrylate esters or a combination of these. The ^{1}H and proton decoupled ^{13}C-NMR spectra of triethylene glycol dimethylacrylate (TRIEGMA) (XIV) recorded on a 270 MHz FT-NMR instrument are shown in Figs. 4.1 and 4.2 respectively. This monomer is used extensively in anaerobic products.

The proton NMR spectrum (Fig. 4.1) shows the methyl substituents from the methacrylate end groups at 1.9 ppm. The methylene groups bonded to ether oxygen appear in the range 3.6–3.8 ppm, and the methylene groups bonded to ester oxygen are found at 4.3 ppm. The terminal olefinic protons give resonances at 5.6 and 6.1 ppm.

The ^{13}C spectrum of the TRIEGMA monomer (Fig. 4.2) shows the following resonances; the methyl substituents at 18 ppm, ester methylene groups at 64 ppm, ether methylene groups at 69 and 71 ppm, olefinic carbon atoms at 136 ppm and 137 ppm (terminal carbon atom) and carboxyl carbon atoms at 168 ppm.

Such multifunctional esters are used to promote the formation of a rigid cross-linked structure on polymerization. These esters are normally determined by GC

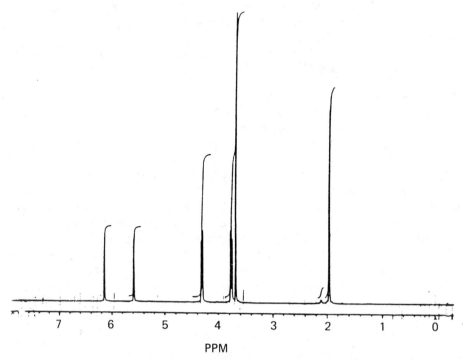

Fig. 4.1 — ¹H NMR spectrum (270 MHz) of triethyleneglycol dimethacrylate in CDCl₃.

analysis with an internal standard. Either narrow or wide bore capillary columns may be used. In the case of ultraviolet cure initiated formulations, the use of higher molecular weight urethane–methacrylate monomers is involved. The 270 MHz ¹H and ¹³C spectra obtained for a typical multifunctional polyether urethane methacrylate resin (XV) are shown in Figs 4.3 and 4.4 respectively.

The proton NMR spectrum of the urethane resin (Fig. 4.3) includes resonances typical of methacrylate esters as outlined above. In addition to these, bands due to alkyl groups (1.0–1.4 ppm), methyl substituents on aromatic rings (2.2 ppm) and aromatic protons (7–8 ppm) are present.

The ¹³C spectrum of the same urethane resin (Fig. 4.4) shows in addition to the methacrylate resonances, resonances associated with alkyl substituents (17 ppm) aromatic rings (110–140 ppm) and the urethane carboxyl moiety (154 ppm).

$$\underset{\displaystyle \overset{\displaystyle O}{\|}}{H_2C=C-C}-O(-CH_2CH_2O)_3-\underset{\displaystyle \overset{\displaystyle O}{\|}}{C-C}=CH_2$$

with CH_3 substituents on the two central carbons.

Triethylene glycol dimethacrylate (XIV)

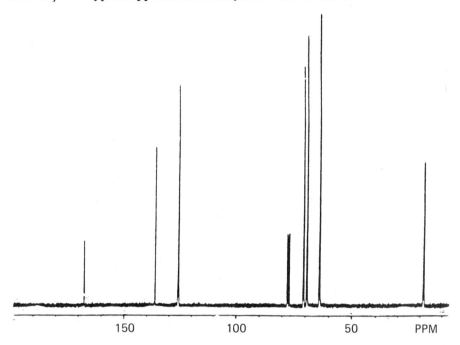

Fig. 4.2 — ^{13}C NMR spectrum (270 MHz) of triethyleneglycol dimethacrylate in CDCl$_3$.

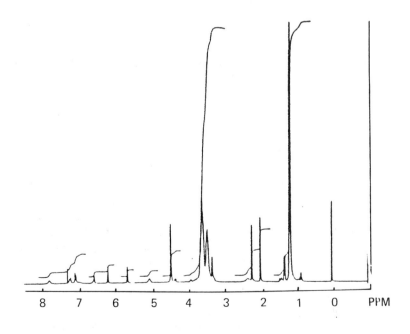

Fig. 4.3 — ^{1}H NMR spectrum (270 MHz) of a urethane–methacrylate resin in CDCl$_3$.

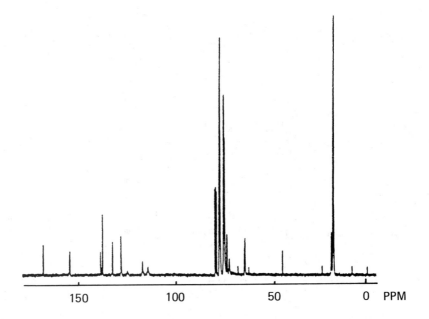

Fig. 4.4 — ^{13}C NMR spectrum (270 MHz) of a urethane–methacrylate resin in $CDCl_3$.

Urethane methacrylate resin (XV)

Such resins may be used in combination with the simpler methacrylate monomers to improve the performance of the cured product. Their relative involatility and the presence of a UV-absorbing chromophore renders HPLC the preferred approach for analysis. Again, the use of an internal standard is favoured for quantification purposes. Typical GC and HPLC traces for an anaerobic sealant and a UV curing adhesive are shown in Figs. 4.5 and 4.6 respectively.

4.3.2 Cure systems
The performance of modern adhesive/sealant formulations is critically dependent on the cure system used. This must be judiciously balanced in order to give the desired performance, especially with respect to cure speed, together with a viable stability

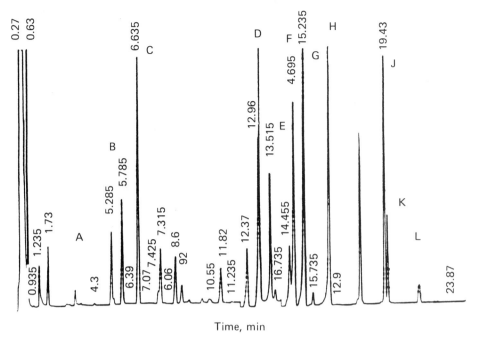

Time, min

Fig. 4.5 — GC trace of a typical anaerobic sealant. Column: 25 m×0.53 mm OV1701 (d_f=1.0 μm). Carrier gas: helium (10 psi). Injector: split 10:1. Detector: F.I.D. at 300°C. Temperature: 60°–180°C, programmed. A, carboxylic acid; B, organic peroxides; C, internal standard(I); D, methacrylate ester; E, aryl hydrazine; F, methacrylate ester; G, internal standard; H, methacrylate ester; I, methacrylate ester; J, plasticizer; K, methacrylate ester; L, methacrylate ester.

for the packaged product. The cure promoters are present at relatively low concentration (approx. 1%). It is essential therefore that appropriate analytical methods and techniques are available to identify and quantify these reactive components.

Thin-layer chromatography (TLC) is a fast, inexpensive technique well suited to adhesive/sealant analysis. Unlike many analytical techniques, little advanced instrumentation is required, apart from an ultraviolet/visible light source for some tests. In addition, sample preparation is generally minimal. Furthermore, in conjunction with specific spray reagents, TLC can rapidly be used to confirm the presence of organic peroxides, reducing agents, organic acids and photoinitiators (if a UV curing formulation is involved). The presence of certain free radical inhibitors can also be established by TLC.

The technique as employed in our laboratory uses commercially available precoated plates (Silica Gel F254) and involves direct application of dilute solutions of the adhesive products followed by development of the plate in an appropriate solvent system. The curatives in the formulation separate from the other major components (and from each other) as the chromatogram is developed. Subsequent analysis of the TLC plate is carried out with the use of specific spray reagents for visual detection of the components. A common curative used in adhesive formulation is the peroxy initiator, cumene hydroperoxide (CHP) (V). A TLC plate

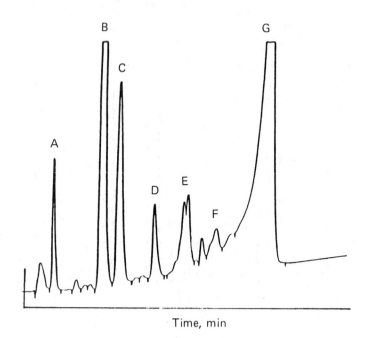

Fig. 4.6 — HPLC trace of a typical UV curing adhesive. Column: Waters M Bondapak, 10 μm, 4.6 mm × 300 mm. Mobile phase: 60–100% THF in water. Flow rate: 1.5 ml/min. Detector: UV at 254 nm. A, adhesion promoter. B, photoinitiator; C, internal standard; D, methacrylate ester; E, urethane–methacrylate oligomers; F, methacrylate ester; G, urethane–methacrylate resin.

containing the substance is sprayed with an aqueous methanolic solution of N,N-dimethyl-p-phenylenediamine dihydrochloride [6] (XVI) and heated gently. The presence of CHP is confirmed by the appearance of purple spots on the plate, corresponding in R_f value with a standard solution of CHP on the same plate.

N,N-dimethyl-p-phenylenediamine dihydrochloride (XVI)

Aromatic amines are also components in adhesive/sealant catalyst systems. The spray reagent used to detect these compounds consists of a dilute aqueous solution of ferric chloride and potassium hexacyanoferrate [7]. The presence and chemical identity of the amines (and other appropriate reducing agents) can be determined by the appearance of variously coloured spots with different R_f values on the chromatogram.

In a similar fashion, acidic species present in some catalyst systems can be determined on TLC plates by the use of a dilute aqueous solution of 2,6-dichlorophenolindophenol sodium salt (XVII) [8].

2,6-dichlorophenolindophenol sodium salt (XVII)

Free radical stabilizers such as hydroquinone, *p*-benzoquinone, *p*-toluquinine, chloranil, and 1,4-naphthoquinone can be detected by reaction with rhodanine (XVIII) and ammonia (spot test or TLC method) to give a coloured complex [9] (XIX).

(XVIII)

(XIX)

The above procedures are useful for qualitative work, but commercially available TLC plate scanners (with varying levels of automation) can be used for quantitative assays, if required.

GC [10,11] and HPLC [12] chromatographic procedures may then be used to confirm the TLC results, as well as being used to quantify the level of each component. These chromatographic techniques moreover have the added advantage that they may also be used to identify and quantify the monomeric and plasticizer components of the formulation (Figs. 4.5 and 4.6). As stated previously, either narrow or wide bore capillary columns may be used for GC analysis. Preferred stationary phases are OV1 (polydimethylsiloxane) and OV101 (polyphenylmethylsi-loxane). Flame ionization (FID) is the usual mode of detection, but where unambiguous confirmation of peak identity is required, a GC coupled to a bench top mass spectrometer is available.

In the case of HPLC analysis [12], the usual approach adopted by our laboratory is to employ the reversed-phase mode using a C18-type stationary phase. The favoured eluant tends to be a binary phase based on methanol or tetrahydrofuran with water. The latter solvent mixture has the advantage that the strong solvent capability of the tetrahydrofuran readily dissolves the polymeric components usually incorporated in these formulations. A variable-wavelength UV/Visible detector is used, and when quantification is required the compound is monitored at its wavelength of maximum absorption.

The peroxide content of sealants/adhesives is normally established by an iodo-metric titration, which involves reducing the peroxy group with iodide ion in an acidic medium and titrating the liberated iodine with standard sodium thiosulphate [13,14]. The end-point is established by potentiometric detection with a platinum–reference electrode combination. Aromatic amine content is determined by titration with perchloric acid in acetic acid as solvent. The end-point is again established by using potentiometric detection with a glass–reference electrode combination. The latter electrode combination is also favoured for determination of the acidic components of the formulation. Dilute sodium hydroxide or tetrabutylammonium hydroxide are used as titrants in aqueous or non-aqueous media respectively.

In the case of the organic peroxides, an important test applied to the incoming materials is an assay procedure based on their measured active oxygen content. This is routinely determined by a titrimetric procedure involving iodometric analysis (Fig. 4.7). This procedure, although well established, is relatively time-consuming. In our laboratory we have compared the accuracy and precision of chromatographic assays [14] (GC or HPLC as appropriate) with the conventional (iodometric) procedure for

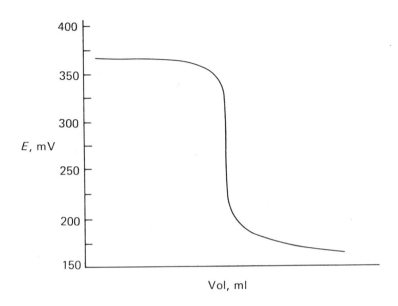

Fig. 4.7 — Representative titration plot obtained from iodometric analysis for organic peroxides.

a number of commercially available organic peroxides. The principal advantage accruing from the use of the chromatographic approaches is the relatively short analysis time (5–6 min) compared with titrimetric assay (15 min). In the case of the GC assay, the use of an on-column injection technique is necessary to avoid thermal decomposition of the peroxide in the injection port.

Representative GC and HPLC traces are reproduced as Figs. 4.8 and 4.9 respectively.

Based on these studies, the following operating procedures may be adopted.

(i) Either GLC and/or HPLC analysis may be substituted for iodometric analysis in the case of benzoyl peroxide and cumene hydroperoxide without any apparent loss in either precision or accuracy.

(ii) GC analysis may be substituted for iodometric analysis in the case of *tert*-butyl hydroperoxide without any apparent loss in accuracy or precision. In addition, di-*tert*-butylperoxide may be readily assayed by GC with acceptable precision. In this case, the alternative iodometric approach is both tedious and problematic.

(iii) HPLC analysis may be readily substituted for iodometric analysis in the case of *tert*-butyl perbenzoate without any apparent loss in either accuracy or precision.

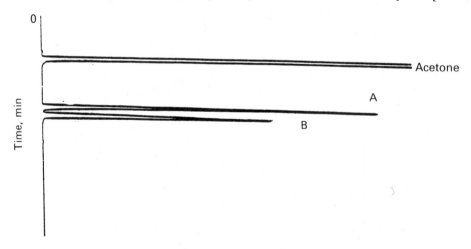

Fig. 4.8 — GC trace of *tert*-butylhydroperoxide. Column: 25 m×0.53 mm, OVI (d_f=1 μm). Injector: on-column. Detector: FID. Temperature: 70–250°C, programmed. A, internal standard; B, *tert*-butylhydroperoxide.

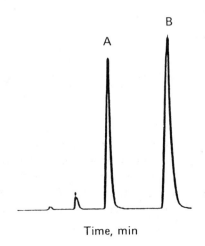

Fig. 4.9 — HPLC trace representing analysis of *tert*-butylperbenzoate. Column: Technopack C_{18}–30 cm×3.9 cm. Mobile phase: acetonitrile:water — 75:25. Flow rate: 1.0 ml/min. Detector: UV at 254 nm. A, *tert*-butylperbenzoate; B, internal standard.

4.3.3 Other compounds

Vicosity modifiers, usually referred to as thickeners, are used in varying concentrations throughout the range of sealants and adhesives. They span a diverse range of chemical structures. Some of the more commonly encountered thickeners are given below:

Poly(alkylmethacrylates) — homo and copolymers
Poly(alkylacrylates)   — homo and copolymers
Polystyrene

Acrylonitrile–butadiene–styrene copolymers
Poly(vinyl acetate)
Fumaric or maleic acid based polyesters

The chemical identity of such materials may readily be confirmed by IR spectroscopy. The initial step involves isolation of the thickener from the formulation matrix. Typically, the adhesive sample is diluted with chloroform to reduce the viscosity. This solution is then subjected to dropwise addition of a polar solvent such as methanol, until the thickener is fully precipitated out of solution. After careful drying to remove all traces of solvent, the sample can be analysed by Fourier transform (FTIR) [15] or dispersive infrared spectroscopy. The IR spectrum of an acrylic thickener identified as a methyl methacrylate/butyl methacrylate copolymer is shown in Fig. 4.10.

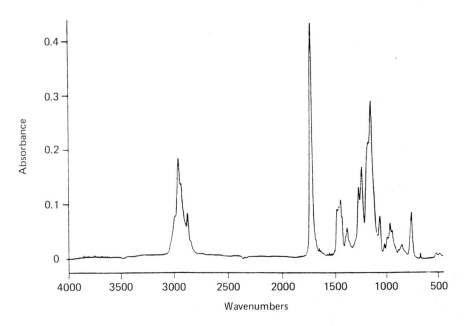

Fig. 4.10 — FTIR spectrum of copolymer of methyl methacrylate and butylmethacrylate.

Fillers and thixotropic agents are used widely throughout the various acrylic sealant and adhesive products. Fillers also form a significant component of epoxy-based adhesives. In a recent example it was desired to identify the filler component used in the hardener component of a competitive epoxy product. The isolation sequence involved centrifugation of the product to isolate the crude filler component. The latter was then washed repeatedly with hexane and dried thoroughly. The fine dry powder was used to prepare a standard KBr disc for infrared analysis. The spectrum obtained for the sample is shown in Fig. 4.11, identifying this particular filler as calcium carbonate.

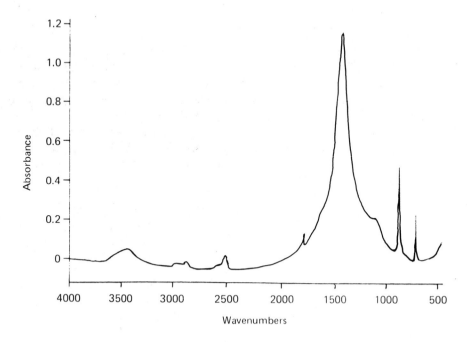

Fig. 4.11 — FTIR spectrum of calcium carbonate filler.

4.3.4 Trace metals

Because of their ability to initiate the polymerization process, there is a requirement to analyse both raw materials and finished formulations for trace levels of the common transition metals (iron(III), copper(II), cadmium(II), nickel(II), cobalt(II), manganese(II) etc.). This is commonly achieved by using atomic-absorption spectrometry (AAS). In recent years there has however arisen the need to distinguish between free and chelated metal ions. This has led us to explore the feasibility of using ion chromatography (IC) [16–18]. At present the dynamically coated reversed-phase approach [19,20], combined with post-column derivatization with PAR [4-(2-pyridylazo)-2-resorcinol] (XX) reagent and visible detection at 500 nm appears to be a very promising technique.

The initial step in the procedure involves the use of a silica-based solid-phase extraction (SPE) cartridge to extract and preconcentrate the metals.

4-(2-Pyridylazo)-2-resorcinol (XX)

After a wash with an appropriate solvent(s) to remove any retained contaminants, the metals are eluted from the SPE with the mobile phase [2mM sodium octanesulphonate/30mM tartaric acid/20mM citric acid, adjusted to pH 3.4 with NaOH]. The eluate is then passed through a C18-type SPE to remove any trace organic contamination prior to chromatographing the extract. This final eluate is then chromatographed on a C18 reverse-phase column with the above mobile phase, and the individual metal ions are quantified after post-column derivatization with PAR reagent and detection at 500 nm.

The analysis of a typical mixture of transition metal ions in a typical organic matrix is shown in Fig. 4.12. A repeat analysis in the presence of an excess of tetrasodium EDTA showed no response for any of the metal ions, indicating that this procedure could be used to detect free metal ions in the presence of complexed metal ions [20].

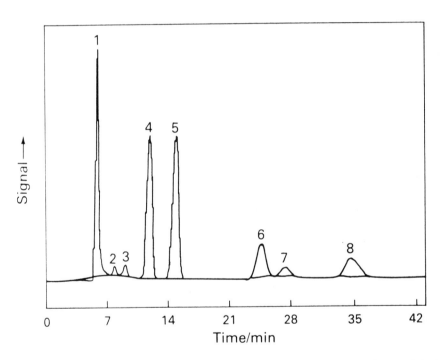

Fig. 4.12 — Separation of: 1, Cu(II); 2, Pb(II); 3, Ni(II); 4, Zn(II); 5, Co(II); 6, Fe(II), 7, Cd(II); 8, Mn(II). (Reproduced from [20] by permission of the Royal Society of Chemistry.)

The technique is essentially restricted to the analysis of bivalent metal ions, but in the case of the metal ions in which we are primarily interested, this represents the state in which they would typically be encountered.

4.4 CASE HISTORIES INVOLVING PROBLEM SOLVING

4.4.1 Trace-metal contamination in a filler component

A problem was identified with a batch of a particular anaerobic formulation immediately after manufacture. The batch in question was found to exhibit evidence of instability (i.e. premature gelation) on accelerated temperature testing. Careful stepwise preparation of a laboratory microbatch, with stability being monitored at each raw material addition stage, led to the identification of a mica-type inorganic filler as the most likely contributor to the instability phenomenon. Investigation revealed that this was the first lot used from a new delivery.

A laboratory microbatch, prepared using a retained portion of the preceding batch, was found to give a normal stability profile. The analytical laboratory was requested to compare the two lots of filler in order to establish the reason for the differing stability behaviour. Accordingly, an experiment was devised to compare the relative concentration of metals that could be extracted from the two mica samples on contact with aqueous mineral acid. The only significant transition metal that was observed in the extracts was iron, and the relative concentrations as determined by atomic absorption spectrometry were as shown in Table 4.2.

Table 4.2 — Analysis of iron in mica samples

Solution	Iron content, ppm	
	Good sample	Bad sample
0.1M HCl	93	906
0.5M HCl	110	1080

4.4.2 Trace metal contamination in pigment component

A further instance where trace metal contamination apparently caused serious product quality problems was traced to the use of a particular batch of a component pigment. A retained sample from the previous shipment of pigment when formulated in the product was found to generate no anomalous effects. Total copper and iron as determined by AAS after dry ashing showed no significant differences. However, there was a considerable difference in the relative amounts of copper solubilized in dilute aqueous acid as detailed in Table 4.3. This particular formulation was known to display an extreme sensitivity to trace/sub-trace contamination by copper. The increased tendency of the later pigment delivery to release souble copper into the acidic formulation undoubtedly had a deleterious influence on the quality (performance/stability) of the sealant batch incorporating that pigment lot.

Table 4.3 — AAS analysis of pigment for trace metal content

Pigment	Observation	Total copper ppm	Total iron ppm	Soluble copper ppm	Soluble iron ppm
Old stock	Excellent stability	1100	86	0.1	56
New delivery	Poor stability	1086	64	18.6	30

4.4.3   Trace chromium contamination of porous metal sealant

A sample from a batch of a heat-curable sealant formulation was received from a customer. The complaint related to significantly reduced cure activity in a particular application. This type of sealant formulation is used to seal the pores of powdered or cast metal parts by filling the pores under vacuum, followed by thermal curing of the sealant. This technique prevents leaks and renders the metal parts more suitable for plating, etc. Contamination with chromium, and specifically chromium(VI), was known to seriously retard the thermal cure. Frequent system cleanout and replacement of the sealant is advised where chomium plated parts are impregnated.

The sealant sample was ashed in a muffle furnace and the residual ash was carefully dissolved in $1M$ sodium hydroxide. An aliquot of this solution was added to a polarographic cell containing $1M$ sodium hydroxide as supporting electrolyte. The potential was scanned by using the differential-pulse polarographic (DPP) mode with a static mercury drop electrode system from -0.60 to $-1.20\,\mathrm{V}$. A peak was obtained at $-0.92\,\mathrm{V}$. When a portion of the ashed residue was scanned over the same potential range in $0.2M$ potassium thiocyanate/$0.2M$ acetic acid, no peak was detected, indicating the absence of chromium(III). When the sample was spiked with 5 ppm each of chromium(III) and chromium(VI), the peak corresponding to chromium(VI) increased. The standard addition algorithm yielded the results for the sample of sealant that are summarized in Table 4.4.

Table 4.4 — Analysis for Cr(VI) and Cr(III) by differential pulse polarography

	Cr(VI), ppm	Cr(III), ppm
Sample	2.7	Not detected
Sample+5 ppm of Cr(VI)+5 ppm of Cr(III)	7.5	4.8

Representative traces for Cr(III) and Cr(VI) are shown in Fig. 4.13.

4.4.4   Varying catalytic effects of ferrous and ferric salts

A further instance where polarography proved to be very useful related to the comparison of a number of oil-soluble iron salts from different sources. These compounds were required as activators for certain anaerobic formulations. However, a carboxylic acid salt specified as containing 6% of iron was found to be considerably less active than a salt containing 9% of iron from another supplier. The difference in activity was far greater than would be expected on the basis of their quoted metal contents. As outlined in Table 4.5, the iron levels specified by the suppliers were confirmed by atomic absorption analysis. However, polarography (in the differential-pulse mode) revealed a considerable difference in the Fe(II)/Fe(III) ratios between sample A and samples B and C (Fig. 4.14). Speciation of iron(II) and iron(III) was achieved by using an ammonium pyrophosphate buffer adjusted to pH 9.0. It was known that Fe(II) was the more active form in terms of speed of cure — thus polarographic analysis rationalized the difference in performance found between the salts [21].

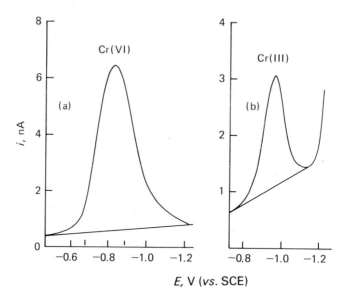

Fig. 4.13 — DPP traces for Cr(VI) and Cr(III). (a) Cr(VI) was dissolved in 0.5M NaOH; (b) Cr(III) was dissolved in 0.2M KSCN–CH$_3$COOH (pH 3.2).

Table 4.5 — Analysis of iron in oil-soluble salt solutions

Iron content, %		Samples		
		A	B	C
Fe	Content (Mfg. Spec.)	6.0	9.0	16.0
Fe	Content — AAS	6.0	8.8	16.0
Fe(II)	Content — DPP	0.0	1.3	2.8
Fe(III)	Content — DPP	6.0	7.5	13.2

4.4.5   Cure retardation arising from the presence of lead in brass

A difficulty associated with the bonding of brass metal, which contains mainly copper and zinc, was traced to the fact that certain brasses can contain up to 5% of added lead. This metal inhibits the formation of free peroxide radicals. For the voltammetric determination of lead in brass, a known mass of the sample was dissolved in concentrated nitric acid and evaporated to dryness. The residue was taken up in 0.1N ammonium citrate (pH 3.0) and scanned by differential-pulse polarography (DPP) over the range -0.1 V to -1.2 V. The traces shown in Fig. 4.15 show the expected peaks due to copper and zinc, and the additional presence of lead was obvious in the trace for the slower curing brass composition [21].

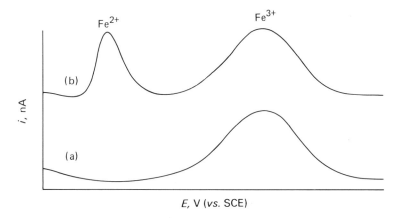

Fig. 4.14 — DPP traces showing speciation of Fe^{2+} and Fe^{3+} in commercial oil-soluble salts. (a) Sample containing Fe^{3+} only; (b) sample containing both Fe^{2+} and Fe^{3+}.

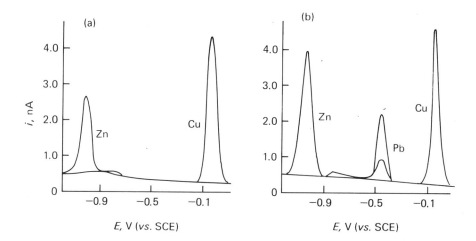

Fig. 4.15 — Typical DPP peaks obtained with brass in $0.1M$ ammonium citrate, pH 3.0: (a) contains no lead; (b) contains 4.5% lead, which was confirmed by re-scanning with a known concentration of lead.

4.4.6 Trace metal profile of a sealant formulation

The occasion sometimes arises where a sealant formulation must be carefully screened for the presence of any adventitious transition metal contamination. The preferred approach is to use differential-pulse anodic stripping voltammetry (DPASV) on a sample prepared as follows: 5.0 g of sealant is ashed for 4 hr along with 0.1 g of *p*-toluenesulphonic acid, in order to retain volatile metals. The resulting ash is dissolved in 20 ml of 0.1M ammonium citrate–citric acid buffer (pH 3.0) and 10 ml of this solution is added directly to the electrochemical cell. After the deposition stage, the metals are stripped from the electrode by scanning from -1.2 V to 0.3 V with a DPV waveform. The trace obtained with a typical sealant in the presence of some typical transition metal contaminants is shown in Fig. 4.16, together with that obtained with 100-ppb standards of the transition metals added to the same volume of supporting electrolyte [21].

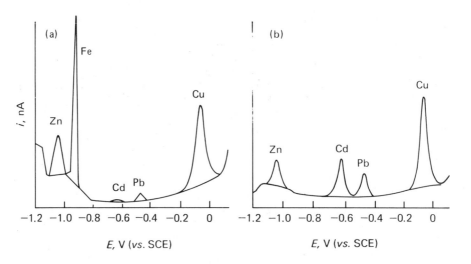

Fig. 4.16 — DPASV of trace metal content of a typical anaerobic sealant in 0.1M ammonium citrate, pH 3.0. (a) In the presence of Zn, Fe, Cd, Pb, and Cu; (b) with standards of 100 ppb of Zn, Cd, Pb, and Cu added to the same volume of supporting electrolyte.

4.4.7 Detection and characterization of sequestering agents

The use of chelating agents in anaerobic sealants has been referred to previously. The stability of many sealants is critically dependent on both the type of chelator and its concentration. Quantifying the level of a chelator in such formulations is a frequent analytical task. For example, it is possible to determine the level of ethylenediaminetetra-acetic acid (EDTA)-type chelators in such formulations by using DPV. This approach takes advantage of the fact that certain metal–EDTA complexes are reduced at more negative potentials than the metal ion itself. If an excess of a metal ion such as copper(II) is added to a suitable supporting electrolyte (e.g. 0.1M acetate buffer, pH 5.0) followed by a known amount of the sealant sample dissolved in methanol, then two separate peaks appear. These peaks represent the free metal ion and the complexed form respectively. The traces for copper(II) alone and that of copper(II) and copper(II)–EDTA mixtures with increasing concentration of chelator using a standard addition method, are shown in Fig. 4.17. If it is

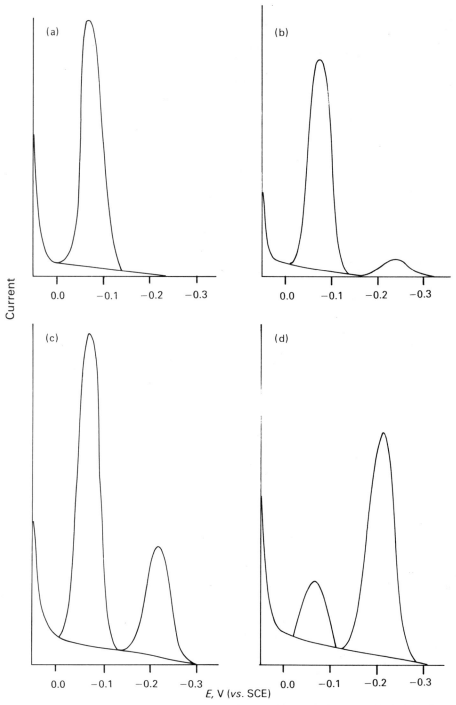

Fig. 4.17 — Determination of EDTA-type chelators in anaerobic sealants in $0.1M$ sodium acetate buffer. (a) Scan of Cu on its own; (b)–(d) scans of increasing concentration of sealant containing the chelator.

required to establish the stoichiometry of a metal–chelate complex, a modified approach may be used. In the case of the copper–EDTPA complex [EDTPA=1,1,1,1-ethylenedinitrilotetra(propan-2-ol)], Job's method of continous variation can be used [22]. The heights of the peaks due to free copper(II) and the copper(II)–EDTPA complex were measured by using DPV, and a Job plot was prepared as shown in Fig. 4.18. Based on the intercept of the two lines, the metal–ligand ratio was established as 1:1.

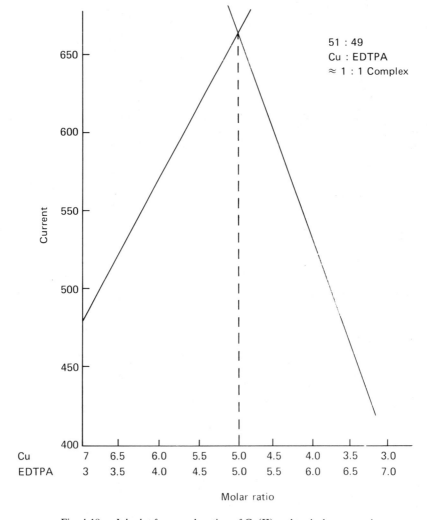

Fig. 4.18 — Job plot for complexation of Cu(II) and typical sequestering agent.

4.4.8 Detection of contaminating residues on printed circuit boards
Fourier–transform infrared spectroscopy (FTIR) [15] is especially useful when applied to the analysis of surface residues or contaminants on parts to be bonded or sealed.

This technique has shown particular benefit in the identification of residues on printed circuit boards (PCB). There is currently a considerable interest in developing adhesives suitable for use in electronic assembly work including solder mask compositions. A key requirement for this latter type of adhesive is that it should be easily removed after the soldering operation and that no contaminating residues should remain on the surface of the PCB.

The presence of such residues can be determined by treating the PCB with several drops of acetonitrile, and analysing the extract by FTIR. The solvent extract is evaporated to dryness on a KRS-5 or ZnSe plate before mounting in an attenuated total reflectance (ATR) accessory. The FTIR spectrum of a residue obtained from a PCB, which had been in contact with a competitive solder mask, is shown in Fig. 4.19. The contaminant was identified as a silicone release agent.

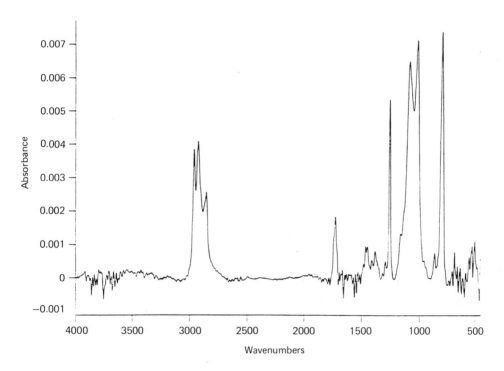

Fig. 4.19 — FTIR spectrum of residue from surface of printed circuit board.

4.4.9 Analysis of packaging materials and lacquers

The highly reactive nature of most sealant/adhesive formulations means that careful attention must be paid to the choice of packaging. The shelf stability, and indeed the eventual reactivity, can be strongly influenced by the choice of packaging. Resin-based adhesives as used within the electronic assembly industry (e.g. chipbonding adhesives) are a case in point. For example, two 'identical' black plastic syringes used for dispensing such a product were shown to have dramatically different effects on product stability. One syringe was identified by FTIR as the expected polypropylene (Fig. 4.20). However, the syringe causing gross product instability was identified as a polyamide (Fig. 4.21).

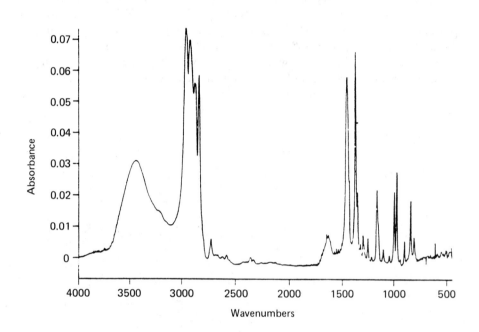

Fig. 4.20 — FTIR spectrum of polypropylene (plastic syringe).

Certain grades of cyanoacrylate esters (XXI) are employed in instant adhesives (superglues) and are packed in aluminium cartridges. The internal surfaces of such cartridges have a lacquer coating. The chemical nature of this coating can be confirmed by FTIR, by one of a number of standard sampling techniques. It is important to establish the identity of such a coating, for product compatibility purposes. Typical coatings include phenolic epoxy-type and styrene–acrylate copolymers. FTIR spectra for these coatings are shown in Figs. 4.22 and 4.23, respectively. Certain lacquer materials have been found to adversely affect the stability of the packaged cyanoacrylate adhesive, and hence it is important to be in a position to confirm the nature of the lacquer material.

$$CH_2 = C \begin{array}{c} CN \\ \diagup \\ \diagdown \\ CO_2R \end{array}$$

α-Cyanoacrylate ester (XXI)

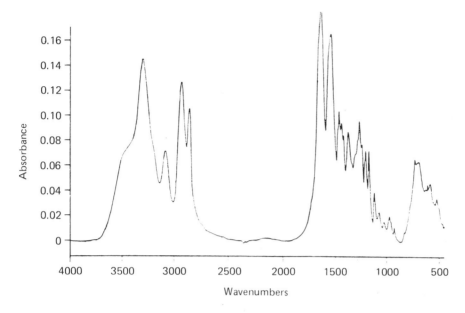

Fig. 4.21 — FTIR spectrum of polyamide (plastic syringe).

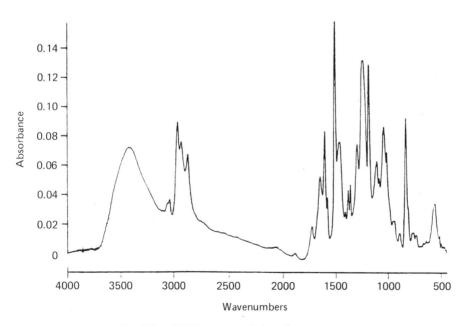

Fig. 4.22 — FTIR spectrum of phenolic–epoxy type lacquer.

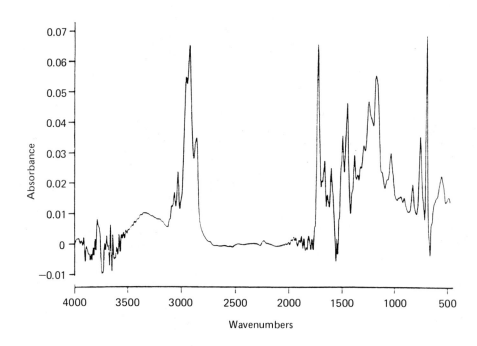

Fig. 4.23 — FTIR spectrum of styrene–acetate type lacquer.

Thermal analysis [23], and in particular differential scanning calorimetry (DSC) is used in the broader area of polymer evaluation and characterization. This technique is also useful for characterizing plastic packaging in terms of glass transition temperature (T_g), melting point, degree of crystallinity, etc.

4.4.10   Identification of curative agent in competitive adhesive

The complementary nature of the various analytical techniques is best illustrated by the compositional analysis of a competitive two-part loudspeaker bonding adhesive. The composition of Part A was established by a combination of techniques including ^{1}H-NMR spectroscopy, IR spectroscopy, TLC and GC to be as follows:

Reactive monomers	— methylmethacrylate
	— 2-hydroxyethylmethacrylate
Thicker	— butadiene–styrene–acrylonitrile–
	acrylate terpolymer
Catalyst	— cumene hydroperoxide
Thixotrope	— fumed silica

In like manner, the composition of the 'B component' was established to be as follows:

Reactive monomers	— methylmethacrylate
	— 2-hydroxyethylmethacrylate
	— trimethylolpropane trimethacrylate
Thickener	— butadiene–styrene–acrylonitrile–acrylate terpolymer
Thixotrope	— fumed silica

Identification of the catalyst in the 'B component' was adjudged to be important, because of the fast curing properties of this two-part product. GC with mass-spectrometric detection was used to analyse the 'B component'. The total ion chromatogram (TIC) is shown in Fig. 4.24. Mass spectra of all the significant

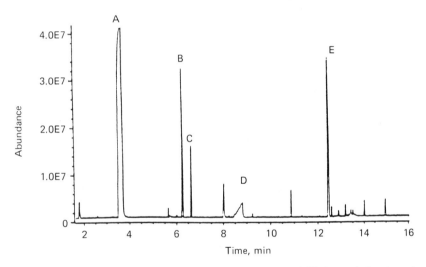

Fig. 4.24 — GC/MS (total ion) scan of competitive adhesive. A, 2-Hydroxyethylmethacrylate; B, methacrylate ester; C, methacrylate ester; D, unknown; E, trimethylolpropane trimethacrylate.

responses were studied in an effort to locate the critical cure promoter. The mass spectrum of the component marked "D" (Fig. 4.25) was adjudged to represent ethylene thiourea, based on both the molecular ion and the associated fragmentation pattern. This identification was confirmed on receipt and analysis of an authentic sample of same. This catalyst component was observed to promote a rapid cure. However, a literature survey established that this compound had certain negative characteristics from a health and safety point of view, and thus our own development programme focused on generating a similar rapid cure profile, but based on an alternative cure chemistry.

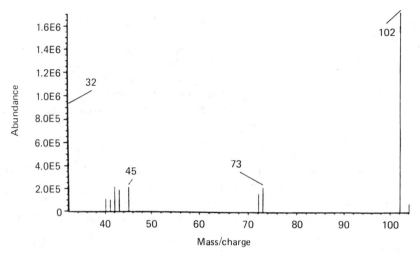

Fig. 4.25 — Mass spectrum of unknown (D).

4.4.11   Contamination originating from a packaging component

In the course of evaluating suitable packaging systems for a recently developed primer, an interesting phenomenon was observed. When the primer was packed in a small aerosol container, its activity or effectiveness was found to decrease sharply with time. A GC trace in the total ion (TIC) mode on the inactive primer revealed a predominant component at an increased retention time as compared to the active component (Fig. 4.26). The mass spectrum (Fig. 4.27) of this peak was attributed to a dibutyl phthalate. Further trials based on contacting the individual packaging

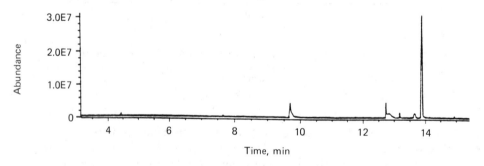

Fig. 4.26 — GC (total ion) trace of problem primer.

components with the primer solvent identified a rubber O-ring as the source of the phthalate plasticizer. The design of the package was amended to overcome this problem.

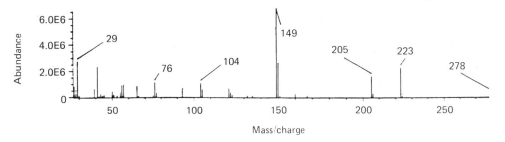

Fig. 4.27 — Mass spectrum of dominant component (from Fig. 4.26).

4.4.12 Determination of soluble oxygen content

Oxygen functions as a stabilizer in anaerobic sealant formulations. The situation sometimes arises where it is necessary to establish the level of dissolved oxygen in a formulation. The best approach based on the matrices involved was determined to be GC analysis with a thermal conductivity detector, and a packed column incorporating molecular sieve 5A and a precolumn packed with 10% OV17 on Chromosorb W [24]. The precolumn functions by retaining the majority of components in the sealant formulation when the temperature is maintained at close to ambient temperatures.

With the separation conditions described above, a linear calibration plot was obtained for peak area *vs.* standard oxygen volumes at STP (Fig. 4.28). The

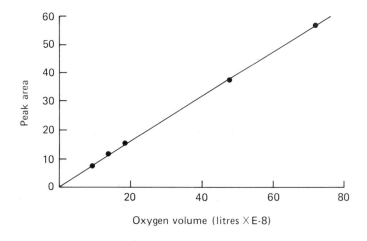

Fig. 4.28 — Graph of peak area *vs.* oxygen volume at STP.

correlation coefficient of this curve was determined to be 0.9996. The method was then applied to the determination of oxygen in typical anaerobic sealant formulations. A typical gas chromatogram for the GC analysis of oxygen in a sealant formulation is shown in Fig. 4.29. Calculation of the oxygen concentrations of

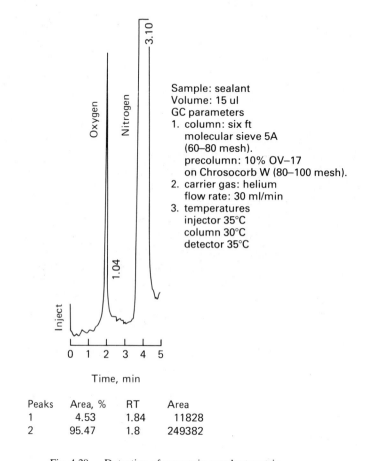

Sample: sealant
Volume: 15 ul
GC parameters
1. column: six ft
   molecular sieve 5A
   (60–80 mesh).
   precolumn: 10% OV–17
   on Chrosocorb W (80–100 mesh).
2. carrier gas: helium
   flow rate: 30 ml/min
3. temperatures
   injector 35°C
   column 30°C
   detector 35°C

Time, min

Peaks	Area, %	RT	Area
1	4.53	1.84	11828
2	95.47	1.8	249382

Fig. 4.29 — Detection of oxygen in a sealant matrix.

samples was obtained from the calibration curve and converting the volume of oxygen to weight by using the ideal gas law. The method was found to have a relative standard deviation of 4.1% and a limit of detection of 0.3 ppm.

4.5 SOME USEFUL ANALYTICAL TECHNIQUES IN THE ANALYSIS OF ANAEROBIC SEALANTS AND RELATED COMPOUNDS

4.5.1 Thermal analysis

Differential scanning calorimetery (DSC) [23] can be used to generate cure profiles of certain adhesive products. For example, in the electronics assembly business, epoxy-based chipboard products are used to hold components on PCBs prior to the automated soldering process. These adhesives are formulated with heat curatives, i.e. they are designed to harden on application of heat. The heat curing profile of these materials can be examined by DSC. The polymerization exotherm detected for a typical epoxy-based chipboard is shown in Fig. 4.30. The heat of reaction, together

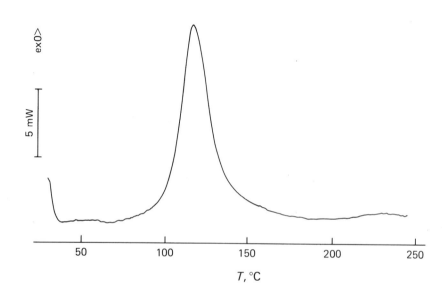

Fig. 4.30 — Cure profile of typical chipbonder product (DSC scan).

with peak temperature and other details, can be obtained from this trace. This type of analysis is useful, not only for product monitoring, but also as an aid to product development work where different cure profiles may be required to match specific applications.

Fully cured adhesive/sealant products can exhibit additional thermal transitions related to individual constituents in the formulation. The DSC trace obtained from a polymerized gasketing product is shown in Fig. 4.31. The pronounced endotherm peaking at 111°C is due to fusion of the partially crystalline polyethylene filler in the product. The process is reversible — recrystallization of the filler within the matrix can be followed during the cooling cycle in the calorimeter.

A further use of DSC in this area is the detection of incomplete (residual) curing in solid samples. A thermally curable epoxy adhesive which has not been fully cured will polymerize as the sample is heated in the calorimeter. This polymerization will give rise to a small exotherm which can be used to estimate the degree of cure in the original sample.

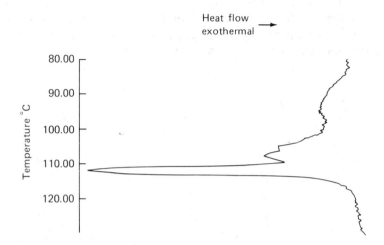

Fig. 4.31 — DSC scan of polymerized gasketing formulation.

Finally, DSC is used in the broader area of polymer evaluation to measure glass transition temperature, degree of crystallinity, decomposition profiles, etc. Current computer control and storage facilities available with thermal analysis equipment allows data to be comprehensively evaluated, re-analysed and compared, reducing the need for repeat analyses.

Thermogravimetric analysis (TGA) is a technique where weight changes in a sample can be continuously monitored with respect to temperature. TGA can be used to look at weight losses associated with the polymerization of an adhesive product, or to determine percentage volatiles released from a cured product during thermal cycling. The technique can also be used to determine mineral filler contents of certain products. In this case, a sample is heated gradually to a high temperature and maintained at this temperature for a period to ensure complete ashing. A typical thermogram obtained for a chipbonding product is shown in Fig. 4.32. The sample was heated from room temperature to 600°C, and held at the final temperature for 30 min. The resultant stable baseline obtained after complete combustion is used in the calculation of filler content for the sample (13.7%).

4.5.2   Ion chromatography
In recent years there has been a steady increase in the number of adhesives developed for particular niches and speciality applications in the industrial market. In the forefront of these are adhesives based on UV curing technology and used for chipbonding and conformal coating applications on printed circuit boards.

For these applications, ionic contamination (in particular the presence of halides) is problematic. Chloride in particular is a pervasive species, and in this context the

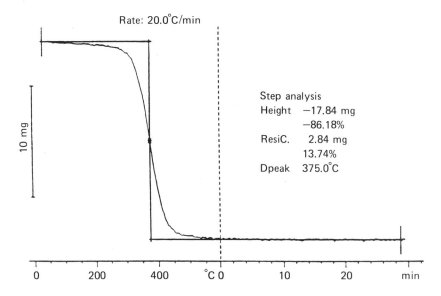

Rate: 20.0°C/min

10 mg

Step analysis
Height   −17.84 mg
             −86.18%
ResiC.   2.84 mg
             13.74%
Dpeak   375.0°C

0   200   400   °C 0   10   20   min

Fig. 4.32 — TG curve (room temp. to 600°C) for a chipbonding product to determine filler content.

hydrolysable or water-soluble chloride content requires to be monitored and carefully controlled. The conventional extraction procedure requires the adhesive to be contacted with deionized water for 16 hr at 80°C. The extract is then analysed for chloride content by chemically suppressed ion chromatography [25] with conductivity detection (Dionex As 4A column, eluent based on $2 \times 10^{-3} M$ NaHCO$_3$: $2 \times 10^{-3} M$ Na$_2$CO$_3$ or $10 \times 10^{-3} M$ sodium tetraborate). A typical IC chromatogram is shown in Fig. 4.33. As an alternative approach, or as a means of confirming the ion chromatogaphic results, polarographic analysis based on differential-pulse cathodic stripping voltammetry (DPCSV) may be used. A typical trace for chloride is depicted in Fig. 4.34. Both of these approaches allow the component raw materials to be individually assessed with regard to extractable chloride, so that a formulation may be developed using raw materials which contribute a negligible level of extractable chloride.

4.5.3   Gel-permeation chromatography
Gel-permeation (GPC) or size-exclusion chromatography [26] (SEC) is a favoured technique for the analysis of polymers and oligomeric compounds. GPC separates sample molecules in the mobile phase on the basis of their effective size. In most cases molecular size can be correlated to molecular weight. For the analysis of organic materials the column(s) is packed with a highly cross-linked spherical polystyrene–divinylbenzene matrix with a tightly controlled pore diameter. Any sample molecules that are smaller than the pore size can diffuse into and out of the pores, whereas sample molecules larger than the pores are effectively excluded. As a

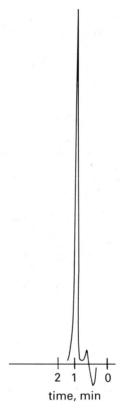

Fig. 4.33 — Ion chromatogram to find chloride content of aqueous extract of adhesive.

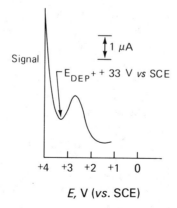

Fig. 4.34 — Differential-pulse polarography of 540 ppb of chloride.

result, larger molecules elute more rapidly than smaller molecules. The separation is thus essentially mechanical in nature, rather than chemical as in other forms of chromatography. All molecules larger than the pores elute first and at the same retention time. The smallest molecule that cannot penetrate the pore defines the exclusion limit of the column. Molecules smaller than a certain size have equal access to the pores and so they elute together at the total permeation volume or dead volume. Molecules beween these extremes are separated in order of their respective molecular weights.

In the analysis of a sample of unknown molecular weight or molecular weight distribution, the first step is to generate a calibration plot (log M.W. *vs.* elution volume or time) for a set of narrow disperse standards of known molecular weights. The function giving the best fit to the plot is then computed and the sample is analysed against this calibration function.

In this manner the number average molecular weight $(Mn=\Sigma N_i M_i/\Sigma N_i)$, the weight average molelcular weight $(Mw=\Sigma N_i M_i^2/\Sigma N_i M_i)$, the peak molecular weight (Mp) and the polydispersity (Mw/Mn) may be calculated.

Polystyrene calibration standards are most frequently used, and with polymers of different chemical structure, molecular weights are reported relative to polystyrene. The GPC trace and computed molecular weights obtained for an acrylate process oligomer are reproduced in Fig. 4.35.

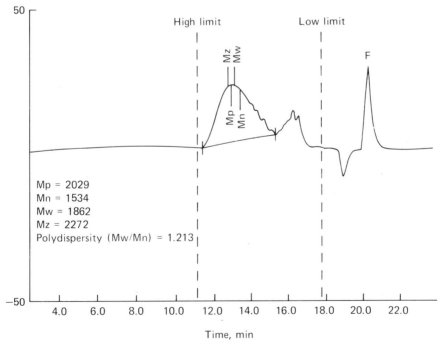

Fig. 4.35 — GPC trace of acrylate process oligomer.

4.5.4 Pyrolysis gas chromatography

Pyrolysis GC [27] is an analytical technique whereby complex involatile materials are broken down into smaller volatile constituent molecules by the use of very high temperatures. Polymeric materials, including cured sealants and adhesives, lend themselves very readily to analysis by this technique. Essentially, a fingerprint uniquely characteristic of the original material is obtained.

In combination with IR analysis, pyrolysis GC is particularly useful in the following areas:

(1) Characterization of polymeric thickeners used in the formulation of sealants and adhesives. The characteristic pyrogram obtained with a styrene–butadiene (SBR) rubber is reproduced as Fig. 4.36. Polymers of this type are frequently used to provide protective coatings — in rust preventative products for example.
(2) Characterization of plastic packaging materials.
(3) Identification of cured adhesives, by generating pyrogram 'fingerprints' for selected adhesives.

Pyrolysis GC is an espcially valuable and informative technique when coupled with mass spectrometric detection which facilitates identification of the pyrolysis products [28].

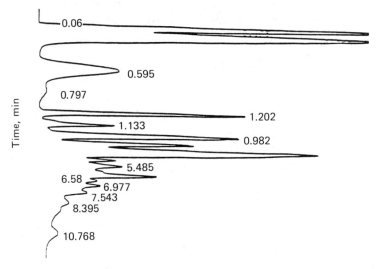

Fig. 4.36 — Pyrogram of styrene–butadiene rubber. Column: 6 ft × ¼″ stainless steel — Tenax GC. Detector: FID — 300°C. Temperature: 80–280°C programmed. Pyrolysis: filament type — 800°C/10 sec.

4.5.5 Supercritical-fluid chromatography

Supercritical-fluid chromatography [29] (SFC) is a relatively new separation technique which uses a fluid above its critical temperature and pressure as the mobile phase. Both capillary and packed column approaches are used. The supercritical fluid possesses the density and solvating power of a liquid, the viscosity of a gas and diffusivity intermediate between a liquid and a gas. In essence, SFC should bridge the gap between GC and HPLC. In our laboratory we have chosen to investigate the potential offered by this technique by adapting one channel of an existing GC with FID detection, to SFC use. The fluid delivery system is a digital syringe-driven pump with twin heads. The fluid we have chosen to work with is supercritical carbon dioxide (approximately 90% of published articles in this field refer to the use of carbon dioxide). Our initial trials have concentrated on a packed column approach using a reverse-phase micro-bore HPLC column.

For most commercial laboratories, the only logical justification for involvement with SFC is to achieve separations which cannot demonstrably be achieved by GC or HPLC. This represents the rationale for our involvement, as we have a requirement to analyse low volatility resins (hence precluding GC) and oligomers which do not possess any significant chromophore (thus difficult to detect with HPLC analysis). Thermolabile compounds with poor chromophores are also potential candidates for SFC separations. We believe that materials such as hydrocarbon-based waxes, surfactants, certain organic peroxides, silicone resins, etc., are worthy of serious evaluation by this approach.

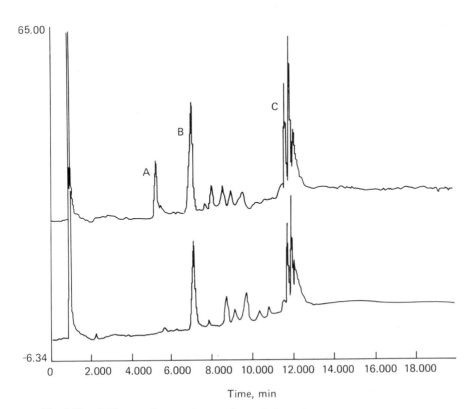

Fig. 4.37 — SFC trace of process intermediates. Column: Deltabond methyl — 25 cm × 1 mm, 5μ. Run conditions: Pressure programme (1500–2500 psi); temperature programme (75–125°C). A, process side-product. B, glutarate ester; C, triaryl phosphate.

One instance we have encountered where SFC appears to offer a unique benefit is in the analysis of a plant residue arising from the vacuum distillation of a reactive acrylic monomer. Spectroscopic (NMR,IR) comparison of two such distillation residues indicated that one residue contained a relatively high level of a process by-product which contained a characteristic functional unit.

However GC, reverse-phase HPLC with UV detection and GPC with refractive index detection were unsuccessful with respect to confirming the difference between the two residues, but SFC revealed the presence of a signifiant level of an additonal component in the residue sample showing the additional responses on NMR (^{1}H, 60MHz) and IR comparison (Fig. 4.37). A third residue, which was shown by NMR and IR to contain a reduced level of this component, was observed to contain a proportionately reduced response at the noted retention time.

4.5.6 Direct-current plasma-emission spectroscopy coupled with flow-injection analysis

A recent development in our laboratory has been the introduction of a computer-assisted metal analyser [30,31] using flow-injection analysis (FIA) coupled with direct-current plasma optical emission spectroscopy. Pneumatic nebulization represents the preferred method of sample introduction for both direct and inductively coupled plasma emission spectroscopy. FIA has the advantage of an improved ability to correct for baseline drift — a problem common to most plasma-based systems. In addition, it has been demonstrated that FIA offers improved precision, accuracy and sample throughput as compared with direct nebulization. Much of the pioneering development work behind this important new combined technique has taken place in our laboratory. As currently developed, the rapid routine determination of more than 70 different elements at trace levels is facilitated in both aqueous and non-aqueous solvents. Sample throughput is of the order of 120 samples/hr.

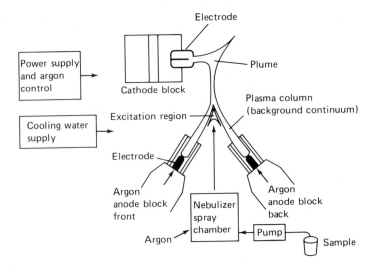

Fig. 4.38 — Typical direct-current plasma three-electrode jet.

A diagram of the DCP plasma jet is given in Fig. 4.38 and a schematic diagram of the automated FIA/DCP-OES system in Fig. 4.39. The signal-measurement approach utilized with the FIA modification is shown in Fig. 4.40. This system is particularly suited to the assay of raw materials for contaminating trace/sub-trace levels of active metals, and to confirm the level of metal salts and complexes in primers used to facilitate the bonding of inactive surfaces.

The complete system is highly automated, and both sample introduction and clean-out of the plasma jet are controlled by a microcomputer. A data-acquisition program collects, stores and processes data generated as line intensities in the spectrometer.

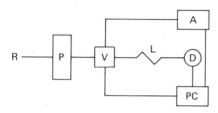

Fig. 4.39 — Automated FIA/DCP–OES system. R=carrier-stream reservoir; P=peristaltic pump; V=flow-injection valve; L=loop; D=DCP detector source; A=autosampler; PC=IBM or compatible personal computer.

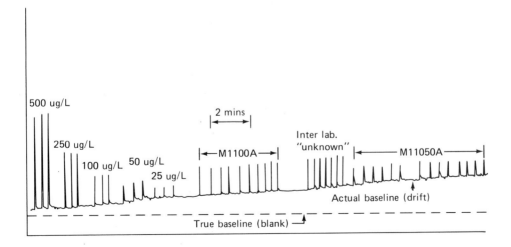

Fig. 4.40 — Reproducibility of signals with a shifting base-line.

4.6   CONCLUDING COMMENTS

The Analytical Laboratory in the speciality chemicals industry may have to have recourse to a variety of modern analytical techniques in the performance of its service role. Specific sample preparation sequences involving matrix elimination and preconcentration may have to be devised which do not alter the integrity of the analyte. The overlying requirement is to provide a fast, accurate and precise response to each query.

There is a further basic requirement for the analyst to be informed and knowledgeable with respect to the technology and the underlying chemistry of the various classes of products. Implicit in this statement is the expectation that analytical chemists will adopt a wider approach towards problem solving and not restrict their involvement to the analysis of samples submitted to the laboratory. The individual analyst should be prepared to participate in and contribute towards the efforts of work-teams seeking to resolve intrinsic technical and process-related problems. This should have the dual benefits of increasing the sense of satisfaction and fulfilment for the analytical chemist concerned, and alerting colleagues to the value of an overall 'Analytical' contribution.

REFERENCES

[1] D. J. Stamper, *British Polymer Journal.*, 1983, **15**, 34.
[2] E. S. Grant, *Drop by Drop*, Loctite Corporation.
[3] G. Hai Lin, R. John, G. G. Wallace, M. Meaney, M. R. Smyth, and R. G. Leonard, *Anal. Chim. Acta*, 1989, **217**, 335.
[4] A. E. Derome, *Modern NMR Techniques for Chemistry Research*, Pergamon Press, Oxford, 1983.
[5] R. M. Silverstein, G. C. Bassler, and T. C. Morill, *Spectrometric Identification of Organic Compounds*, 4th edn, John Wiley & Sons, New York, 1980.
[6] E. Knappe and D. Peteri, *Z. Anal. Chem.*, 1962, **190**, 386.
[7] M. Gillio-Tos, S. A. Previtera and A. Vimercati, *J. Chromatogr.*, 1964, **13**, 571.
[8] C. Passera, A. Pedrotti and G. Ferrari, *J. Chromatogr.*, 1964, **14**, 289.
[9] F. Feigl, *Spot Tests in Organic Analysis*, Elsevier, Amsterdam, 1978.
[10] L. S. Ettre, *Introduction to Open Tubular Columns*, Perkin Elmer, 1978.
[11] R. R. Freeman, *High Resolution Gas Chromatography*, Hewlett-Packard, 1981.
[12] L. R. Snyder, J. L. Glajch and J. J. Kirkland, *Practical HPLC Method Development*, Wiley–Interscience, New York, 1988.
[13] L. Horner and C. Jurgen, *Angew. Chemie*, 1958, **70**, 266.
[14] D. F. Heatley, M.Sc. Thesis, Dublin City University, 1988.
[15] P. R. Griffiths and J. A. De Haseth, *Fourier Transform Infrared Spectrometry*, John Wiley, New York, 1986.
[16] J. P. Mooney, M. Meaney, M. R. Smyth and R. G. Leonard, *Analyst*, 1987, **112**, 1555.
[17] P. O'Dea, M. Deacon, M. R. Smyth and R. G. Leonard, *Anal. Proc.*, 1991, **28**, 82.
[18] R. M. Cassidy and S. Elchuk, *Anal. Chem.*, 1982, **54**, 1558.
[19] J. Krol, *Waters Ion Chromography Notes*, 1988, **2**, 1.
[20] M. Deacon, M. R. Smyth and R. G. Leonard, *Analyst*, 1991, in press.
[21] M. C. Brennan and G. Svehla, *Anal. Proc.*, 1989, **26**, 343.
[22] M. C. Brennan and R. G. Leonard, *Electrochemistry, Sensors and Analysis*, Elsevier, Amsterdam, 1986, p. 169.
[23] W. W. Wendlandt, *Thermal Analysis*, 3rd Ed., John Wiley, New York, 1985.
[24] T. O'Shea, M. Meaney, M. R. Smyth and R. G. Leonard, *Anal. Lett.*, 1991, **24**, 103.
[25] J. F. Lawrence, *Liquid Chromatography in Environmental Analysis*, Humana Press, New Jersey, 1984.
[26] K. H. Altgelt, and L. Segal, *Gel Permeation Chromatograhy*, Marcel Dekker, New York, 1971.
[27] M. P. Stevens, *Characterization and Analysis of Polymers by Gas Chromatogrphy*, Marcel Dekker, New York.

[28] G. Montaudo, *British Polymer Journal*, 1986, **18**, 231.
[29] R. M. Smith, *Supercritical Fluid Chromatography*, Royal Society of Chemistry, London, 1988.
[30] M. C. Brennan, R. A. Simons, G. Svehla, and P. B. Stockwell, *Journal of Automatic Chemistry*,
      1990, **12**, 183.
[31] M. C. Brennan and G. Svehla, *Irish Chemical News*, 1990, **35**.

5

Air pollution analysis

Imelda Shanahan
Peat Research Centre, Bord na Mona, Newbridge, Co Kildare, Ireland

5.1 INTRODUCTION

Concern for the quality of our air is not a new phenomenon, although the recent upsurge in interest in environmental matters has certainly heightened our awareness of the risks which threaten our environment.

In the early thirteenth century, coal began to replace wood for domestic heating and industrial uses in London, resulting in a catastrophic deterioration in air quality [1]. There was renewed interest in air pollution and its effects in the 20th century when a series of so-called killer smogs (a combination of the words smoke and fog) occurred, the worst of which in December 1952 resulted in 4000 fatalities. Following this episode, Britain passed a Clean Air Act to reduce emissions to the atmosphere from coal-burning activities, and subsequent smog episodes were less severe and caused fewer deaths than the 1952 incident.

In Dublin, serious winter smog episodes became prevalent during the late 1980s — the situation grew so bad that the authorities placed a ban on the sale of smokey coal during the autumn of 1990, but stopped short of declaring parts of the city as totally smoke-free zones. While it is too early to assess the impact on air quality from this action, some radical action was absolutely essential and further action may be required in the future.

An air pollution of a related kind, photochemical smog, was first described in 1951, following an observation of the harmful effects of such smog on plants in Los Angeles County during the late 1940s. A major constituent of photochemical smog is ozone, formed during the reaction of organic compounds and oxides of nitrogen in air in the presence of sunlight, in contrast to the sulphur dioxide and suspended particulate matter which are major constituents of sulphurous smog.

With the recognition of the severity of the many smog episodes in London, and the identification of photochemical smog in Los Angeles in the 1940s, came the need to monitor a wide range of primary and secondary pollutants which can contribute to the deterioration of air quality, and which may have other harmful effects as well. In this chapter, a series of common techniques for sampling and analysis of a wide range of gas-phase pollutants over relatively short ranges are described. Discussions of remote sensing techniques to measure the column abundance of pollutants (i.e. the integrated concentration of pollutants in a column of air from the surface of the earth through the atmosphere) are not covered — the interested reader is referred to alternative sources [2–4] for a discussion of such techniques.

Air-pollution monitoring is an expensive business, so it is essential to examine the objectives carefully before any attempt is made to establish an air-monitoring programme. Although it may be tempting to design a system that could serve a multitude of different objectives, it is likely that the cost and effort involved would be prohibitive. In practice, it appears that only certain objectives are realizable with a given network. It is not the purpose of this chapter to deal comprehensively with the design of air monitoring programmes — the interested reader is referred to the WHO/WMO monograph on this subject [5] for further information.

Some of the main objectives of air monitoring programmes are given below:

  (i)   to observe long-term trends in air quality;
 (ii)   to judge compliance with air-quality standards;
(iii)   to evaluate control strategies;
(iv)   to evaluate risk to human health and to the environment;
(iiv) to investigate specific complaints arising from pollution episodes.

Consideration of these objectives and the types and amount of data required are essential if the appropriate information is to be collected with minimum effort and cost. Having established the reasons for conducting the study, and the type and amount of information required, the analyst must choose a representative sampling site and sampling point for the study, suitable analytical techniques, and procedures for evaluating and interpreting the data collected. Since the objectives of the proposed monitoring programme must be established independently for each study, it is appropriate to start this discussion with a review of factors which will influence the choice of sampling site, and the choice of suitable sampling methods and analytical procedures. Consequently, it is first necessary to consider the physical and chemical characteristics of the atmosphere that influence these choices.

5.2 THE ATMOSPHERE

5.2.1 Physical characteristics

The earth's atmosphere is an envelope of gases extending to a height of approximately 2000 km, with the density of these gases decreasing rapidly with increasing altitude. Temperature also varies with altitude, and this characteristic variation is used to divide the atmosphere into different layers. The temperature profile is shown in Fig. 5.1, where average atmospheric temperature for mid-latitudes is plotted as a function of height above sea level.

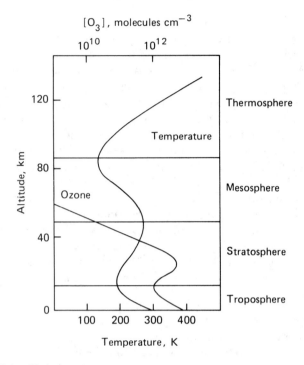

Fig. 5.1 — Variation of temperature and ozone concentration with altitude.

The troposphere is characterized by a steady decrease in temperature with increasing altitude from about $-15°C$ at the earth's surface to about $-60°C$ at the tropopause. It is maintained as a distinct layer of the atmosphere by the cooler air of the stratosphere which lies directly above. The decrease in temperature within the troposphere is due to the strong heating effect at the earth's surface from the absorption of visible and ultraviolet radiation. Accompanying this, there is strong circulation and vertical mixing of atmospheric constitutents, because warmer air from near ground level rises and is replaced by cooler air from above. Thus, any pollutants released at the earth's surface can move to the top of the troposphere in a few days or less, depending on meteorological conditions. Essentially all of the water vapour, clouds and precipitation in the earth's atmosphere are found in this region, so removal of water-soluble pollutants by precipitation scavenging is an important atmospheric cleansing mechanism.

The stratosphere shows a reversal of the temperature gradient found in the troposphere with the temperature increasing with increasing altitude to about 5°C at the stratopause. Relatively little vertical mixing occurs within the stratosphere, since the cooler air is at lower altitudes and does not readily rise. However, transport in the horizontal direction is quite rapid, and a tracer may be dispersed around the world on a line of latitide from East to West in 1 to 4 weeks. There is a relatively slow rate of transport of material across the tropopause from the troposphere into the stratosphere. Upward moving air from the troposphere passes through the coldest region of the tropopause and this results in all but a few parts per million of water vapour being precipitated out, so the stratosphere is very dry and no precipitation scavenging occurs in this region. Consequently, massive injections of particles, for example from volcanic eruptions, often produce layers of particles in the stratosphere which persist for long periods of time.

In the mesosphere, from about 50 to 85 km, the temperature again falls rapidly with increasing altitude and vertical mixing again occurs in this region. At about 85 km, in the thermosphere, the temperature starts to rise again because of increased absorption of solar radiation with wavelengths less than 200 nm by O_2 and N_2 as well as by atomic species.

Once a species is emitted at the surface of the earth into the troposphere, it can undergo three overall processes:

(a) chemical reaction in the troposphere, e.g. reaction with the OH radical and subsequent oxidation;
(b) deposition at the surface of the earth, e.g. acid rainfall;
(c) transport into the stratosphere.

Precisely which fate befalls primary pollutants emitted into the troposhere will be dependent on their solubility in water and on their reactivity towards other species present in the troposphere. It is essential that such factors are also considered in choosing sampling and analytical procedures for monitoring atmospheric pollutants.

5.2.2 The ozone layer

Ozone is an important trace constituent of the stratosphere; it is continually being formed and destroyed in a series of photochemically initiated reactions. The basic mechanism for the ozone equilibrium was established in 1930 by Chapman, who suggested that reactions (1)–(4), now known as the Chapman Cycle, were important:

$$O_2 + h\nu \ (\lambda < 220\,nm) \ \rightarrow \ 2O \tag{1}$$

$$O + O_2 + M \qquad\quad \rightarrow \ O_3 + M \tag{2}$$

$$O + O_3 \qquad\qquad\quad \rightarrow \ 2O_2 \tag{3}$$

$$O_3 + h\nu \ (\lambda < 320\,nm) \ \rightarrow \ O(^1D) + O_2 \tag{4}$$

Ozone is formed when oxygen atoms produced from the photo-dissociation of molecular oxygen combine with oxygen, the excess energy being removed by a third molecule, M. Although the maximum rate of photolysis of molecular oxygen occurs above the stratopause, the region of maximum ozone formation is at the lower stratospheric levels where the higher pressures ensure that third-body collisions are more frequent. Ultraviolet radiation in the wavelength range 240–320 nm can dissociate ozone, and it is the resulting absorption that shields the earth's surface from harmful ultraviolet radiation, and also contributes to the characteristic temperature profile of the stratosphere.

The relatively slow reactions (1) and (3) control the amounts of ozone and atomic oxygen in the stratosphere, whereas the rapid processes (2) and (4) determine their relative abundances. This reaction scheme predicts considerably more ozone than is actually present, so other ozone destruction pathways must be considered.

It is now generally accepted that the Chapman cycle accounts for some 20% of the natural ozone destruction rate.

The contribution of catalytic cycles involving hydrogen- and nitrogen-containing species has also been recognised [6–8]. The electronically excited oxygen atoms, $O(^1D)$, formed in the photodissociation of ozone [reaction (3)], are responsible for the formation of odd hydrogen and nitrogen radicals. The hydroxyl radical is produced by the reaction of excited oxygen atoms with water vapour and methane, both of which are present at trace levels in the natural stratosphere:

$$O(^1D) + H_2O \rightarrow 2OH \tag{5}$$
$$O(^1D) + CH_4 \rightarrow OH + CH_3 \tag{6}$$

while NO is produced as a result of the reactions of $O(^1D)$ atoms with nitrous oxide which has a number of important natural and ozone anthropogenic sources:

$$O(^1D) + N_2O \rightarrow 2NO \tag{7}$$

The OH radical and NO then participate in catalytic cycles which account for the remaining 10% and 70% respectively of the natural ozone destruction rate:

$$OH + O_3 \rightarrow HO_2 + O_2 \tag{8}$$
$$HO_2 + O \rightarrow OH + O_2 \tag{9}$$

$$O + O_3 \rightarrow 2O_2 \text{ (net)}$$
$$NO + O_3 \rightarrow NO_2 + O_2 \tag{10}$$
$$NO_2 + O \rightarrow NO + O_2 \tag{11}$$

$$O + O_3 \rightarrow 2O_2 \text{ (net)}$$

Because the atomic oxygen that reacts in the second reaction in each of these two cycles would otherwise have formed ozone, the net effect of each cycle is the destruction of two ozone molecules.

If these natural ozone destruction cycles were the only processes contributing to atmospheric ozone loss, significantly larger concentrations of ozone would be found in the atmosphere. The importance of another catalytic cycle involving chlorine atoms in the destruction of ozone has new been well established, following the pioneering work of Rowland and Molina [9] and other research groups [10–12]. Reactions (12) and (13) constitute a catalytic chain leading to net ozone destruction:

$$Cl + O_3 \rightarrow ClO + O_2 \tag{12}$$
$$ClO + O \rightarrow Cl + O_2 \tag{13}$$

$$O + O_3 \rightarrow 2O_2 \text{ (net)}$$

There are several chlorine-containing compounds from natural and (primarily) human sources which serve as sources of chlorine for propagation of the catalytic cycle described above. Sources of these and other atmospheric pollutants are discussed in the following sections.

5.3   THE AIR POLLUTION SYSTEM

The air pollution system comprises all anthropogenic and natural emissions into the atmosphere, their sources, and all processes to which they are subjected in the atmosphere (Fig. 5.2). The species emitted directly into the atmosphere from

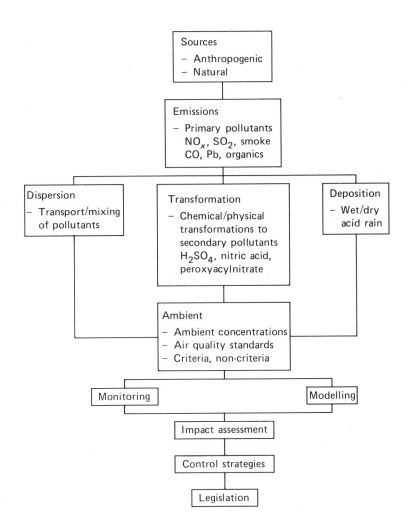

Fig. 5.2 — The air pollution system.

identifiable sources are termed **primary** pollutants, e.g. SO_2, NO_x, CO, Pb, volatile organic compounds and particulate matter from combustion sources, whereas those that are produced in the atmosphere as a result of interactions between primary pollutants and/or natural species are called **secondary** pollutants; e.g. sulphuric acid derived from atmospheric oxidation of sulphur dioxide is a secondary pollutant.

Once emitted into, or formed in, the atmosphere, pollutants may be subjected to the removal or transport processes referred to earlier. Mathematical models have been developed which, together with data on measured concentrations of atmospheric pollutants and data relating to potential chemical and/or physical transformations in the atmosphere, may be used to describe the variation in concentration of selected pollutants within, for example, a plume. Combining this information with the documented impact assessment on receptors such as plants, animals and human beings, allows development of suitable control measures which may be enshrined in relevant legislation.

In the next section, we look at each element of the air pollution system with a view to assessing the potential relevance of each element to the air pollution monitoring objective.

5.3.1 Emissions

In devising a programme for monitoring air quality, it is necessary to consider the relative importance of all natural and anthropogenic sources of primary emissions, together with the chemical and physical transformations which these pollutants may undergo in a given air mass.

Atmospheric pollutants may usually be characterized according to the physical state in which they exist, i.e. gaseous pollutants, aerosols and particulates. It is also convenient to classify natural and anthropogenic emissions according to their chemical nature rather than their physical state. Thus, one such classification of atmospheric pollutants might be:

- hydrocarbons;
- halogenated compounds;
- nitrogen-containing compounds;
- sulphur-containing compounds;
- oxidants;
- aerosols;
- metals;
- particulate matter;
- polycyclic aromatic hydrocarbons.

Although such a classification is also very useful, it is more usual to combine the two systems, since there are many overlapping features. One further addition to the classification is odours, since odorous compounds represent an important class of emissions in their own right, primarily due to the public nuisance effect which generally accompanies such emissions.

In the following sections, the sources, both anthropogenic and natural, and strengths of emissions of a wide variety of gaseous pollutants into the atmosphere will be discussed. The principle natural and man–made emissions into the troposphere are summerised in Table 5.1, and a brief discussion of their relative significance follows. In terms of bulk emissions, by far the greatest sources of atmospheric pollutants are combustion sources, although other sources often result in much more harmful emissions into the atmosphere.

Table 5.1 — Natural and man-made sources of the minor trace gases of the troposphere

Compound	Natural sources	Man-made sources
Carbon-containing trace gases		
Carbon monoxide (CO)	Oxidation of natural methane, natural C_5, C_{10} hydrocarbons; oceans; forest fires	Oxidation of man-made hydrocarbons; incomplete combustion of wood, oil, gas, and coal, in particular motor vehicles, industrial processes
Carbon dioxide (CO_2)	Oxidation of natural CO; destruction of forests; respiration by plants	Combustion of oil, gas, coal and wood
Methane (CH_4)	Enteric fermentation in wild animals; emissions from swamps, bogs, etc., natural wet land areas, oceans	Enteric fermentation in domesticated ruminants; emissions from paddies, natural gas leakage, sewage gas; combustion sources
Light paraffins, C_2–C_6	Aerobic biological source	Natural gas leakage; motor vehicle evaporation emissions; refinery emissions
Olefins, C_2–C_6		Motor vehicle exhaust; diesel engine exhaust
Aromatic hydrocarbons		Motor vehicle exhaust; evaporative emissions; paints, petrol, solvents
Terpenes	Vegetation	
Nitrogen-containing trace gases		
Nitric oxide (NO)	Forest fires; anaerobic processes in soil; electric storms	Combustion of oil, gas and coal
Nitrogen dioxide (NO_2)	Forest fires; electric storms	Combustion of oil, gas and coal; atmospheric transformation of NO
Nitrous oxide (N_2O)	Emission from denitrifying bacteria in soil; oceans	Combustion of oil, gas and coal
Peroxyacetyl nitrate (PAN)	Degradation of hydrocarbons	Degradation of hydrocarbons
Ammonia (NH_3)	Aerobic biological source in soil; Breakdown of amino acids in organic waste material	Coal and fuel oil combustion; waste treatment
Amines		Fish-meal processing, decay of fish
Sulphur-containing trace gases		
Sulphur dioxide (SO_2)	Oxidation of H_2S; volcanic activity	Combustion of oil and coal; roasting sulphide ores
Hydrogen sulphide (H_2S)	Anaerobic fermentation; volcanoes	Oil refining; animal manure; Kraft paper mills, rayon production; coke-oven gas

Continued next page

Table 5.1 (*Continued*)

Compound	Natural sources	Man-made sources
Carbon disulphide (CS$_2$)	Anaerobic fermentation	Viscose rayon plants; brick making; fish-meal processing
Methyl mercaptan (CH$_3$SH)	Anaerobic biological sources	Animal rendering; animal manure; pulp and paper mills; brick manufacture; oil refining
Dimethyl sulphide (CH$_3$SCH$_3$)	Aerobic biological sources	Animal rendering; animal manure; pulp and paper mills; sewage treatment
Dimethyl disuphide (CH$_3$SSCH$_3$)		Animal rendering; fish-meal processing; sewage treatment
Other organic sulphur compounds, C$_2$–C$_4$ mercaptans, alkyl disulphides, dimethyl trisulphide, alkyl thiophenes, benzothiophenes	Anaerobic biological sources	Animal rendering, fish-meal processing; brick making; sewage treatment
Chlorine-containing trace gases		
Hydrogen chloride (HCl)	Volcanoes; degradation of CH$_3$Cl	Coal combustion; degradation of chlorocarbons
Methyl chloride (CH$_3$Cl)	Slow combustion of organic matter; marine environment; algae	PVC and tobacco combustion
Methylene dichloride (CH$_2$Cl$_2$)		Solvent
Chloroform (CHCl$_3$)		Pharmaceuticals; solvent; combustion of petrol; bleaching of wood pulp; degradation of C$_2$HCl$_3$
Carbon tetrachloride (CCl$_4$)		Solvent; fire extinguishers; degradation of C$_2$Cl$_4$
Methyl chloroform (CH$_3$CCl$_4$)		Solvent; degreasing agent
Trichloroethylene (C$_2$HCl$_3$)		Solvent, dry cleaning agent; degreasing agent
Tetrachloroethylene (C$_2$Cl$_4$)		Solvent; dry cleaning agent, degreasing agent
Other chlorofluorocarbons (CCl$_3$F, CCl$_2$F$_2$, C$_2$Cl$_3$F$_3$, C$_2$Cl$_2$F$_4$, C$_2$ClF$_5$)		Aerosol propellants; refrigerant; foam blowing
Other minor trace gases		
HF	Volcanoes	
Ozone	Stratosphere: natural NO–NO$_2$ conversion	Man-made NO–NO$_2$ conversion
CH$_3$Br	Aerobic biological source	Fumigation of soil and grain
CH$_3$I	Aerobic biololgical source	Insignificant

5.3.1.1 Hydrocarbons

The characteristic pleasant smell associated with pine forests, shrubs and flowers indicates that volatile organic compounds must be emitted naturally by living plants.

Of course, the decay of plants by microbiological action also produces emissions, although the odour is usually considerably less pleasant!

The single naturally produced organic compound present in the greatest quantity in ambient air throughout the world is methane. It is produced by anaerobic bacterial fermentation processes in water which contains substantial quantities of organic matter, such as swamps, marshes, rice fields, and lakes. In addition, CH_4 is produced by enteric fermentation in animals as well by other species — such processes account for about 20–25% of the natural methane emissions (there are an estimated 1.5 billion cows worldwide, each producing an average 10 lb (4.5 kg) of methane per day ... think about it!). Smaller amounts are emitted into the air from seepage of natural gas from the earth and from forest fires, some of which are due to natural causes.

The background concentration of methane is about 1.3–1.6 ppm in the northern hemisphere and slightly lower in the southern hemisphere, although the concentration appears to be rising slowly at 1–2% per annum. Anthropogenic sources contribute about 10% of the global atmospheric burden per annum. Methane has been identified as a greenhouse gas, so-called because of its potential contribution to warming of the earth's atmosphere as a result of absorption of infrared radiation emitted from the surface of the earth.

Natural sources are responsible for emission of larger organic compounds in addition to methane — one review cites emission of 367 different organic species from vegetation [13]. These include simple alkanes from natural gas seepage and bacterial fermentation, and from plants, trees and shrubs, and higher molecular weight compounds from these sources and from combustion. A comprehensive review of organics observed in the troposphere has been published [14].

At the ppb concentrations normally encountered in the atmosphere, these species are relatively harmless to mammals. However, a serious problem arises in the phenomenon of photochemical smog which is produced under certain meteorological conditions by the reactions of nitrogen oxides with other air pollutants including hydrocarbons, leading to the formation of secondary pollutants such as ozone, aldehydes, peroxyacylnitrate and alkyl nitrates. These reaction products can cause irritation of the eyes and mucous membranes, and visibility may also be reduced as a result of formation of aerosols from polymer molecules originating from photochemical reactions of the hydrocarbons.

5.3.1.2 Halogenated compounds

Many varied halogenated organic and inorganic compounds are found in the troposphere, some of which are emitted from natural sources. The halogenated hydrocarbon gases include a wide variety of chlorinated and fluorinated compounds, and a much lesser quantity of bromine-containing compounds. The halogenated species identified in the trosphere, their sources (both natural and man-made), and their reactions in the atmosphere have been reviewed by Cicerone [15].

Methyl chloride, methyl bromide and methyl iodide are all produced naturally, predominantly in the oceans, with ambient concentrations in the sub-ppb range. Several halogenated species are used as insecticides (DDT, lindane etc.), and concern has arisen about the potential toxic hazards of such species. An additional problem is their possible contribution to stratospheric ozone depletion owing to release of chlorine atoms as a result of chemical transformations in the atmosphere (Section 5.3.3).

Chlorofluorocarbons (CFCs) are compounds containing chlorine, fluorine, carbon, and possibly hydrogen, which have been used extensively in industrialized countries, primarily as aerosol propellants, refrigerants and as blowing agents in the production of, for example, polyurethane foam. In the early 1970s, 75% of all CFCs were in aerosol sprays, but in the past decade this figure has been reduced to about 25%. The dominant uses now are foam blowing and refrigeration systems. They are ideally suited to such uses since they are generally non-toxic, non-flammable and chemically inert.

The dominant CFCs released from the surface of the earth are CCl_3F, CCl_2F_2 and $CHClF_2$, often referred to as CFC-11, CFC-12 and CFC-22, with the numbers referring to the numbers of halogen and hydrogen atoms present. The first number is the number of hydrogen atoms plus one, while the second number gives the number of fluorine atoms. For CFCs containing two or more carbon atoms, a three digit numbering system is used. The first digit gives the number of carbon atoms minus one, the second digit gives the number of hydrogens plus one and the third digit gives the number of fluorine atoms; the balance of atoms is made up of chlorine atoms. Thus CCl_2FCClF_2, a solvent extensively used in the manufacture of electronic components, is CFC-113.

In addition to the CFCs, other halogenated species are also emitted from the earth's surface. These include methyl chloroform (CH_3CCl_3), carbon tetrachloride (CCl_4), dichloromethane (CH_2Cl_2), perchloroethene (C_2Cl_4) and trichloroethene (C_2HCl_3). The main source of these species in the atmosphere is their use as solvents and degreasing agents, with carbon tetrachloride also used as a reagent in the manufacture of CFC-11 and CFC-12.

The principal atmospheric significance of these and other chlorinated species is their potential contribution to stratospheric ozone depletion (Section 4.2.2).

5.3.1.3 Sulphur-containing compounds

The gaseous compounds of sulphur which are of interest in air pollution studies fall into three main categories — oxides, hydrides and organic compounds.

Of the oxides of sulphur, only SO_2 and SO_3 are important air pollutants. Sulphur dioxide is emitted primarily from combustion sources, with some minor sources including metallurgical processes and miscellaneous chemical industries. SO_3 is also produced from combustion sources, but in much smaller amounts. Owing to its reactivity — it is converted into sulphuric acid aerosol rapidly on contact with water vapour — it has never been observed freely in atmospheric air.

Hydrogen sulphide is released mainly from pulp and paper manufacture and also from certain refining and coking activities. It is oxidized rapidly to sulphuric acid in air and this, coupled with the unpleasant odour associated with this species, is the main concern from an air-pollution point of view.

Emissions of organic sulphur compounds are, in volume terms, relatively minor. They are, however, of growing importance as a source of odour nuisance, a field that is receiving a growing amount of attention. The main sources are industries which process natural products such as wood pulp, paper and animal offal, sewage treatment plants and miscellaneous chemical processes. The main compound types of interest are sulphides, thiols and mercaptans.

Many of these species also have natural sources, e.g. H_2S from swamps, and organic sulphides from a variety of aerobic and anaerobic biological processes.

5.3.1.4 *Nitrogen-containing species*

The oxides, hydrides and organic compounds of nitrogen are the categories of emissions of great importance from an atmospheric-pollution point of view. The most important oxides are nitric oxide (NO) and nitrogen dioxide (NO_2) — the higher oxides are present at insignificant concentrations. The major source of nitrogen oxides (NO_x) is combustion, when fixation of nitrogen from the atmospheres occurs at high flame temperatures. The oxides are emitted mainly as NO which is converted rapidly into nitrogen dioxide. Motor vehicle exhaust contributes a sizeable fraction of total emissions, while fertilizer manufacture also releases NO and NO_2.

Ammonia is a pollutant of secondary importance, but severe local problems may be experienced. Deleterious effects are mainly those associated with its role in the formation of atmospheric particulate matter.

Organic nitrogen compounds, like the organic sulphur compounds, are emitted primarily from fish-processing industries, decay of fish, etc. Their main atmospheric significance is the public nuisance effect of odorous emissions — primarily attributed to organic amines such as trimethylamine and triethylamine. Natural sources of ammonia and certain amines include animal urine and chicken dung.

5.3.1.5 *Odours*

Because of the growing interest in the field of odorous emissions, this section looks briefly at the characteristics of such emissions. The components occur primarily in the gas phase, although the contribution of particulate matter and aerosols to the overall odour burden cannot be overlooked.

The dominant species associated with malodours are hydrogen sulphide, the organic sulphides, mercaptans, thiols, and amines such as trimethylamine and triethylamine. It is seldom possible to identify a single constituent as being responsible for a malodour — the synergistic effect of the mixture must always be considered. Thus, while odorous gas streams may be characterized primarily by specific constituents such as sulphur compounds from sewage treatment plants, amines from fish-processing, such gases are usually complex mixtures which may contain hundreds of different constituents, all of which contribute to the perceived odour.

Not all odorous gas streams are unpleasant, although even a pleasant smelling gas may become overpowering if the concentration and frequency of emission is high enough. Volatile esters, aldehydes and ketones are characterized by very pleasant smells, but at very high concentrations, these can create as much a public nuisance as malodorous emissions.

5.3.1.6 Aerosols and particulate matter

An aerosol is defined as a suspension or dispersion of solid or liquid particles in a gaseous medium, whereas particulate matter refers to either suspended or deposi-table particulates, depending on the size ranges. Although these are obviously not characterized as gaseous pollutants, it is important to consider their presence, and the presence of any other pollutants adsorbed on these particulates, in any air-pollution monitoring scheme.

5.3.2 Dispersion

The concentrations of pollutants in ambient air and their potential impacts are determined not only by their rates of emissions, chemical transformations and/or wet/dry deposition, but also to a significant extent by meteorology. During the severe London smog episodes of the early 1950s, and more recently during similar episodes in Dublin in the late 1980s, local meteorological conditions resulted in pollutants being trapped in relatively small volumes of air by temperture inversion, thus resulting in high pollutant concentrations.

In the lowest 10 km of the earth's atmosphere (the troposphere), air temperature generally decreases with altitude at an approximate rate of 7°C per km. As described earlier, warm air from near ground level rises and is replaced by cooler air from above, resulting in good vertical mixing within the troposphere under normal conditions.

Occasionally, however, the temperature of the air within the troposphere may start to rise with increasing altitude before reverting to the normal positive lapse rate (Fig. 5.3), resulting in the formation of an **inversion layer**. This layer of warm air may then act as a lid on the cooler, denser air below, preventing the normal rapid mixing of pollutants throughout the troposphere. Thus, much higher than normal ground-level concentrations of pollutants are observed.

The formation of temperature inversions of the type described above is one of the most important meteorological factors contributing to air pollution in urban areas. The inversion characteristic of London/Dublin smog is caused by the rapid cooling of the earth's surface and the layer above it due to emission of infrared radiation immediately after sunset. If air turbulence is low on calm nights, the cooling effect may be sufficiently rapid to allow formation of an inversion which can persist until dawn, when sufficient heating of the earth's surface and the air above it causes the inversion to break. The formation of inversions also contributes to the photochemi-cal smog episodes which occur in Los Angeles and Mexico City, but the reasons for formation of the inversion layers are different.

The role of meteorological transport in establishing the contribution of various sources to downwind ambient pollutant concentrations is not limited to pollutants *within* a given air basin — the long-range transport of pollutants to remote regions is also affected. For example, tracer studies have shown that air masses originating along the coast in southern California in the morning travel long distances during the course of the day [16].

An important current problem of atmospheric research is the contribution of meteorological processes to the phenomenon of acid deposition. Meteorological

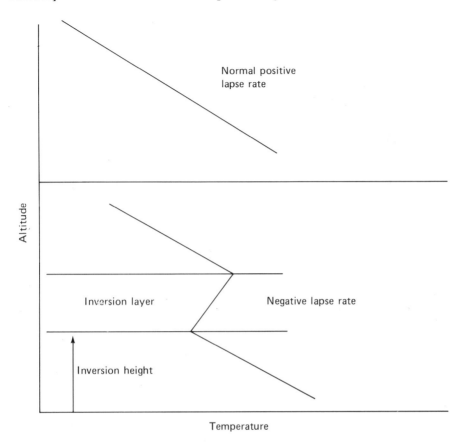

Fig. 5.3 — Variation of temperature with altitude in the troposphere showing: (a) normal positive lapse rate; (b) change from positive to negative lapse rate, characteristic of a thermal inversion.

processes transport the acid precursors, during which time they react to form acids. As a result, acid deposition is believed to be characterized by long-range transport, say 1000 km, from source to receptor, although recent studies suggest that a significant portion of acid deposition in some locations comes from local sources [17] rather than by long-range transport from distant ones.

In addition to the direct dispersion and transport of pollutants, other meteorological factors also play an important role in atmospheric chemistry. For example, fog water droplets may provide an aqueous medium for the liquid phase conversion of SO_2 into sulphate. Furthermore, the cycle of condensation of water vapour on aerosols at low temperatures, followed by evaporation during the day, is thought to be a major factor controlling pollutant concentrations within the droplets [18]. The intensity of sunlight is another major factor in the formation of secondary pollutants such as acids, owing to the importance of photochemical reactions. Temperature is, of course, important because of its effects on the rates of chemical reactions, the solubility of gases in water, and the volatility of water droplets in the atmosphere.

In summary, meteorological parameters play an important role in determining the dispersion and transport of pollutants, and also in their atmospheric chemistry.

5.3.3 Chemical transformations

Within the past twenty years, there has been a revolution in our knowledge of the chemistry of urban atmospheres. A combination of laboratory work and computer modelling studies has led to the quantitative characterization of the rates and mechanisms of atmospheric reactions of primary pollutants such as organic compounds, NO_x and SO_x. Complementing the fundamental kinetic and mechanistic studies has been the development and application of sensitive and selective analytical techniques for the identification of trace atmospheric pollutants.

Many varied chemical processes within the atmosphere transform the relatively small number of inorganic primary pollutants, and the much larger number of organic compounds emitted into the atmosphere, into a diverse range of secondary air pollutants. The main driving force for atmospheric chemistry comes in the form of energy from the sun to promote gas-phase photochemical reactions, although heterogeneous processes also play an important, if largely uncharacterized, role in atmospheric chemistry.

The temperature at the surface of the sun is approximately 6000°C, so that solar radiation is largely in the ultraviolet and visible region of the spectrum. About 47% of the total radiation directed at the earth from the sun actually reaches the earth's surface; the remainder is lost either by reflection or absorption by atmospheric gases — this absorption results in a heating effect on the earth's atmosphere in certain regions. The potentially lethal ultraviolet radiation with wavelengths between 200 and 300 nm is absorbed in the stratosphere by ozone and, to a smaller extent, by molecular oxygen (Fig. 5.4). It is this absorption of solar radiation by atmospheric gases that drives the photochemical transformations of primary pollutants in the atmosphere.

The foundations of air pollution photochemistry were laid down in 1961 by Leighton [19] and were further established during the 1960s by several research groups [20–28]. An important development in our understanding of the photochemistry of polluted atmospheres came with the recognition that the hydroxyl radical (OH) is, with several important execptions, the key reactive intermediate in the photo-oxidation of most organic and many inorganic atmospheric pollutants [29–36]. To date, the rates of OH reactions with more than 400 organic compounds of atmospheric signficance have been measured [37–39].

The major sources of the OH radical in the atmosphere are photolysis of nitrous acid (HONO), formaldehyde (HCHO) and O_3/H_2O systems. The relative importance of these formation routes for OH radicals as a function of time of day have been calculated [40], with HONO photolysis being the dominant source in the early morning, and ozone photolysis becoming more important in the early afternoon. The OH radical then initiates chain reactions by attack on organic pollutants which are oxidized under ambient atmospheric conditions (Fig. 5.5).

As we have seen, the atmosphere is a very complex air system which contains hundreds if not thousands of pollutants, and a range of highly reactive species such as the OH and nitrate radicals and ozone. Since the number of possible chemical

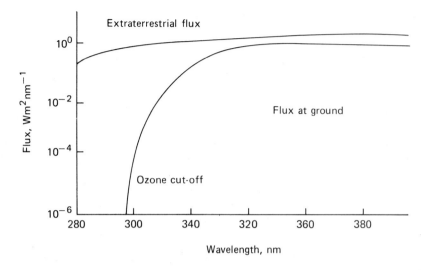

Fig. 5.4—This plot compares the solar UV spectrum at the top of the atmosphere with the solar energy that reaches the ground.

$$OH^\bullet + C_3H_8 \quad \rightarrow \quad C_3H_7^\bullet + H_2O$$

$$C_3H_7^\bullet + O_2 \quad \rightarrow \quad C_3H_7O_2^\bullet$$

$$C_3H_7O_2^\bullet + NO \quad \rightarrow \quad C_3H_7O^\bullet + NO_2$$

$$C_3H_7O^\bullet + O_2 \quad \rightarrow \quad C_2H_5CHO + HO_2^\bullet$$

$$HO_2^\bullet + NO \quad \rightarrow \quad OH^\bullet + NO_2$$

Fig. 5.5 — Oxidation of propane initiated by the OH radical, simplified by showing only the abstraction of a primary hydrogen, although secondary hydrogens are abstracted more rapidly.

reactions between these species is so large, a precise evaluation of the chemistry of a given air mass is a formidable task. However, the vast database of kinetic and mechanistic information on reactions of these pollutants allows us to estimate with reasonable confidence the relative contributions of each of the oxidizing species to the net removal rates of most pollutants by homogeneous gas-phase processes. This, of course, narrows down significantly the number of reactions which must be evalutaed when considering the chemistry of a given air mass.

To illustrate the point, consider the calculation of tropospheric lifetimes (τ) of some selected organic compounds of anthropogenic origin. For a second order reaction between an organic compound, A, and an oxidizing species, Ox:

A + Ox $\rightarrow$ products

Rate = $k_{ox}[Ox]$

$\tau = 1/k_{ox}[Ox]$

Thus, to evaluate τ, we need to know the absolute value of the rate constant, k_{ox}, and the concentration of the specific oxidant involved. By using the measured value of k_{ox} for reactions of OH radical with n-butane and the maximum observed concentration of OH radicals in air, the calculated atmospheric lifetime of n-butane with respect to OH radicals is 11 hours. Similar calculations of lifetimes with respect to other oxidizing species indicate that reaction with OH radicals is a major removal mechanism for n-butane — the same result is obtained for a large number of organic pollutants.

The significance of such lifetime determinations is clearly illustrated by considering the tropospheric lifetimes of chlorofluorocarbons. These species have very long tropospheric lifetimes owing to the fact that they do not absorb light of wavelengths greater then 290 nm and do not react at significant rates with either ozone or OH radicals. In addition to the lack of chemical reactivity, they are not very soluble in water and hence are not removed by precipitation. As a result, CFCs reside in the troposphere for years, slowly diffusing up into the stratosphere. The actual lifetimes are in the range 50–150 years.

In contrast to the CFCs, the lifetimes of which in the troposphere are determined by their rate of transport into the stratosphere, other halogenated species such as 1,1,1-trichloroethane have finite lifetimes with respect to OH radicals, owing to the presence of abstractable hydrogen atoms. Thus some are removed in the troposphere and the balance are transported into the stratosphere where photodissociation leading to chlorine atoms may occur. The released chlorine atoms may then participate in catalytic ozone destruction cycles.

While the OH radical is the major driving force for the daytime chemistry of both polluted and clean atmospheres, other reactions also play an important role in providing sinks for primary pollutants and in the formation of secondary pollutants. Organic pollutants may also be degraded by reaction with ozone and by direct photolysis.

The chemistry of the polluted atmosphere is continually under investigation, and significant new reaction pathways are still being identified and characterized. Relatively recently (*ca.* 1980 — 1983) the important role of the nitrate (NO_3) radical and HONO in the night-time chemistry of polluted urban atmospheres has been established [41–51]. In 1977, these species had not even been identified in urban atmospheres; HONO was first identified and measured by long-path spectroscopic techniques in clean continental air [41,42], and later in polluted urban atmospheres by Platt [43].

A comprehensive discussion of air pollution chemistry is beyond the scope of this chapter — the reader is referred to the many excellent texts and review articles which discuss this topic. The purpose of this limited discussion is to highlight the implications of chemical transformations of primary air pollutants for ambient air analysis. In particular, it is important to identify secondary pollutants and their significance before attempting to devise an air monitoring programme. This will identify the species which must be determined in a polluted atmosphere. Potential interferences for selected analyses may be identified prior to commencing the monitoring programme, and a decision may be taken to remove potential interferences during sampling. Furthermore, secondary pollutants are frequently more hazardous than the primary pollutants from which they were formed (e.g. peroxyacyl nitrate (PAN) and phosgene (CCl_2O)), so that it is important to include such pollutants in any monitoring programme.

One air-monitoring consequence of the presence of significant ambient concentrations of secondary air pollutants such as PAN, HNO_3 and other nitrogenous species is their interference with measurements of NO_2 by chemiluminescent analysers, since these are known to respond quantitatively to such species when operated in the NO_2 mode [52]. The presence of a range of aldehydes as secondary air pollutants derived from oxidation of hydrocarbons may interfere with analysis for HCHO since many of the methods of analysis used for this species are based on reactions of the aldehyde functionality in general, and are not specific for HCHO.

5.3.4 Deposition

Removal of atmospheric pollutants by deposition at the surface of the earth and by precipitation in the form of rain or snow is an important atmospheric cleansing mechanism, but the processes also contribute to the phenomenon of acid rain. Deposition processes may be classified as either wet or dry depending on the phase in which the pollutant strikes the earth's surface and is taken up. Dry deposition occurs when gases or particles are taken up at the surface of the earth, while wet deposition occurs when pollutants dissolved in rain or snow are taken up at the earth's surface. When the pollutants are dissolved within the cloud, the process is referred to as rainout, whereas when dissolution take place below the cloud, the process is referred to as washout. A further process, occult deposition, occurs when cloud droplets impact on solid surfaces at the surface of the earth.

Because of the highly variable nature of precipitation events, quantitative evaluation of wet deposition of pollutants is difficult. In addition to meteorological factors, the solubility of the pollutant in snow, ice and rain and how this varies with pH, the size of the water droplets, and the number present must also be considered; for example, snow may scavenge HNO_3 much more efficiently than air [53]. As an approximation, the rate of wet deposition of a pollutant is sometimes taken as λC where C is the pollutant concentration and λ is known as a washout coefficient, and is proportional to the precipitation intensity [54].

Dry deposition can also be a very important removal mechanism for pollutants, especially in arid regions. The process is usually characterized by a deposition velocity, V_g, which is defined as the flux (F) of the species (S) to the surface, divided by the concentration, [S], at some reference height, h. Further details on the process are contained in the literature [55].

5.3.5 Ambient concentrations and air quality standards for selected pollutants

Several pollutants have well-documented effects on people, plants or materials at concentrations approaching those found in ambient air. In the USA, seven of these are known as **criteria pollutants** and national ambient air quality standards (NAAQS) have been set for them (Table 5.2). The recommended values or limits set for these criteria pollutants by other countries and/or the World Health Organisation

(WHO) are given in Table 5.3. These limits are subject to periodic review — those listed in Table 5.2 are at present under review, but no changes are expected.

Table 5.2 — US National Ambient Air Quality Standards (NAAQS) for some criteria pollutants

Pollutant	Primary		Secondary	
	Concentration, ppm	Time, hr	Concentration, ppm	Time, hr
CO	9.0	8[†]	9.0	8
	35.0	1	35.0	1
SO$_2$	0.03	Annual mean	—	—
	0.14	24	0.5	3
O$_3$	0.12	1	0.12	1
NO$_2$	0.05	Annual mean	0.05	Arithmetic mean
NMHC[‡]	0.24	3	0.24	3

† CO concentration should not exceed 9.0 ppm for more than 8 hours.
‡ NMHC = non-methane hydrocarbons, expressed as ppm of carbon.

Table 5.3 — Recommended values or limits[†] for various pollutants set by various countries and by the World Health Organisation [56]

Pollutant	Country	Concentration, ppm	Time, hr
CO	WHO	25	24
		30	8
		100	1
	W. Germany	26	0.5
	USSR	1.3	24
SO$_2$	WHO	0.038–0.057	24
		0.015–0.023	Annual
	W. Germany	0.06	24
	USSR	0.02	24
O$_3$	WHO	0.10	1

† **Limits** are differentiated from **recommendations** in that limits are defined by legislation.

The USA sets two types of NAAQS — **primary** air standards are designed to protect public health with an adequate margin of safety, while **secondary** air standards are designed to protect public welfare, e.g. from economic losses due to damage to plants, materials, etc.

In addition to the criteria pollutants, there are many trace gases present in the atmosphere, background concentrations of which are given in Table 5.4, together with ambient concentrations of criteria pollutants. Clearly, the presence of such a wide variety of trace gases at such significant concentrations requires careful assessment when planning any air monitoring programme.

Table 5.4 — NIOSH recommended sampling procedures and methods of analysis for volatile compounds in air

Airborne contaminant	OSHA standard, ppm		Sample volume, litres		Sample rate, ml/min		Collection device	Desorbing solvent	Analytical method (Key at end of table)
	TWA	Ceil.	Ceil.	TWA	Ceil.	TWA			
Acetaldehyde	200	—	—	60	—	200	Bubbler	—	HPLC
Acetic acid	10	—	—	25	—	50	Charcoal	Formic acid	GC
Acetic anhydride	5	—	—	100	—	1000	Bubbler	—	Colour
Acetone	1000	—	2	2	100	20	Charcoal	Carbon disulphide	GC
Acetone cyanohydrin	—	—	3	10	200	20	Porapak QS	Ethylacetate	GC
Acetonitrile	40	—	—	10	—	20	Charcoal	Benzene	GC
Acetylene tetrabromide	1	—	100	100	1000	200	Silica gel	THF	GC
Acrolein	0	—	1.5	48	—	100	Coated XAD-2	Toluene	GC
Acrylonitrile	20	—	—	12	—	25	Charcoal	Carbon disulphide	GC
Allyl alcohol	2	—	2	10	200	20(50)	Charcoal	Carbon disulphide + 5% 2-PrOH	GC
Allyl chloride	1	—	100	100	1000	200	Charcoal	Benzene	GC
Allyl glycidyl ether	—	10	3	—	200	—	Tenax	Ethyl ether	GC
2-Aminoethanol	3	—	96	—	200	—	Silica gel	—	GC
2-Aminopyridine	0	—	12	—	—	100	Tenax	—	GC
n-Amyl acetate	100	—	2	10	200	20(50)	Charcoal	Carbon disulphide	GC
sec-Amyl acetate	125	—	2	10	200	20(50)	Charcoal	Carbon disulphide	GC
Aniline	5	—	6	48	200	100	Silica gel	Ethyl alcohol	GC
o-Anisidine	0.5 mg/m³	—	—	240	—	500	XAD-2 resin	Ethyl alcohol	HPLC
p-Anisidine	0.5 mg/m³	—	—	240	—	500	XAD-2 resin	Ethyl alcohol	HPLC
Anthracene	—	—	—	600	—	2000	Filter	Benzene	GC
Arsenicals (particulate)	0.2 mg/m³	—	—	300	—	1500	Filter	Carbonate/bicarbonate/borate buffer	IC–AA
Arsenic, inorganic	0.2 mg/m³	—	—	400	—	2000	Filter, treated	—	AA
Arsine	0.05	—	2	10	200	20(50)	Charcoal	0.01M Nitric acid	AA, GF
Azelic acid	—	5	—	480	—	2000	Filter	Ethyl alcohol	GC
Benzene	1	—	2	10	200	20(50)	Charcoal	Carbon disulphide	GC
Benzidine	—	—	—	500	—	1700	Filter	Distilled water	HPLC
Benzo(c)acridine	—	—	—	600	—	2000	Filter	Benzene	GC
Benzo(a)anthracene	—	—	—	600	—	2000	Filter	Benzene	GC
Benzo(a)anthrene	—	—	—	600	—	2000	Filter	Benzene	GC
Benzo(b)fluoranthene	—	—	—	600	—	2000	Filter	Benzene	GC
Benzo(j)fluoranthene	—	—	—	600	—	2000	Filter	Benzene	GC
Benzo(k)fluoranthene	—	—	—	600	—	2000	Filter	Benzene	GC
Benzo(g,h,i)perylene	—	—	—	600	—	2000	Filter	Benzene	GC
Benzo(a)pyrene	—	—	—	600	—	2000	Filter	Benzene	GC
Benzo(e)pyrene	—	—	—	600	—	2000	Filter	Benzene	GC
Benzoylperoxide	5 mg/m³	—	—	90	—	1500	Filter	Ethyl ether	HPLC
Benzyl chloride	1	—	2	10	200	20(50)	Charcoal	Carbon disulphide	GC
Bibenzyl	—	—	—	600	—	2000	Filter	Benzene	GC
Bisphenol A (BPA)	—	—	—	288	—	1600	Filter	Acetonitrile	HPLC

Table 5.4 (*continued*) — NIOSH recommended sampling procedures and methods of analysis for volatile compounds in air

Airborne contaminant	OSHA standard, ppm TWA	OSHA standard, ppm Ceil.	Sample volume, litres Ceil.	Sample volume, litres TWA	Sample rate, ml/min Ceil.	Sample rate, ml/min TWA	Collection device	Desorbing solvent	Analytical method (Key at end of table)
Boron carbide	—	—	—	500	—	1700	Filter	—	XRD
Bromoform	0.5	—	2	10	200	20(50)	Charcoal	Carbon disulphide	GC
Butadiene	1000	—	1	1	50	20	Charcoal	Carbon disulphide	GC
2-Butanone	200	—	2	10	200	20(50)	Charcoal	Carbon disulphide	GC
2-Butoxyethanol	50	—	2	10	200	20(50)	Charcoal	Carbon disulphide + 1% methanol	GC
Butylacetate	150	—	2	10	200	20(50)	Charcoal	Carbon disulphide	GC
sec-Butyl acetate	200	—	2	10	200	20(50)	Charcoal	Carbon disulphide	GC
tert-Butyl acetate	200	—	2	10	200	20(50)	Charcoal	Carbon disulphide	GC
Butyl alcohol	100	—	2	10	200	20(50)	Charcoal	Carbon disulphide + 1% 2-propanol	GC
sec-Butyl alcohol	150	—	2	10	200	20(50)	Charcoal	Carbon disulphide + 1% 2-propanol	GC
tert-Butyl alcohol	100	—	2	10	200	20(50)	Charcoal	Carbon disulphide + 1% 2-propanol	GC
Butyl amine	5	—	6	48	200	100	Silica gel	—	GC
Butyl glycidyl ether	50	—	—	10	—	25	Charcoal	Carbon disulphide	GC
n-Butyl glycidyl ether	50	—	2	10	200	20(50)	Charcoal	Carbon disulphide	GC
n-Butyl mercaptan	10	—	1.5	—	—	20	Chromosorb 104	Acetone	GC
p-tert-Butyltoluene	10	—	2	10	200	20(50)	Charcoal	Carbon disulphide	GC
Camphor	2	—	2	10	200	20(50)	Charcoal	Carbon disulphide + 1% MeOH	GC
Carbon disulphide	20	30	6	12	200	200	Charcoal	Benzene	GC
Carbon tetrachloride	10	25	5	15	1000	50	Charcoal	Carbon disulphide	GC
Chlordane	0.5 mg/m³	20 mg/m³	—	120	—	1000	Filter	Toluene	GC
Chlorinated camphene (Toxaphene)	0.5 mg/m³	—	—	15	—	1000	Filter	Petroleum ether	GC
Chlorine	1	—	—	30	—	1000	Bubbler	—	Colour, ISE
Chloroacetaldehyde	—	1	3	—	200	—	Silica gel	50% MeOH/deionized water	GC
a-Chloroacetophenone	—	—	—	12	—	25	Tenax	—	GC
2-Chloroaniline	—	—	—	480	—	1000	Filter	—	HPLC
Chlorobenzene	75	—	10	10	200	20(50)	Charcoal	Carbon disulphide	GC
1,2,3,5-Tetrachlorobenzene	—	—	3	12	200	25	Filter	Hexane	GC
o-Chlorobenzylidene malononitrile	—	—	—	90	—	1.5	Filter/Tenax	—	HPLC
Chlorobromomethane	200	—	2	5	200	20(50)	Charcoal	Carbon disulphide	GC
Chlorodiphenyl (42% Cl)	1 mg/m³	—	—	270	—	1500	Filter + bubbler	Acetone	GC
Chlorodiphenyl (54% Chlorine)	0.5 mg/m³	—	—	100	—	1500	Filter	Petroleum ether	GC
Chloroform	—	50	2	10	200	20(50)	Charcoal	Carbon disulphide	GC
bis(Chloromethyl)ether	—	—	—	10	—	20	Chromosorb 101	—	GC

Table 5.4 (*continued*) — NIOSH recommended sampling procedures and methods of analysis for volatile compounds in air

Airborne contaminant	OSHA standard, ppm		Sample volume, litres		Sample rate, ml/min		Collection device	Desorbing solvent	Analytical method (Key at end of table)
	TWA	Ceil.	Ceil.	TWA	Ceil.	TWA			
Chloromethyl methylether	—	—	—	60	—	500	Chromosorb 101	—	GC
1-Chloro-1-nitropropane	20	—	—	12	—	50	Chromosorb 108	Ethylacetate	GC
p-Chlorophenol	—	—	—	24	—	50	Silica gel	Acetonitrile	HPLC
Chloroprene	25	—	—	3	—	20	Charcoal	Carbon disulphide	GC
Chlorotoluene	50	—	—	10	—	20	Charcoal	Carbon disulphide	GC
Chromium, hexavalent	0.05 mg/m³	0.1 mg/m³	—	600	—	2000	Filter	3% Na₂CO₃/2% NaOH	Colour
Chrysene	—	—	—	600	—	2000	Filter	Benzene	GC
Cresol (all isomers)	5	—	—	20	—	40	Silica gel	Acetone	GC
Crotonaldehyde	2	—	—	25	—	50	Bubbler	—	POL
Cumene	50	—	2	10	200	20(50)	Charcoal	Carbon disulphide	GC
Cyclohexane	300	—	2	2.5	200	20	Charcoal	Carbon disulphide	GC
Cyclohexanol	50	—	2	10	200	20(50)	Charcoal	Carbon disulphide + 5% 2-PrOH	GC
Cycloheaxanone	50	—	2	10	200	20(50)	Charcoal	Carbon disulphide	GC
Cyclohexene	300	—	2	5	200	20(50)	Charcoal	Carbon disulphide	GC
Cyclohexylamine	5	—	—	96	—	200	Silica gel coated	—	GC
Cyclopentadiene	—	—	—	3	—	20	Chromosorb 104	—	GC
2,4-D	10 mg/m³	—	—	100	—	1000	Filter	Ethyl acetate	GC
DDT	1 mg/m³	—	—	90	—	1500	Filter	Ethyl alcohol	GC
Decane	—	—	2	4	200	20(50)	Charcoal	Iso-octane	GC
Demeton	0.1 mg/m³	0.3 mg/m³	—	480	—	1000	Filter	Toluene	GC
Diacetone alcohol	50	—	2	10	200	20(50)	Charcoal	Carbon disulphide + 5% 2-PrOH	GC
o-Dianisidine	—	—	—	500	—	1700	Filter	Distilled water	HPLC
Diazomethane	0.2	—	—	10	—	200	Coated R	Carbon disulphide	GC
Dibenzo(a,h)anthracene	—	—	—	600	—	2000	Filter	Benzene	GC
Diborane	0.1 mg/m³	—	—	120	—	1000	Impregnated charcoal	3% hydrogen peroxide	PES
1,2-Dibromoethane	20	—	—	180	—	20	Charcoal	Benzene	GC
Dibutyl phosphate	—	—	—	30	—	1.5	Filter	Acetonitrile	GC
Dibutylphthalate	5 mg/m³	—	—	10	—	1000	Filter	Carbon disulphide	GC
o-Dichlorobenzene	50	—	—	10	—	20	Charcoal	Carbon disutphide	GC
p-Dichlorobenzene	75	—	—	3	—	10	Charcoal	Carbon disulphide	GC
3,3'-Dichlorobenzidine	—	—	—	48	—	200	Filter	—	HPLC
Dichlorodifluoromethane (F-12)	1000	—	—	1.2	—	10	Charcoal	—	GC

Table 5.4 (*continued*) — NIOSH recommended sampling procedures and methods of analysis for volatile compounds in air

Airborne contaminant	OSHA standard, ppm		Sample volume, litres		Sample rate, ml/min		Collection device	Desorbing solvent	Analytical method (Key at end of table)
	TWA	Ceil.	Ceil.	TWA	Ceil.	TWA			
1,1-Dichloroethane	100	—	2	10	200	20(50)	Charcoal	Carbon disulphide	GC
1,2-Dichloroethane	200	—	2	3	200	20(50)	Charcoal	Carbon disulphide	GC
Dichloroethyl ether	—	15	—	12	—	25	Charcoal	—	GC
Dichloromonofluoromethane (F-21)	1000	—	—	1.2	—	10	Charcoal	—	GC
1,1-Dichloro-1-nitroethane	—	10	15	—	1000	—	Charcoal	Carbon disulphide	GC
1,2-Dichloropropane	75	110	—	4	—	50	Charcoal	15/85 Acetone/cyclohexane	GC
Dichlorotetrafluoroethane (F-114)	1000	—	—	3	—	50	Charcoal	—	GC
Dichlorvos	—	—	—	120	—	500	XAD-2 resin	Toluene	GC
Diethyl carbamoyl chloride	—	—	—	10	—	20(50)	Porapak P	Ethyl acetate	GC
Diethylamine	25	—	50	50	1000	100	Silica gel	0.2N H$_2$SO$_4$ in 10% MeOH	GC
2-Diethylaminoethanol	10	—	—	96	—	200	Silica gel	20% water in MeOH	GC
Dieldrin	0.25 mg/m^3	—	—	180	—	1500	Filter	Iso-octane	GC
Diethylenetriamine	10 mg/m^3	—	—	96	—	200	Silica gel	—	GC
Di(2-ethylhexyl) phthalate	5 mg/m^3	—	—	30	—	1000	Filter	Carbon disulphide	GC
1,1-Difluorodibromomethane	100	—	2	10	200	20(50)	Charcoal	Isopropyl alcohol	GC
Difluorodibromomethane	100	—	2	10	200	20	Charcoal	Isopropyl alcohol	GC
Di-isobutyl ketone	50	—	—	12	—	25	Charcoal	Carbon disulphide	GC
Di-isopropylamine	5	—	6	48	200	100	Silica gel	1N H$_2$SO$_4$, neutralized	GC
Dimethylacetamide	10	—	—	50	—	100	Silica gel	Carbon disulphide + 2% acetone	GC
Dimethylamine	10	—	50	50	1000	100	Silica gel	0.2N H$_2$SO$_4$ in 10% MeOH	GC
4-Dimethylaminobenzene	—	—	—	72	—	200	Filter	Isopropyl alcohol	GC
Dimethylaniline	5	—	—	10	—	20	Charcoal	Carbon disulphide	GC
N,N-Dimethylaniline	—	—	—	—	—	—	Charcoal	—	GC
Dimethylformamide	10	—	—	15	—	50	Silica gel	Carbon disulphide + 2% acetone	GC
1,1-Dimethylhydrazine	0.5	—	—	100	—	1000	Bubbler	—	GC, Colour
Dimethyl sulphate	1 mg/m^3	—	—	12	—	50	Porapak P	Ethyl ether	GC
N,N-Dimethyl-p-toluidine	1 mg/m^3	—	—	96	—	1500	Silica gel	Ethyl ether	GC
Dinitrobenzene	0.2 mg/m^3	—	—	90	—	1500	Filter	—	HPLC
Dinitro-o-cresol	1.5 mg/m^3	—	—	180	—	1.5	Filter/bubbler	—	HPLC
Dinitrotoluene	—	—	—	90	—	1500	Filter	—	HPLC
Dioxane	100	—	2	10	200	20(50)	Charcoal	Carbon disulphide	GC
Diphenyl	—	—	—	—	—	—	Tenax	Carbon tetrachloride	GC
p,p-Diphenylmethane di-isocyanate	—	0.2 mg/m^3	20	—	1000	—	Bubbler	—	Colour

Table 5.4 (*continued*) — NIOSH recommended sampling procedures and methods of analysis for volatile compounds in air

Airborne contaminant	OSHA standard, ppm TWA	OSHA standard, ppm Ceil.	Sample volume, litres Ceil.	Sample volume, litres TWA	Sample rate, ml/min Ceil.	Sample rate, ml/min TWA	Collection device	Desorbing solvent	Analytical method (Key at end of table)
Dipropylene glycol methyl ether	100	—	2	10	200	20(50)	Charcoal	Carbon disulphide	GC
Endrin	0.1 mg/m^3	0.3 mg/m^3	—	120	—	1000	Filter	Toluene	GC
Epichlorohydrin	5	—	20	20	200	50	Charcoal	Carbon disulphide	GC
EPN (*O*-Ethyl-*O*-*p*-nitrophenyl phenylphosphorothioate)	0.5 mg/m^3	—	—	120	—	1500	Filter	Iso-octane	GC
2-Ethoxyethanol (Cellosolve)	200	—	—	5	—	10	Charcoal	5% MeOH in dichloromethane	GC
2-Ethoxyethyl acetate	100	—	2	10	200	20(50)	Charcoal	Carbon disulphide	GC
Ethyl acetate	400	—	2	6	200	20(50)	Charcoal	Carbon disulphide + 1% 2-BuOH	GC
Ethyl alcohol	1000	—	—	—	—	50	Charcoal	Carbon disulphide + 1% 2-BuOH	GC
Ethylamine	10	—	6	48	200	100	Silica gel	1N H$_2$SO$_4$, neutralized	GC
Ethyl benzene	100	—	2	10	200	20(50)	Charcoal	Carbon disulphide	GC
Ethyl bromide	200	—	2	4	200	20(50)	Charcoal	Isopropyl alcohol	GC
Ethyl butyl ketone	50	—	2	10	200	20(50)	Charcoal	Carbon disulphide + 1% MeOH	GC
Ethyl chloride	1000	—	—	10	—	20	Charcoal	Carbon disulphide	GC
Ethylene chlorohydrin	5	—	2	10	200	20(50)	Charcoal	Carbon disulphide + 5% 2-PrOH	GC
Ethylenediamine	10	—	—	96	200	200	Silica gel	—	GC
Ethylene dibromide	20	30	2	10	200	200	Charcoal	Carbon disulphide	GC
Ethylene dichloride	50	100	3	10	200	20(50)	Charcoal	Carbon disulphide	GC
Ethylene glycol	50	—	3	48	200	200	Filter/silica gel	2% PrOH in water	GC
Ethylene oxide	50	—	—	5	—	20	Charcoal	Carbon disulphide	GC
Ethyl ether	400	—	2	3	200	20(50)	Charcoal	Ethyl acetate	GC
Ethyl formate	100	—	2	10	200	20(50)	Charcoal	Carbon disulphide	GC
n-Ethylmorpholine	20	—	—	10	—	20	Silica gel	1N H$_2$SO$_4$ neutralized	GC
Fluoranthene	600	—	—	2000	—	5	Filter	Benzene	GC
Fluorotrichlormethane (F-11)	1000	—	—	10	—	25	Charcoal	—	GC
Formaldehyde	2	—	—	6	—	200	Coated Supelpak 20	Carbon disulphide + 1% 2-BuOH	Colour
Formaldehyde	—	—	—	6	—	200	Alumina	Carbon disulphide + 1% 2-BuOH	Colour
Formaldehyde		—	—	96	—	200	Impregnated charcoal	Hydrogen peroxide	LG/IC
Formic acid	9 mg/m^3	—	—	100	—	2500	Bubbler/filter 20	—	GC
Furfural	5	—	—	25	—	50	Bubbler	—	—
Furfural	5	—	—	120	—	500	Bubbler	—	HPLC
Furfural alcohol	50	—	—	6	—	50	Porapak Q	Acetone	GC

Table 5.4 (*continued*) — NIOSH recommended sampling procedures and methods of analysis for volatile compounds in air

Airborne contaminant	OSHA standard, ppm TWA	OSHA standard, ppm Ceil.	Sample volume, litres Ceil.	Sample volume, litres TWA	Sample rate, ml/min Ceil.	Sample rate, ml/min TWA	Collection device	Desorbing solvent	Analytical method (Key at end of table)
Glycidol	50	—	2	10	200	20(50)	Charcoal	Tetrahydrofuran	GC
Heptachlor	0.5 mg/m^3	—	—	60	—	200	Chromosorb 102	Toluene	GC
Heptane	500	—	2	4	200	20(50)	Charcoal	Carbon disulphide	GC
Hexachlorobutadiene	—	—	—	100	—	200	XAD-2 resin	Hexane	GC
Hexachloro-cyclopentadiene	—	—	—	100	—	200	Porapak T	Hexane	GC
Hexachloroethane	1	—	—	—	—	100	Charcoal	Carbon disulphide	GC
Hexachloronaphthalene	0.2 mg/m^3	—	—	30	—	1000	Filter	Hexane	GC
Hexamethylenetetramine	—	—	—	—	—	—	Filter		GC
Hexane	500	—	2	4	200	20(50)	Charcoal	Carbon disulphide	GC
2-Hexanone	100	—	10	10	200	20	Charcoal	Carbon disulphide	GC
Hexanone (MEK)	100	—	2	10	200	20(50)	Charcoal	Carbon disulphide	GC
Hydrazine	1	—	—	100	—	1000	Bubbler		GC
Hydrogen bromide	3	—	—	100	—	1000	Silica gel	—	ISE, IC
Hydrogen chloride	—	5	15	—	1000	—	Silica gel	—	ISE, IC
Hydrogen cyanide	10	—	—	12	—	200	Filter/bubbler	—	ISE, IC
Hydrogen fluoride	3	—	—	45	—	1500	Silica gel	—	ISE, IC
Hydrogen sulphide	10	20	5	—	200	17	Molecular sieve	Thermal	GC
Hydroquinone	2 mg/m^3	—	—	90	—	1500	Filter		HPLC
Inorganic acids	100	—	—	48	—	200	Filter/silica gel	Carbonate/bicarbonate buffer	HPLC
Isoamyl acetate	100	—	2	10	200	20(50)	Charcoal	Carbon disulphide	GC
Isoamyl alcohol	100	—	2	10	200	20(50)	Charcoal	Carbon disulphide + 5% 2-PrOH	GC
Isobutyl acetate	150	—	2	10	200	20(50)	Charcoal	Carbon disulphide	GC
Isobutyl alcohol	100	—	2	10	200	20(50)	Charcoal	Carbon disulphide + 1% 2-PrOH	GC
Iso-octane	400	—	—	4	—	20	Charcoal	Carbon disulphide	GC
Isophorone	25	—	—	12	—	25	Charcoal	Carbon disulphide	GC
Isopropyl acetate	250	—	2	9	200	20(50)	Charcoal	Carbon disulphide	GC
Isopropyl alcohol	400	—	2	3	200	20(50)	Charcoal	Carbon disulphide + 1% 2-BuOH	GC
Isopropylamine	5	—	6	48	200	100	Silica gel	1N H$_2$SO$_4$, neutralized	GC
Isopropyl ether	500	—	—	3	—	10	Charcoal	Carbon disulphide	GC
Isopropyl glycidyl ether	50	—	2	10	200	20(50)	Charcoal	Carbon disulphide	GC
Kepone	—	—	—	120	—	1000	Filter		GC
Ketene	0.9 mg/m^3	—	—	50	—	1000	Impinger		Colour
Lead, inorganic	0.15 mg/m^3	—	—	600	—	2000	Filter	Acid digestion	AA
Lead sulphide	—	0.15 mg/m^3	—	500	—	1700	Filter		XRD
Lindane	0.5 mg/m^3	—	—	90	—	1500	Filter/bubbler	Iso-octane	GC
Malathion	15 mg/m^3	—	—	120	—	1000	Filter	Iso-octane	GC

Table 5.4 (*continued*) — NIOSH recommended sampling procedures and methods of analysis for volatile compounds in air

Airborne contaminant	OSHA standard, ppm TWA	OSHA standard, ppm Ceil.	Sample volume, litres Ceil.	Sample volume, litres TWA	Sample rate, ml/min Ceil.	Sample rate, ml/min TWA	Collection device	Desorbing solvent	Analytical method (Key at end of table)
Maleic anhydride	—	—	—	360	—	1.5	Bubbler	—	HPLC
Mercury	0.1 mg/m³	—	—	48	—	20	Filter/atomic absorp.	—	AA
Mercury, in air	0.5 mg/m³	0.1 mg/m³	3	12	200	200	—	Thermal	AA
Mercury, inorganic and organic	0.05 mg/m³	0.1 mg/m³	3	—	100	—	—	Thermal	AA
Mercury, inorganic only	0.05 mg/m³	0.1 mg/m³	3	—	100	—	—	Thermal	AA
Mercury (organic)	0.1 mg/m³	0.04 mg/m³	3	12	200	200	Carbosieve B	Thermal	AA
Mercury, organic (allyl)	0.01 mg/m³	0.14 mg/m³	3	12	200	200	—	Thermal	AA
Mesityl oxide	25	—	2	10	200	20(50)	Charcoal	Carbon disulphide + 1% MeOH	GC
Methoxychlor	15 mg/m³	—	—	100	—	1500	Filter	Iso-octane	GC
Methyl acetate	200	—	2	7	200	20(50)	Charcoal	Carbon disulphide	GC
Methyl acetylene	1000	—	—	5	—	50	Sampling bag	—	GC
Methyl acetylenepropadiene	1000	—	—	3.6	—	20	Sampling bag	—	GC
Methyl acrylate	10	—	2	5	200	20(50)	Charcoal	Carbon disulphide	GC
Methylal (dimethoxymethane)	1000	—	2	2	200	20(50)	Charcoal	Hexane	GC
Methyl alcohol	200	—	2	5	200	20(50)	Silica gel	Distilled water	GC
Methylamine	10	—	6	48	200	100	Silica gel	1N H₂SO₄, neutralized	GC
Methyl-n-amylketone	100	—	2	10	200	20(50)	Charcoal	Carbon disulphide + 1% MeOH	GC
Methyl bromide	—	20	11	—	1000	—	Charcoal	Carbon disulphide	GC
Methyl butylketone	100	—	—	10	—	20	Charcoal	Carbon disulphide	GC
Methyl cellosolve	25	—	2	10	200	20(50)	Charcoal	5% MeOH in methylene chloride	GC
Methyl cellosolve acetate	25	—	2	10	200	20(50)	Charcoal	Carbon disulphide	GC
Methyl chloride	100	200	0.5	1.5	80	25	Charcoal	Methylene chloride	GC
Methylene chloride	500	1000	1	2.2	200	20(50)	Charcoal	Carbon disulphide	GC
Methyl chloroform	350	—	2	6	200	20(50)	Charcoal	Carbon disulphide	GC
Methyl cyclohexane	500	—	2	4	200	20(50)	Charcoal	Carbon disulphide	GC
Methyl cyclohexanol	100	—	—	12	—	20	Charcoal	Methylene chloride	GC
Methyl cyclohexanone	100	—	—	3	—	50	Charcoal	Methylene chloride	GC
4,4'-Methylenebis phenylisocyanate	—	0.02	—	—	—	—	Porapak Q	Acetone	GC
	—	—	—	—	—	1000	Filter, impreg.	Methylene chloride	HPLC
Methyl ethyl ketone (see 2-butanone)	—	—	—	—	—	—	Charcoal	—	—

Table 5.4 (*continued*) — NIOSH recommended sampling procedures and methods of analysis for volatile compounds in air

Airborne contaminant	OSHA standard, ppm TWA	OSHA standard, ppm Ceil.	Sample volume, litres Ceil.	Sample volume, litres TWA	Sample rate, ml/min Ceil.	Sample rate, ml/min TWA	Collection device	Desorbing solvent	Analytical method (Key at end of table)
Methyl ethyl ketone peroxide	0.2	—	—	240	—	1700	Impinger, midget	Dimethylphthalate	Colour
Methyl formate	100	—	—	5	—	10	Carbosieve B	Carbon disulphide	GC
5-Methyl-3-heptanone	25	—	2	10	200	20(50)	Charcoal	Carbon disulphide + 1% MeOH	GC
Methyl iodide	5	—	2	10	200	20(50)	Charcoal	Toluene	GC
Methyl isoamylacetate	50	—	—	10	—	20	Charcoal	Carbon disulphide	GC
Methylisobutyl carbinol	25	—	2	10	200	20(50)	Charcoal	Carbon disulphide + 5% 2-PrOH	GC
Methyl isobutyl ketone (MIBK)	100	—	—	5	—	10	Charcoal	Carbon disulphide	GC
Methyl methacrylate	100	—	—	5	—	10	XAD-2	Carbon disulphide	GC
α-Methylstyrene	100	—	—	12	—	25	Charcoal	Carbon disulphide	GC
Monochloroacetic acid	—	—	100	48	200	100	Silica gel	Distilled water	LC/IC
Monomethyl aniline	2	—	—	100	—	1000	Bubbler	—	GC
Monomethylhydrazine	0.2	—	—	22.5	—	1500	Bubbler	—	GC, Colour
Morpholine	20	—	—	20	—	50	Silica gel	1N H_2SO_4, neutralized	GC
Naphtha, coal tar	100	—	2	10	200	20(50)	Charcoal	Carbon disulphide	GC
Naphthalene	10	—	—	12	—	25	Charcoal	Carbon disulphide	GC
Naphthylamines	—	—	—	96	—	200	Filter silica gel	—	GC
α-Naphthylthiourea	0.3 mg/m³	—	—	480	—	1500	Filter	Ethyl alcohol	HPLC
Nickel carbonyl	0.05	—	250	72	200	150	Charcoal acid wash	—	AA
Nickel, metal and inorganic	1 mg/m³	—	—	400	2500	2000	Filter	HF/nitric acid	AA
Nicotine	0.5 mg/m³	—	—	100	—	200	XAD-2 resin	Ethyl acetate	GC
p-Nitroaniline	1	—	90	96	1000	200	Silica gel	Ethyl alcohol	GC
Nitrobenzene	1.0	—	—	50	—	100	Silica gel	Carbon disulphide + 2% acetone	GC
4-Nitrobiphenyl	—	—	—	48	—	200	Filter/silica gel	Isopropyl alcohol	GC
p-Nitrochlorobenzene	1 mg/m³	—	—	50	—	100	Silica gel	Carbon disulphide + 2% acetone	GC
Nitroethane	100	—	—	3	—	50	Supelpak 20F	Ethyl acetate	GC
Nitroglycerin	0.2	—	—	15	—	50	Tenax	Ethyl alcohol	GC
Nitromethane	100	—	—	5	—	10	Chromosorb 106	—	GC
1-Nitropropane	25	—	—	12	—	25	Chromosorb 106	—	GC
2-Nitropropane	25	—	—	12	—	25	Chromosorb 106	—	GC
N-Nitrosodimethylamine	—	—	—	72	—	200	Tenax	—	GC
Nitrotoluene	5	—	—	20	—	40	Silica gel	Carbon disulphide + 2% acetone	GC
Nonane	400	—	—	4	—	20	Charcoal	Carbon disulphide	GC
Octachloronaphthalene	0.1 mg/m³	—	—	30	—	1000	Filter	Hexane	AA
Octane	500	—	2	4	200	20(50)	Charcoal	Carbon disulphide	GC

Table 5.4 (*continued*) — NIOSH recommended sampling procedures and methods of analysis for volatile compounds in air

Airborne contaminant	OSHA standard, ppm — TWA	OSHA standard, ppm — Ceil.	Sample volume, litres — Ceil.	Sample volume, litres — TWA	Sample rate, ml/min — Ceil.	Sample rate, ml/min — TWA	Collection device	Desorbing solvent	Analytical method (Key at end of table)
Paraquat	0.5 mg/m³			90		1500	Filter	Distilled water	HPLC
Parathion	0.11 mg/m³		3	12	200	25	Filter	Iso-octane	GC
Pentachlorobenzene	—		3	12	200	25	Filter	Hexane	GC
Pentachloroethanol				10	200	50	Porapak R	Hexane	GC
Pentachloronaphthalene	0.5 mg/m³			250		1300	Filter/bubbler	Iso-octane	GC
Pentachlorophenol	0.5 mg/m³			180		1500	Filter/bubbler	—	HPLC
Pentane	1000		2	2	50	20	Charcoal	Carbon disulphide	GC
2-Pentanone	200		2	10	200	20(50)	Charcoal	Carbon disulphide	GC
Petroleum distillate	500		2	4	200	20(50)	Charcoal	Carbon disulphide	GC
Phenol	5		2	25	200	50	Bubbler	—	GC
Phenyl ether (vapour)	1		2	10	200	20(50)	Charcoal	Carbon disulphide	GC
Phenyl ether-biphenyl	1		2	10	200	20(50)	Silica gel	Benzene	GC
Phenyl glycidyl ether	10		2	10	200	20(50)	Charcoal	Carbon disulphide	GC
Phenylhydrazine	5			—		—	Bubbler	—	GC
Phosdrin	0.1 mg/m³	0.3 mg/m³		240		1000	Supelpak 20F	Toluene	GC
Phosgene	0.4 mg/m³			50		1000	Bubbler	—	Colour
Phosphine	0.3			16		50	Coated silica gel	—	Colour
Phosphorus pentachloride	1 mg/m³			48		200	Filter/bubbler	—	Colour
Phosphorus trichloride				24		200	Bubbler	—	Colour
Phthalic anhydride	2			100		1500	Filter	—	HPLC
Picric acid	0.1 mg/m³			100		1500	Filter	—	HPLC
Platinum, inorganic	0.002 mg/m³			720		2000	Filter	Acid digestion	AA
n-Propyl acetate	200		2	5	200	10	Charcoal	Carbon disulphide	GC
n-Propyl acetate	200		2	10	200	20(50)	Charcoal	Carbon disulphide	GC
Propyl alcohol	200			10		20(50)	Charcoal	Carbon disulphide + 1% 2-PrOH	GC
Propylene dichloride	75		2	10	200	20(50)	Charcoal	Carbon disulphide	GC
Propylene oxide	100		2	5	200	20(50)	Charcoal	Carbon disulphide	GC
n-Propyl nitrate	25			70		150	Charcoal	Carbon disulphide	GC
Pyrene	—			600		2000	Filter	Benzene	GC
Pyrethrum	5 mg/m³	10 mg/m³		120		1000	Filter	Acetonitrile	HPLC
Pyridine	5			25		50	Charcoal	Carbon disulphide	GC
Quinone	0.4 mg/m³			24		100	XAD-2 resin	—	HPLC
Ronnel	10 mg/m³			120		1000	Filter	Toluene	GC
Rotenone	5 mg/m³			100		1500	Filter	Acetonitrile	GC

Table 5.4 (*continued*) — NIOSH recommended sampling procedures and methods of analysis for volatile compounds in air

Airborne contaminant	OSHA standard, ppm		Sample volume, litres		Sample rate, ml/min		Collection device	Desorbing solvent	Analytical method (Key at end of table)
	TWA	Ceil.	Ceil.	TWA	Ceil.	TWA			
Selenium compounds	0.2 mg/m³	—	—	600	—	2000	Filter	Acid digestion	AA
Silica, amorphous	—	—	—	400	—	1700	Filter	—	XRD
Stoddard solvent	500	—	2	3	200	20(50)	Charcoal	Carbon disulphide	GC
Styrene	100	200	2	10	200	20(50)	Charcoal	Carbon disulphide	GC
Styrene oxide	—	—	—	13	—	50	Tenax	Ethyl acetate	GC
Sulphur hexafluoride	1000	—	—	3	—	50	Sampling bag	—	GC
Sulphuric acid	1 mg/m³	—	—	1500	—	2	Silica gel	—	Colour
Sulphuric fluoride	5	10	—	3	—	50	Sampling bag	—	GC
2,4,5-T	10 mg/m³	—	—	100	—	1000	Filter	Ethyl alcohol	HPLC
Terphenyls (orthoterphenyl)	—	1	15	—	1000	—	Filter	Carbon disulphide	GC
1,1,2,2-Tetrachloro-1,2-difluorothane	500	—	2	2	50	20	Charcoal	Carbon disulphide	GC
1,1,1,2-Tetrachloro-2,2-difluoroethane	500	—	2	2	35	20	Charcoal	Carbon disulphide	GC
1,1,2,2-Tetrachloroethane	5	—	10	10	200	20(50)	Charcoal	Carbon disulphide	GC
Tetrachloroethylene	100	200	3	10	200	20(50)	Charcoal	Carbon disulphide	GC
Tetrachloronaphthalene	2 mg/m³	—	—	104	—	1300	Filter/bubbler	Iso-octane	GC
Tellurium	0.1 mg/m³	—	—	104	—	1300	Filter	Acid digestion	GC
Tetraethylpyrophosphate	—	—	—	48	—	200	Supelpak 20P	Toluene	GC
Tetrahydrofuran	200	—	2	9	200	20(50)	Charcoal	Carbon disulphide	GC
Tetramethylsuccinonitrile	0.5	—	—	50	—	100	Charcoal	Carbon disulphide	GC
Tetranitromethane	1	—	—	250	—	1000	Bubbler	Ethyl acetate	GC
Tetryl (2,4,6-trinitrophenyl methylinitramine)	1.5 mg/m²	—	—	100	—	1000	Filter	—	Colour
Thiophene	—	—	—	48	—	100	Charcoal	Toluene	GC
Titanium diboride	—	—	—	500	—	1700	Filter	—	XRD
Toluene	200	300	2	10	200	20(50)	Charcoal	Carbon disulphide	GC
2,4-Toluene di-isocyanate	—	0.02	20	—	1000	—	Bubbler	—	Colour
o-Toluidine	5	—	—	50	—	100	Silica gel	Ethyl alcohol	GC
Trace elements	—	—	—	500	2500	1500	Filter	Acid digestion	ICP
Tributyl phosphate	5 mg/m³	—	—	100	—	1500	Filter	Petroleum ether	GC
1,2,4-Trichlorobenzene	—	5	3	12	200	25	Filter-Supelpak 20F	Hexane	GC
1,2,2-Trichlorethane	10	—	2	10	200	20(50)	Charcoal	Carbon disulphide	GC
Trichloroethylene	100	200	3	10	200	20(50)	Charcoal	Carbon disulphide	GC
Trichloronaphthalene	5 mg/m³	—	—	104	—	1300	Filter	Iso-octane	GC
1,2,3-Trichloropropane	50	—	—	10	—	20	Charcoal	Carbon disulphide	GC

Table 5.4 (*continued*) — NIOSH recommended sampling procedures and methods of analysis for volatile compounds in air

Airborne contaminant	OSHA standard, ppm		Sample volume, litres		Sample rate, ml/min		Collection device	Desorbing solvent	Analytical method (Key at end of table)
	TWA	Ceil.	Ceil.	TWA	Ceil.	TWA			
1,1,2-Trichloro-1,2,2-trifluoroethane	1000	—	1.5	1.5	50	20	Charcoal	Carbon disulphide	GC
Triethylamine	25	—	6	48	200	100	Silica gel	1N sulphuric acid, neutralized	GC
Trifluoromono-bromomethane	1000	—	1	1	50	50	Charcoal	Methylene chloride	GC
Trimellitic anhydride	0.005	—	—	400	—	1700	Filter	Methanol	GC
2,4,7-Trinitro-9-fluorenone	—	—	—	500	—	2000	Filter	Toluene	HPLC
Triorthocresyl phosphate	0.1 mg/m^3	—	—	100	—	1500	Filter	Petroleum ether	GC
Triphenyl phosphate	3.0 mg/m^3	—	—	100	—	1500	Filter	Petroleum ether	GC
Turpentine	100	—	2	10	200	20(50)	Charcoal	Carbon disulphide	GC
Vinyl acetate	10	—	—	3	—	20	Chromosorb 107	—	GC
Vinyl bromide	5	—	—	6	—	50	Charcoal	Ethyl alcohol	GC
Vinyl chloride	1	—	—	5	—	20	Charcoal	Carbon disulphide	GC
Vinyl toluene	100	—	2	10	200	20(50)	Charcoal	Carbon disulphide	GC
Warfarin	0.1 mg/m^3	—	—	400	—	1500	Filter	Methanol	HPLC
Welding and brazing fumes	—	—	—	120	—	1000	Filter	—	XRD
Xylene	100	150	2	10	200	20(50)	Charcoal	Carbon disulphide	GC
2,4-Xylidine	—	—	—	—	—	—	—	—	GC

Key: GC, gas chromatography; AA, atomic absorption spectrophotometry; HPLC, high pressure liquid chromatography; LC, liquid chromatography; IC, ion chromatography; ICP, inductively coupled plasma atomic-emission spectroscopy; XRD, X-ray diffraction; POL, polarography; ISE, ion-selective electrodes; AA, GF, AA/graphite furnace; PES, plasma-emission spectroscopy; Colour, colorimetry.

The remaining elements of the air pollution system have been discussed in detail in the literature, and will not be discussed further here since they do not impinge directly on the subject of immediate interest.

5.4   EXPRESSION OF GAS CONCENTRATIONS

Irrespective of the goals of an air-monitoring programme, it will be necessary to calculate gas-phase concentrations of pollutants in air from experimental measurements. Also, it is usually necessary to determine the efficiency of certain sampling procedures, and to calibrate the quantitative response of a given instrument to the species of interest. Such calculations require a good working knowledge of the behaviour of gases in relation to temperatures and pressure, and a clear understanding of the fundamental principles governing such behaviour. Furthermore, it is frequently necessary to convert flow-rate, volume, density, pressure and concentration measurements from one system of expression to another. The purpose of this section is to present the fundamental relationships that exist in pure and mixed gases, and to illustrate the principles involved in expressing gas concentrations in a variety of different units.

5.4.1   Ideal gases

The relationship between the pressure and volume of a gas at a constant temperature is given by Boyle's Law, which states that, at a fixed temperature, the volume of a fixed mass of gas varies inversely with the pressure:

$$PV = \text{constant}, \; K$$
$$V \;\; = K(1/P)$$

where K is a constant of proportionality. Alternatively:

$$P_1V_1 = P_2V_2$$

When pressure remains constant, the relationship between volume and temperature is given by Charles' Law which states that, at a fixed pressure, the volume of a given mass of gas varies directly with absolute temperature:

$$V = KT$$

When a mixture of gases is considered, each constituent of the mixture obeys Boyle's and Charles' laws. Furthermore, Dalton's law of partial pressures states that the total pressure exerted by a mixture of gases in a container is equal to the sum of the partial pressures. The partial pressure is the pressure which would be exerted by each gas if it were the only gas in the container.

$$P_{\text{TOTAL}} = P_1 + P_2 + \ldots + P_n$$

where P_1, P_2, $\ldots$ P_n are the partial pressures of components 1, 2, $\ldots$ n. Partial pressures of gases are not affected by the presence of other gases.

Under the same conditions of temperature and pressure, equal volumes of gases contain the same number of molecules. At STP (0°C and 1 atmosphere), one mole of an ideal gas contains 6.023×10^{23} molecules and occupies a volume of 22.414 l. Thus a given volume of different gases will contain the same number of moles or molecules, but a different weight.

A more useful equation which combines both Boyle's and Charles' Laws is the ideal gas law:

$$PV = KT$$

$$\frac{P_1 V_1}{T_1} = \frac{P_2 V_2}{T_2}$$

If K is proportional to the number of moles of gas, then:

$$PV = nRT$$

where R is the gas constant ($R = 8.314$ mol^{-1}K^{-1} = 0.082056 l.atm mol^{-1}K^{-1}).

5.4.2 Deviations from ideal behaviour
Most gases adhere closely to the ideal gas law at room temperature and atmospheric pressure, but marked deviations from ideality occur at high pressures, and correction factors related to the compressibility of the gas must be applied to achieve even reasonable accuracies of $\pm 20\%$. The compressibilities of many pure gases are known, but not those of gas mixtures. Great care must be taken in calculating concentrations of gases in pressurized systems since marked deviations from ideality cannot always be compensated.

Van der Waals proposed a modification to the ideal gas law in 1873 which considered the attractive forces that exist between the atoms or molecules of the gas and the volume occupied by the molecules themselves, which reduces the effective volume. The van der Waals equation of state for one mole of a real gas is:

$$[P + (n^2 a/V^2)][V - nb] = nRT$$

At low pressures, V is large and the term a/V^2 is also small while b is small relative to V. Thus the equation approximates to the ideal gas law and the gas behaves ideally. At higher pressures, b becomes significant, and deviations from ideal behaviour result.

5.4.3 Concentration units
For gas-phase species, the most commonly used units of concentration are parts per million (ppm, or ppm v/v), parts per hundred million (pphm), parts per billion (ppb) and parts per trillion (ppt). These units express the number of molecules of gaseous constituent found in a million (10^6), a hundred million (10^8), a billion (10^9 — an American billion) or a trillion (10^{12}) molecules of air, respectively. Because numbers of molecules, or moles, are proportional to their volumes according to the ideal gas law, these units may also be thought of as the numbers of volumes of gaseous constituents found in 10^6, 10^8, 10^9 or 10^{12} volumes of air, respectively. For ideal gases, the concentration in ppm, for example, is calculated from:

$$C_{\text{PPM}} = \frac{10^6 V_a}{V_d + V_a} = \frac{10^6 P_a}{P_d + P_a}$$

where V_d and V_a are the volumes of the diluent gas and constituent a, respectively, and P_d and P_a are the partial pressures of the diluent and constituent gas, respectively. The V_a and P_a terms in the denominators may be neglected when dealing with concentrations less than 5000 ppm since their contribution to the total volume or pressure is neglible. However, these terms must be included at higher concentrations since their contribution to total volume and pressure is significant.

Concentrations may also be calculated when a known weight or volume of liquid is vaporized in a known volume of diluent gas. If the weight of liquid is W, the molecular weight is M, and if it is evaporated at constant temperature and pressure, the volume of vapour produced is:

$$V_L = WRT/MP$$

and the concentration on dilution is:

$$C_{\text{PPM}} = \frac{10^6 V_a}{V_d} = \frac{10^6 WRT/MP}{V_d}$$

It is usual to express W in terms of the volume (V_L) and density (ρ) of the liquid:

$$C_{\text{PPM}} = \frac{10^6 V_L \rho RT}{V_d MP}$$

At room temperature and atmospheric pressure (25°C and 760 mm Hg), this reduces to:

$$C_{\text{PPM}} = \frac{24.5 \times 10^6 V_L \rho}{V_d M}$$

$$C_{\text{w/v}} \frac{10^{-6} C_{\text{PPM}} MP}{RT}$$

Concentrations of pollutants in ambient air are usually sufficiently small that ppm is the largest unit in use. However, pollutant concentrations in stack or exhaust trains prior to dilution and mixing with air are much higher, and percentage (parts per hundred) is often used as an expression of concentration. Thus:

$$C(\%) = \frac{10^2 V_a}{V_a + V_b + \ldots + V_n} = \frac{10^2 P_a}{P_a + P_b + \ldots + P_n}$$

where $V_{a, b \ldots n}$ and $P_{a, b \ldots n}$ are the volumes and partial pressures of constituents a, b $\ldots$ n at constant temperature. For concentrations greater than 0.1% (1000 ppm) this concentration term is the one most frequently used.

When gaseous constituents are present at sub-ppt levels (e.g. free radicals such as the OH radical in tropospheric air), the most suitable expression of concentration is the number of molecules, atoms or free radicals in a given volume of air, usually a cubic centimetre (cm^3).

At one atmosphere pressure and 25°C (298K), the volume occupied by 1 mole of an ideal gas is given by

$$V = \frac{nRT}{P} = \frac{1\,\text{mol} \times 0.08205\,\text{l.atm}\,\text{mol}^{-1}\,\text{K}^{-1} \times 298\text{K}}{1\,\text{atm}}$$

$$= 24.453\,\text{litres} .$$

Since 1 mole contains 6.023×10^{23} molecules, we can calculate that $1\,\text{cm}^3$ of ideal gas contains $6.023 \times 10^{23} \times 0.001\,\text{l}/24.453\,\text{l} = 2.463 \times 10^{19}$ molecules.

From the definition of ppm as the number of molecules per million molecules of air, 1 ppm corresponds to $(2.46 \times 10^{19} \times 10^{-6}) = 2.46 \times 10^{13}$ molecules/cm^3 at 25°C and one atmosphere total pressure. It follows that a concentration of, for example, the OH radical in air of 0.1 ppt is $(2.46 \times 10^{19} \times 10^{-12} \times 0.1) = 2.46 \times 10^6$ molecules/cm^3 at 25°C and one atmosphere total pressure. In clean atmospheres, the OH concentration may be an order of magnitude smaller than this, so the use of fractional ppt units would be most cumbersome. Clearly, corrections must be made if the temperature or pressure varies significantly from those used in this sample calculation.

The basis of the concentration units used to express gas concentrations is the number of molecules, or volumes, of a given gaseous species. In aqueous systems, mass rather than volume is the basis for expressing concentrations in ppm, etc. Thus 1 ppm in an aqueous solution is 1 mg/l. (since the density of water is approx 1 g/ml). It is important to note that these expressions are quite different.

Occasionally, mass per unit volume is used to express gas phase concentrations, usually 10^{-6}g per cubic metre (μg/m^3). Since one atmosphere at 25°C contains 4.09×10^{-2} mol/l, one ppm must contain $(4.09 \times 10^{-2} \times 10^{-6}) = 4.09 \times 10^{-8}$ mol/l or 4.09×10^{-5} mol/m^3. If the molecular weight of the gaseous constituent of interest is expressed as MW g/mol, then one ppm in units of mass per m^3 is $(4.09 \times 10^{-5} \times$ MW) g/m^3 or $40.9 \times$ MW μg/m^3.

The conversion between μg/m^3 and ppm, pphm, ppb and ppt can be summarized as follows:

$$\begin{aligned}
\mu\text{g/m}^3 &= \text{ppm} \times 40.9 \times (\text{MW}) \\
&= \text{pphm} \times 0.409 \times (\text{MW}) \\
&= \text{ppb} \times 0.0409 \times (\text{MW}) \\
&= \text{ppt} \times 4.09 \times 10^{-5} \times (\text{MW})
\end{aligned}$$

Concentrations of suspended particulate matter (SPM) in air are usually expressed in units of mass per volume, usually μg/m^3, or in the number of particles per unit volume, usually number per cm^3. This expression of concentration is also frequently used to express the concentrations of criteria pollutants such as sulphur dioxide in air. In cases where emission into the atmosphere from a given source are monitored, the concentrations are frequently expressed as mass per unit time of emissions.

5.4.4 Vapour pressures of pure substances

A substance which is volatile is one which vaporizes readily. The volatilities of liquids are readily explained by means of the kinetic molecular theory which considers the Maxwell–Boltzmann distribution of kinetic energies of molecules in the liquid. A certain fraction of molecules will have sufficient energy to escape from the surface of the liquid, and, if the temperature remains constant, the process may be repeated until all of the liquid evaporates.

If the liquid is contained in a closed container, the molecules having sufficient energy to escape from the condensed phase will escape into the vapour phase, and, since the temperature is constant, the evaporation continues at a constant rate. However, in a closed container, as the number of molecules in the vapour phase increases, these molecules also achieve a Maxwell–Boltzmann distribution of kinetic energies. Some of these molecules will have such low kinetic energies that they will condense to liquid when they collide with the surface of the container and of the liquid. Ultimately, evaporation and condensation will reach equal rates, and an equilibrium vapour composition is reached, with no net transfer of molecules between the two phases. The pressure exerted by the molecules in the vapour phase under these conditions is called the vapour pressure of the liquid. The vapour pressure is a function of the nature of the substance and the temperature, and is independent of the volume above the liquid. Vapour pressure always increases with temperature, and the experimentally observed variation is described by an equation of the form:

$$\log P_{vap} = -A(1/T) + B$$

where A and B are constants. This experimental variation in vapour pressure with temperature is only partially described by a simple equation of the type shown. The equation described above is a simplified form of the Clausius–Clapeyron equation, which, in its differential form is:

$$\frac{d \ln P_{vap}}{d(1/T)} = -\frac{\Delta H}{R}$$

Alternatively:

$$\log p_2 - \log p_1 = -(\Delta H/2.303R)[1/T_2 - 1/T_1]$$

With the simplifying assumption that ΔH is constant (true over narrow ranges in temperature), integrating this equation yields an equation of the type:

$$\log P_{vap} = -A(1/T) + B$$

where $A\ (= -\Delta H/2.303R)$ and B are constants.

The dependence of vapour pressure on temperature applies only over a narrow range of temperature, so alternative equations must be used to describe the empirical dependence on temperature — several modifications have been attempted [58]. One particularly suitable equation of this type is the Antoine equation [59] which represents the temperature dependence of most substances' vapour pressures over large temperature intervals [60]:

$$\log p_{vap} = A - B/(T + C)$$

where p_{vap} is the vapour pressure at the required temperature, T is the temperature, in °C, and A, B and C are constants which are characteristic of the given substance and the temperature range of interest. Thus for methanol, $A = 7.20660$, $B = 1582.698$ and $C = 239.765$; and the calculated vapour pressure at 30°C is 21.86 kPa.

5.4.5 Vapour pressures of constituents in a mixture

As mentioned earlier, Dalton's law of partial pressures states that the sum of the partial pressures of the gaseous components in a mixture equals the total pressure — this is also true for the partial pressures of vapours over a liquid mixture. However, depending on the nature of the liquids in the mixture, the vapour pressures of the components may not be as expected. As a general rule, the vapour pressure of a component in a liquid mixture depends on the nature of the liquids present in the mixture, and on the relative amounts of each constituent present. This is quite different from the case of pure liquids.

In an ideal solution, the vapour pressure of a component depends only on the vapour pressure component and the mole fraction of the component in solution — this relationship is known as Raoult's Law and is true over the entire composition range:

$$p_a = x_a p_a^*$$

where p_a and p_a^* are the vapour pressures of the component a in solution and in the pure state, respectively, and x_a is the mole fraction of component a in solution. Real solutions show marked deviations from ideal behaviour, but they do approximate to Raoult's Law when solute concentrations are very low. As a general rule, when the components of a solution are similar, e.g. ethanol and water, strong positive deviations from Raoult's Law are observed, that is, the measured partial pressures are greater than the calculated vapour pressures.

As with calculations involving the ideal gas law, care must be taken when calculating expected vapour compositions from Raoult's Law to avoid the serious errors that could arise when there are strong positive or negative deviations from the law. In general, only chemically similar liquids obey Raoult's law reasonably well over a narrow concentration range — at high solute concentrations, deviations from ideal behaviour should be expected. To estimate vapour compositions for mixtures of components, particularly for multicomponent dissimilar liquids, it is generally best to measure the gas phase composition by a suitable technique rather than to rely on calculations which assume ideal behaviour. When the mutual solubilities of the components are very small, as occurs with toluene and water, a situation arises where the vapour composition is dependent only on the temperature, and is independent of the relative amounts of each constituent present.

In ideal solutions, the solute also obeys Raoult's law, but in real solutions, the relationship does not hold over the entire composition range. Although vapour pressure is proportional to mole fraction at low concentrations, the linear dependence is expressed by Henry's law, which states that

$$p_{vap} = x_a K_a$$

where K_a is some constant, having the dimensions of pressure, chosen such that the plot of vapour pressure of the component *vs.* its mole fraction is tangential to the experimental curve at $x_a = 0$. Henry's law constants for many species are known [61], but they are rarely known for multicomponent mixtures. Again, there is no substitute for experimental measurement, although many systems, particularly gases dissolved in liquids, are very adequately described by Henry's law.

5.5 GAS SAMPLING TECHNIQUES

5.5.1 General sampling considerations

Gas-phase organic pollutants are generally present at very low concentrations in large volumes of air, and are usually present together with large numbers of other pollutants which may interfere with representative sampling and analysis of the pollutant of interest. Although there are many analytical instruments available for use in the field, the low concentrations and the large number of potential interferences usually ensure that the analyses must be carried out in the laboratory with the full range of analytical instrumentation available. Thus, the precise and accurate determination of the levels of these species present in air depends on the collection and transport of a representative sample to the laboratory, with unwanted reactions or losses during sampling and storage minimized.

Because gases mix freely with ambient air, the sampling of organic vapours in air is much easier than the sampling of aerosols, dusts, etc. However, the reactive nature of many organic pollutants and the susceptibility of such species to absorptive and/or adsorptive losses during storage ensures that the collection and storage of representative samples of organic pollutants in air presents many challenges to the atmospheric chemist.

Although this chapter deals primarily with the sampling and analysis of gaseous organic pollutants in air, some reference will be made to the techniques used for collection of aerosol and dust samples. It is particularly important to remember that many organic pollutants are soluble in water and may be found dissolved in aqueous aerosols, and organic material is frequently found sorbed on particulate matter — consequently, any monitoring programme which seeks to establish the levels of a particular pollutant in air must consider all forms in which the pollutant may be found.

Where high concentrations of a gas may be present in a stagnant or slow moving air mass, an uneven distribution of the gas within the air mass may arise. This is usually overcome by sampling from a well mixed air mass or a turbulent flow of air.

Particular care must be taken to ensure that the materials from which the sampling system is manufactured do not interfere with the collection of a representative sample through unwanted reactions with constituents of the sampling system with respect to possible reactivity towards gaseous constituents of interest. Frequently, the reactivity of the species of interest requires that the sample be pretreated to remove unwanted interferences, particulates or moisture. In such cases, care must also be taken to analyse the material collected in the pretreatment devices for the content of the gaseous constituent of interest.

Many gaseous organic pollutants are photochemically unstable, so light is generally excluded during storage to prevent losses. All possible precautions should be taken against inward or outward leakage of gas samples, particularly during storage. The use of sealing lubricants to prevent leaks is not recommended since these frequently absorb or dissolve gaseous constituents, thus interfering with the sampling process. Condensation of atmospheric moisture within the sampling system should also be avoided for the same reason, and also because moisture may allow certain unwanted reactions to occur.

5.5.2 Gas sampling systems

Air sampling equipment, intended for evaluation of airborne exposure, has evolved markedly during recent decades towards increased miniaturization and automation. This trend towards miniaturization has been greatly facilitated by enormous improvements in the limits of detection of modern analytical instrumentation; this has permitted satisfactory analyses at the microgram, and occasionally nanogram, quantities of airborne contaminants.

Despite the evolutionary trends in the development of modern air sampling equipment, the basic principles of air sampling remain the same, the main improvements arising from the manner in which these basic principles are implemented. The basic components of an air sampling system are:

(i)   air intake system;
(ii) collection device;
(iii) flow measurement device;
(iv) air moving component.

Since the performance of the entire system depends on satisfactory performance of each individual component, it is essential that the behaviour of each component is evaluated under precisely the same conditions as those to be used during sampling. The preliminary objective in evaluating the performance of the components of the sampling system is to ensure that a representative sample is taken.

5.5.2.1 Air inlet system

The nature of the intake system varies according to the overall objective of the sampling programme, from thin-walled probes used for aerosol source sampling to free vertical access for dustfall deposit gauges. The detailed description of such devices for all possible sampling objectives is beyond the scope of this text, but there are a number of general considerations which apply to the design of systems to meet stated objectives.

The first requirement is that the inlet be designed to allow a representative sample of the desired pollutant to be collected at the sampling point. Although this issue is of fundamental importance, the evaluation of performance for specific atmospheric pollutants of individual intake devices commonly used in air sampling has received relatively little attention. This factor assumes over-riding significance when considering sampling of aerosols and particulate matter, but is of lesser significance in sampling of gases.

The materials from which the intake system and the transfer lines to the collection device are fabricated must be carefully chosen in order to minimize the occurrence of phenomena that will modify the properties and/or quantity of the pollutants of interest. Particular attention must be addressed to the prevention or minimization of such phenomena as adhesion to tube walls, condensation of volatile components within transfer lines, reaction of gaseous constituents with transfer system materials, and adsorption and reaction with collected particulates [62–69]. When specific pollutants are of interest, further problems uniquely associated with these species may be identified; for example, special care must be exercised in handling gas samples which contain sulphur compounds such as SO_2 because they are easily oxidized, particularly in contact with stainless steel.

5.5.2.2 *Collection devices*

The choice of collection device is determined by the specific constituent(s) of interest. Gaseous sampling usually incorporates one or more of grab sampling, adsorption, absorption, and cryogenic sampling, descriptions of which systems and their applications are given in the next section. One factor common to all collection systems is the need to determine the collection efficiency of the device for the specific airborne pollutant(s) of interest. Although 100% collection efficiencies are highly desirable, in practice lower collection efficiencies may be used since it is so difficult to attain 100% efficiency for all components of interest. Collection efficiencies of 90% or better are generally recognized as being acceptable [67], but the lowest collection efficiency that can be tolerated usually depends on the significance of the quantitative results. Regardless of the precise collection efficiency which is acceptable for each component, it is essential that the collection efficiency can be precisely and reproducibly measured for each component of interest.

Verification of collection efficiencies of selected collection devices requires preparation of standard test atmospheres [67,70]. For gaseous constituents, this may be achieved by introducing known quantities of the test gas into a container providing a known dilution volume. Collapsible containers such as Tedlar or Teflon bags are frequently used, since they offer the advantage of allowing no dilution or pressure changes within the test chamber as the standard mixture is withdrawn for analysis. This technique is the one used exclusively for sampling odorous gas mixtures for subsequent analysis. An alternative method of generating a standard gas mixture involves the use of permeation tubes. The pollutant, in liquid form, is enclosed in a tube which is placed in an oven at a known, constant temperature. The pollutant permeates through the plastic tube at a known rate, and gas mixtures of known composition are generated by the passage of a known flow-rate of air past the tube. These and other methods of generating standard gas mixtures for calibration purposes have been reviewed by Barratt [70].

Regardless of the apparent efficiency of the collection device, great care must be taken to ensure that losses within the transfer lines do not contribute to a reduction in the overall efficiency of the collection system. Furthermore, the use of two or more collection devices in series is frequently practised to identify overloading or poor efficiency of the first collection device in the series. For example, charcoal adsorption tubes which are commonly used in air sampling contain two different sections separated by a plastic plug for this purpose, i.e. identification of overloading or breakthrough from the first section.

When collection devices are used in series, specific evaluation of the effect of upstream components on the constituents reaching subsequent collection components is usually necessary. For example, particulate matter collected on an upstream filter may contain adsorbed organic material that was intended to be collected at a subsequent collection step. Alternatively, a screening filter may be used upstream of a second collection component in order to remove selectively constituents that may interfere with subsequent collection and analysis of a component of interest. It may not be possible to avoid all of the problems mentioned here, but a knowledge of such potential interferences will allow modifications to the chosen analytial techniques, and may affect the interpretation of the quantitative analytical results.

5.5.2.3 *Flow-measurement devices*

The accurate determination of the concentration of an airborne pollutant in an air sample is fundamentally dependent on the accurate measurement of the volume of air sampled. Flow-measuring components of air sampling systems usually fall into one of two categories — volume meters and rate meters [63,71–73].

Volume meters measure the total volume of air sampled in a fixed period of time. The most commonly used type of volume meter is the dry gas meter, capable of measuring to an accuracy of 2–4% under optimum conditions. Rate meters measure the instantaneous volume flow-rate through the sampling system, and therefore require frequent checks to ensure that the instantaneous readings are representative of the average flow-rates through the system. The commonest systems in use include rotameters, pitot tubes, orifice meters, venturi meters and hot-wire anemometers.

Flow-measuring devices of the type mentioned here tend to be inaccurate at low flow-rates, i.e. less than $10\,cm^3/min$, but sampling rates for most sampling systems tend to be higher than this limiting value, thus ensuring reasonable accuracy. Frequent calibration of all measuring devices is recommended under the conditions expected during sampling, taking care to ensure that the inlet system and collection device are in place so that the calibration will be valid for field measurements. Finally, measurement of air pressure and temperature at the inlet to the flow measuring device is essential, since air volume is dependent on these two parameters.

5.5.2.4 *Air-moving devices*

The final component is essential if air is to be either drawn or forced through the air-sampling system. In order to minimize the possiblity of sample contamination during sampling, the preferred air-moving device is a pump or fan which draws air through the entire sampling system. As with all components of the air-sampling system, the air-moving device must be calibrated under field conditions with all components of the system in place. Care must also be taken to ensure that the device delivers constant performance throughout the sampling period.

5.6 GAS-COLLECTION DEVICES

Although every component of the gas sampling system plays a crucial role in the collection of truly representative gas samples, the component which allows greatest scope in ensuring that this objective is met is the collection device. The collection device may allow for collection of an instantaneous sample by direct methods such as grab sampling in a variety of different containment devices, or for concentration of the gaseous constituents of interest by indirect processes such as adsorption, absorption, and cryogenic enrichment. A brief resumé of such devices and their applications follows.

5.6.1 Grab sampling

Grab sampling of gases from the atmosphere involves the direct collection of the test gas in an impermeable container — the technique is obviously limited in application to those gaseous constituents present at high enough concentrations for subsequent analysis. Instantaneous gas samples may be collected in rigid glass or metal flasks, or in flexible plastic bags made of Tedlar® or Teflon®, or combinations of these and other materials with aluminium foil. In all cases, the containers must be evaluated for reactivity with and/or adsorption of gaseous organic pollutants prior to use for field measurements. The technique is most widely used for the collection of gas samples for subsequent analysis by olfactometry to determine the odour threshold of the sample.

Instantaneous gas sampling may also be achieved with direct-reading analytical instruments that have a response time measured in seconds. These instruments tend to be relatively expensive, and are generally restricted in application to those cases where relatively high concentrations of pollutants are present.

5.6.2 Adsorption

Adsorption of gases from the atmosphere is a surface phenomenon whereby intermolecular forces of attraction between the gas molecules and those on the surface lead to concentration of the gas molecules at the surface. The process is dependent on the surface area of the adsorbent material, and is particularly useful as a preconcentration technique since large volumes of air can be sampled.

Materials commonly used as adsorbents for air sampling include activated charcoal, silica gel, molecular sieve and gas chromatographic supports such as Tenax®. Among the factors which must be considered in the selection of an adsorbent for a particular application are high surface area and the relative affinity of the adsorbent for polar and non-polar phases. A large contact area for the adsorption is maintained with a high relative surface area, while certain adsorbents may selectively adsorb polar or non-polar species depending on their own polarity. Activated charcoal, for example, is non-polar and therefore has an affinity for non-polar organic compounds. Silica gel is polar and the opposite effect is observed.

The adsorption/desorption capacity of the adsorbent must be known for each species of interest, and care must be taken to ensure that breakthrough effects resulting from exceeding the adsorption capacity of the adsorbent are not observed. Depending on the objectives of the monitoring programme, selective adsorption may be desirable, in which case it will be possible to exploit the different affinities of adsorbents for polar and non-polar compounds.

The adsorbent is usually packed in glass or stainless steel tubes. For example, in NIOSH Method P&CAM 127 (see below), a tube 7-cm long and 6-mm o.d. containing 150 mg of coconut charcoal is recommended for collecting solvent vapours. The 150 mg of charcoal is divided into a 100-mg front section and a 50-mg back-up section. The front bed traps the airborne contaminant, and the backup section is monitored to determine if any contaminant has broken through the front section. This design is common to many adsorbent tubes for collecting airborne contaminants.

To sample airborne vapours, the sealed ends of the tube are broken and it is connected to the inlet of a pump. The tube is normally kept in a vertical position to prevent shifts in the adsorbent during sampling. When the recommended volume of air has been collected, the tube is capped and may be refrigerated for subsequent analysis. Desorption is often achieved by using solvents such as carbon disulphide or methanol, although thermal desorption is now becoming increasingly popular for selected contaminants.

Although adsorbent tubes are extremely reliable and widely used for collecting airborne contaminants, certain limitations must be acknowledged. These include:

(a) adsorbents have saturation limits for each contaminant sampled; when these are exceeded, breakthrough occurs, and the results are unreliable;
(b) breakthrough can occur if the recommended sample volume is exceeded;
(c) the contaminant of interest may be displaced by another contaminant which is more strongly adsorbed by the adsorbent;
(d) high humidity may severely reduce the breakthrough volumes of adsorbents such as charcoal and silica gel, although some adsorbents are unaffected by moisture;
(e) reactive components may be converted into other species during sampling, hence determination may be inaccurate;
(f) if two or more contaminants are of interest, it may be necessary to choose a solvent for one which differs from that recommended for all others.

Methods for monitoring approximately 500 airborne contaminants have been published by the US National Institute for Occupational Safety and Health (NIOSH) [74] in a seven-volume series. The methods are continually revised and updated as collection efficiencies and/or exposure limits to various species change, and a fourth edition of this comprehensive database is currently in preparation. One of the most widely used methods is P&CAM 127 which describes techniques for collecting and analysing organic solvents in air.

Table 5.4 summarizes the techniques recommended by NIOSH for sampling and analysis of volatile organic compounds in air. Also included are the OSHA worker's exposure limits to selected species. The US Occupational Safety and Health Administration (OSHA) has restricted worker's exposure to airborne contaminants [76]. The OSHA limits are expressed as Time-Weighted Averages (TWA) over a specific period of continuous exposure. The ceiling limit is the maximum concentration that a worker can be exposed to for short periods of time. Also included in this table are the recommended methods of analysis for each compound — these will be discussed in Section 5.7.

The US Environmental Protection Agency (EPA) also publishes a comprehensive set of standards for the measurement of a variety of organic compounds in air. A particularly useful source of information is the *Compendium of Methods for the Determination of Toxic Organic Compounds in Ambient Air* [75]. In most cases, NIOSH and OSHA are interested in workplace exposure and eight-hour sampling periods, whereas the EPA emphasizes outside environmental exposure and longer exposure periods. At present, the EPA routinely use fourteen methods for sampling and analysis — these are summarized in Table 5.5.

5.6.3 Absorption
The process involved in collecting airborne contaminants by absorption in a liquid collection phase is a solubility phenomenon. Since the efficiency of the process is limited by the vapour pressure of the gas over its solution, efficiency is greatly improved by choosing an absorbing medium that reacts chemically with the species of interest. Sulphur dioxide in ambient air is routinely monitored by using a hydrogen peroxide solution as the absorbing medium. Then, sulphur dioxide is oxidized to sulphuric acid during the sampling process.

The efficiency of the process is acutely dependent on the contact area available for the gas molecules, so collection devices such as bubblers, impingers etc. are designed to maximize the interfacial area between the gas molecules and the absorbing solution. This is usually achieved by ensuring that the sampled gas stream is effectively bubbled through a significant volume of absorbing solution. The efficiency is also enhanced at lower temperatures.

As with all collection devices, the actual efficiency of collection during sampling must be established to avoid over-sampling and subsequent misinterpretation of results.

5.6.4 Cryogenic sampling
The efficiency of most collection devices is greatly improved at lower temperatures, but the method of lowering the temperature below the boiling or freezing points of the gases to be collected without any other collection system is also a useful sampling technique. The main limitation is the bulky nature of the equipment, and the difficulty of transporting sufficient quantities of cryogenic liquids for routine sampling purposes.

One of the principal problems encountered in the use of this technique in the field is the large volume of water collected during sampling. This is frequently overcome by using a series of traps cooled to successively lower temperatures so that the water will be selectively removed at an early stage.

Although cryogenic sampling systems are not suitable for routine use, they do have the advantage or providing stabilized bulk atmospheric samples for subsequent analysis.

Table 5.5 — EPA recommended methods for the determination of toxic organic compounds in air

METHOD TO1

is for the determination of volatile organics. Air sample is drawn through Tenax sample tube. After collection it is heat desorbed into a cold trap and subsequently analysed by capillary GC/MS.

METHOD TO2

is for the determination of volatile organic compounds. Air sample is drawn through a carbon molecular sieve sample tube. The sample is then heat desorbed into a cold trap and subsequently analysed by GC/MS with capillary column.

METHOD TO3

is for determination of volatile organic compounds. Ambient air is drawn through collection trap submerged in liquid argon. Once the collection is complete the sample is analysed by GC using temperature programming techniques with FID and EC detection.

METHOD TO4

is for the determination of organochlorine pesticides and polychlorinated biphenyls. Air sample is taken by collection on a glass-fibre filter and polyurethane foam. Materials are recovered by Soxhlet solvent extraction and then subjected to GC/EC analysis.

METHOD TO5

is specific for the determination of aldehydes and ketones. Air is sampled through midget impinger with appropriate solution and then analysed by HPLC.

METHOD TO6

is specific for phosgene. The air sample is drawn through midget impinger containing appropriate solution. After collection the solution is reduced to dryness, dissolved in acetonitrile. Results are determined by HPLC.

METHOD TO7

is specific for *N*-nitrosodimethylamine. Air is drawn through a proprietary sorbent cartridge. Sorbent is then solvent extracted with dichloromethane and analysed by GC/MS using capillary column.

METHOD TO8

is specific for phenols and methylphenols (cresols). Air is drawn through two midget impingers using NaOH solution. Phenols are trapped as phenolates. Phenols are determined by using reversed-phase HPLC.

METHOD TO9

is specific for polychlorinated dibenzo-*p*-dioxins (PCDVE). Air is drawn through an inlet filter followed by a polyurethane foam cartridge. Filters and foam are extracted with benzene followed by GC/MS.

METHOD TO10

is specific for a variety of organochlorine pesticides. Air sample is drawn through a polyurethane foam cartridge. Pesticides are solvent extracted and then analysed by GC with EC detection.

METHOD TO11

is specific for formaldehyde. Air is drawn through a coated silica gel cartridge to form a stable derivative of formaldehyde. The sample is solvent eluted and then determined by HPLC.

METHOD TO12

is for the determination of non-methane organic compounds. Air is drawn through a glass bead trap at − 186°C (liquid argon). It is then back-flushed with helium to remove methane. Temperatures are then raised to 90°C and sample is flushed directly into a flame-ionization detector giving total non-methane organic compounds.

METHOD TO13

is specific for PNAs. Air sample is drawn through sorbent and polyurethane foam cartridges. Sample is solvent (Soxhlet) extracted and then analysed either by GC/FID, GC/MS or HPLC.

METHOD TO14

is for the determination of volatile organics. Air sample is drawn into a passivated pre-evacuated canister. The sample is then concentrated in a cryogenic trap and analysed by GC with one or more appropriate GC detectors including MASS, FID, etc.

NOTE: In most cases, NIOSH and OSHA methods are focused on workplace exposure measurements for eight-hour periods, whereas the EPA emphasizes outside environmental exposure and longer exposure times.

5.7 CHOICE OF ANALYTICAL TECHNIQUE

Because ambient air normally contains hundreds of different constituents, by far the most widely used technique in air pollution monitoring is gas chromatography with detectors chosen to suit the particular analysis. Thus, separation and measurement are performed simultaneously. However, some analyses require selective and sensitive techniques for identification and determination of selected pollutants at low concentrations in the presence of large numbers of potential interferents. Frequently, the choice of technique is determined by the concentration range expected, although costs also prohibit use of certain techniques for routine monitoring purposes. This section considers briefly the principal techniques used for monitoring volatile air pollutants.

5.7.1 Chemical methods of analysis

Chemical techniques were applied to air pollution monitoring long before the relatively recent upsurge in interest in atmospheric chemistry. For example, the presence of nitric acid in air was inferred in the 1800s by its reaction with calcium carbonate. Many chemical methods of analysis are still used for routine monitoring purposes — the most common examples are those in which the pollutant of interest is trapped in an absorbing solution in which a colour-forming reaction takes place; the change in colour indicates the presence of the pollutant of interest, and the intensity of colour at a pre-determined wavelength is proportional to the concentration of pollutant present. Other methods are based on properties such as acidity, conductivity or electrochemical properties of the species of interest.

Acidimetric techniques involving the determination of hydrogen ions produced by absorption of SO_2 in an oxidizing solution such as hydrogen peroxide have been widely used for the determination of sulphur dioxide in air. The acid may be determined by titration, or by conductivity or pH measurements. Any other gas that is readily absorbed to give an acidic solution interferes with the determination. Alkaline pollutants such as ammonia can neutralize the acid and give a negative interference.

A wide variety of colorimetric methods has been applied to routine pollutant monitoring, especially for nitrogen- and sulphur-containing pollutants. The technique usually involves reaction of the pollutant of interest (or its hydrolysis or oxidation products) with a colour-forming reagent, followed by spectrophotometric analysis. Of the chemical methods of analysis, the colorimetric methods are the most sensitive and versatile for routine monitoring purposes — they are also the most cost-effective.

Colour-indication tubes such as those marketed by Draeger are a cheap but effective means of identifying and quantifying certain pollutants in air. They are based on the adsorption of a pollutant of interest on a solid surface with the simultaneous occurrence of a colour reaction. The tubes are made of glass and usually contain a layer of colour-forming reagents along the gas stream. Air samples are sucked through the tube by a small hand-pump, allowing the pollutant of interest to react with the colour-forming reagent. The length of the coloured layer is then proportional to the amount of pollutant. The standard deviation of the results is around 5–20%.

5.7.2   Gas chromatography

GC with a suitable detection system is the most widely used technique for monitoring complex ambient gas mixtures. The flame-ionization detector (FID) is very sensitive to hydrocarbons, so that this is the most widely used detector for monitoring hydrocarbons in ambient air. A continuous monitoring instrument is the dual FID, used without a gas chromatograph, to measure total hydrocarbon concentration. Mass spectrometry interfaced to GC is frequently used to confirm the identities of compounds tentatively identified by comparison of their retention times on the GC column with those of authentic samples.

A second type of detector particularly suited to hydrocarbon monitoring is based on the principle of photoionization. Hydrocarbons are ionized on exposure to ultraviolet light with an energy of 10.2eV, and the ions are detected by conventional methods. The photoionization detector (PID) may be used as a detector with a GC, or on its own as a hand-held direct-reading instrument. While its main advantage is the absence of any requirement for fuel gases such as hydrogen, the detector is insensitive to C_2–C_4 hydrocarbons, so that the total hydrocarbon content reported is generally lower than that reported by FID.

The electron-capture detector (ECD) is particularly sensitive to halogenated compounds and is widely used for measuring ambient concentrations of such species. Sulphur-containing compounds are the main species to which the flame photometric detector (FPD) responds, so the technique is widely used for the detection and quantification of such species in ambient air. The only other detector routinely used for air pollutant monitoring is the thermal-conductivity detector (TCD) although this detector is rarely used for such studies.

5.7.3   Spectroscopic monitoring techniques

Chemiluminescence and fluorescence are the principal emission spectrometry techniques routinely used for air-pollutant monitoring. Chemiluminescent analysers are the recommended instruments for analysis of ozone and the oxides of nitrogen in air. The flame-photometric detector is also an emission-spectrometic technique — sulphur compounds burnt in a hydrogen-rich flame form excited S_2 molecules which emit light at 394 nm. The most notable feature of these analysers is their selectivity and sensitivity.

Fluorescence is the technique employed in some commercial sulphur dioxide monitors, but the most important application of this technique is in the determination of the hydroxyl radical in ambient air. For this purpose, laser induced fluorescence (LIF) is widely used — the technique may also be adapted for analysis of NO in air.

Infrared absorption spectrometry in its conventional form is of limited value as a technique for monitoring atmospheric pollutants since they are usually present at such small concentrations that the fraction of light absorbed is generally very low. However, consideration of the Beer–Lambert law, which states that the intensity of light absorbed by a polluting species is proportional to the path–length as well as to the pollutant concentration, clearly indicates the potential of the technique. While the experimenter has no control over the concentrations of the analytes which are to be determined, it is possible to increase the path-length through which the analysing radiation is passed, and this has the effect of allowing much lower concentrations of gases to be determined. Long-path cells are widely used for this purpose, although their use is severely limited by the presence of such strongly absorbing species as carbon dioxide, water and hydrocarbons.

The development of Fourier-transform infrared spectrometry (FTIR) in the early 1970s led to a marked increase in the potential use of infrared spectrometry for routine monitoring purposes. The technique offers a number of advantages over conventional IR spectrometers including sensitivity, speed, resolution and improved data processing.

5.7.4 Olfactometric analysis

The basic principle of olfactometric analysis is that a sample of odorous gas is diluted with odour-free air and presented to a panel of trained observers in increasing concentrations in order to determine the odour threshold, which is the concentration at which 50% of the observers can perceive the odour. The olfactometer used for the analysis is esentially a dilution apparatus which usually operates within the range 1–250 000 dilutions.

Although this sensory technique is essentially subjective, careful choice and training of panellists, together with stringent operating conditions, can lead to results with a confidence level of ± 100%. There are two basic methods — the forced-choice method and the yes–no method. In the former, two or three samples of air, only one of which is odorous, are presented to each observer who is forced to choose which sample is odorous even if no odour is detectable. The other method uses a single gas sample stream with a simple yes/no choice of odour perception or not.

5.8 CONCLUSIONS

In this chapter, the important considerations associated with any air monitoring study have been summarized, with particular reference to the range of techniques available for sampling and analysis of volatile atmospheric pollutants. The chapter does not seek to cover the subject exhaustively, but rather to give a flavour of the complex nature of the air pollution system. It is hoped that the reader's interest in the subject has been stimulated, and that the references cited will act as a useful source of information for those interested in pursuing the subject in greater depth.

REFERENCES

[1] J. P. Lodge, Jr., Selections of *The Smoake of London, Two Prophecies*, Maxwell Reprint Co., Elmsford, New York, 1969.
[2] P. L. Hanst, *Adv. Environ. Sci. Technol.*, 1971, **2**, 91.
[3] E. D., Hinkley, R. T. Ku and P. L. Kelley, *Laser Monitoring of the Atmosphere*, E. D. Hinkley, ed., Springer-Verlag, New York, 1976.
[4] K. D. Killinger and A. Mooradian eds., *Optical and Laser Remote Sensing*, Springer-Verlag, New York, 1983.
[5] World Health Organisation, *Air Monitoring Programme Design for Urban and Industrial Areas*, WHO Offset Publication 33, WHO, Geneva, 1977.
[6] D. R. Bates and M. Nicolet, *Geophys. Res.*, 1950, **55**, 301.
[7] P. J. Crutzen, *Quart. J. Roy. Meteorol. Soc.*, 1970, **96**, 320.

[8] H. S. Johnson, *Science*, 1971, **173**, 517.

[9] M. J. Molina and F. S. Rowland, *Nature*, 1974, **249**, 810.

[10] R. S. Stolarski and R. J. Cicerone, *Can. J. Chem.*, 1974, **52**, 1510.

[11] S. C. Wofsy and M. B. McElroy, *Can. J. Chem.*, 1974, **52**, 1582.

[12] P. J. Crutzen, *Can. J. Chem.*, 1974, **52**, 1569.

[13] T. E. Graedel, *Rev. Geophys. Space Phys.*, 1979, **17**, 937.

[14] R. A. Duce, V. A. Mohnen, P. R. Zimmerman, D. Grosjean, W. Cautreels, R. Chatfield, R. Jaenicke, J. A. Ogren, E. D. Pellizari and G. T. Wallace, *Rev. Geophys. Space Phys.*, 1983, **21**, 921.

[15] R. J. Cicerone, *Rev. Geophys. Space Phys.*, 1981, **19**, 123.

[16] Metronics Associates, Inc., *Field Studies of Air Pollution Transport in the South Coast Basin*, Technical Report No. 186, 1973.

[17] R. W. Shaw, *Atmos. Environ.*, 1982, **16**, 337.

[18] J. W. Munger, D. J. Jacob, J. M. Waldman and M. R. Hoffman, *J. Geophys. Res.*, 1983, **88**, 5109.

[19] P. A. Leighton, *Photochemistry of Air Pollution*, Academic Press, New York, 1961.

[20] H. Buckberg, K. W. Wilson, M. H. Jones and K. G. Lindth, *Int. J. Air Water Pollut.*, 1963, **7**, 257.

[21] M. W. Korth, A. H. Rose, Jr. and R. C. Stahman, *J. Air Pollut. Control Assoc.*, 1964, **14**, 168.

[22] S. W. Nicksic, J. Harkins and L. J. Painter, *Int. J. Air Water Pollut.*, 1966, **10**, 15.

[23] W. J. Hamming and J. E. Dickinson, *J. Air Pollut. Control Assoc.*, 1066, **16**, 317.

[24] J. C. Romanovsky, R. M. Ingels and R. J. Gordon, *J. Air Pollut. Control Assoc.*, 1967, **17**, 454.

[25] B. Dimitraides, *J. Air Pollut. Control Assoc.*, 1967, **17**, 460.

[26] W. A. Glasson and C. S. Tuesday, *J. Air Pollut. Control Assoc.*, 1970, **20**, 239.

[27] A. P. Altshuller, S. L. Kopzynski, D. L. Wilson, W. A. Lonemann and F. D. Sutterfield, *J. Air Pollut. Control Assoc.*, 1969, **19**, 787.

[28] B. Dimitraides, *Environ. Sci. Technol.*, 1972, **6**, 253.

[29] J. Heicklen, K. Westberg and N. Cohen, *Cent. Air. Environ. Stud. Pa. State Univ.*, *Report No.* 115–69, 1969.

[30] H. Niki, E. E. Daby and B. Weinstock, *Adv. Chem. Ser.*, 1972, **113**, 16.

[31] T. A. Hecht, J. H. Seinfeld and M. C. Dodge, *Environ. Sci. Technol.*, 1974, **8**, 327.

[32] K. L. Demerjian, J. A. Kerr and J. G. Calvert, *Adv. Environ. Sci. Technol.*, 1974, **4**, 1.

[33] H. Levy, II, *Adv. Photochem.*, 1974, **9**, 369.

[34] K. R. Darnall, A. C. Lloyd, A. M. Winer and J. N. Pitts, Jr., *Environ. Sci. Technol.*, 1976, **10**, 692.

[35] A. C. Lloyd, K. R. Darnall, A. M. Winer and J. N. Pitts, Jr., *J. Phys. Chem.*, 1976, **80**, 789.

[36] B. J. Finlayson and J. N. Pitts, Jr., *Science*, 1976, **192**, 111.

[37] R. Atkinson, K. R. Darnall, A. C. Lloyd, A. M. Winer and J. M. Pitts, Jr., *Adv. Photochemistry*, 1979, **11**, 375.

[38] R. Atkinson and A. C. Lloyd, *J. Phys. Chem. Ref. Data*, 1984, **13**, 315.

[39] R. A. Atkinson, *J. Phys. Chem. Ref. Data, Monograph* 1, 1989.

[40] A. M. Winer, in *Handbook of Air Pollution Analysis*, 2nd Ed., R. M. Harrison and R. Perry, eds., Chapman and Hall, London, 1985.

[41] J. F. Noxon, R. B. Norton and W. R. Henderson, *Geophys. Res. lett.*, 1978, **5**, 679.

[42] J. F. Noxon, R. B. Norton, and E. Marovich, *Geophys. Res. Lett.*, 1980, **7**, 125.

[43] V. Platt, D. Perner, A. M. Winer, G. W. Harris and J. N. Pitts, Jr., *Geophys. Res. Lett.*, 1980, **7**, 89.

[44] V. Platt and D. Perner, *J. Geophys. Res.*, 1980, **85**, 7453.

[45] V. Platt, D. Perner, G. W. Harris, A. M. Winer and J. N. Pitts, Jr., *Nature*, 1980, **285**, 312.

[46] W. P. L. Carter, A. M. Winer and J. N. Pitts, Jr., *Environ. Sci. Technol.*, 1981, **15**, 829.

[47] V. Platt, D. Perner, J. Schroder, C. Kessler and A. Toennisen, *J. Geophys. Res.*, 1981, **86**, 11, 965.

[48] G. W. Harris, W. P. L. Carter, A. M. Winer, J. N. Pitts, Jr., V. Platt and D. Perner, *Environ. Sci. Technol.*, 1982, **16**, 414.

[49] W. R. Stockwell and J. G. Calvert, *J. Geophys. Res.*, 1983, **88**, 6673.

[50] V. Platt, A. M. Winer, H. W. Biermann, r. Atkinson and J. N. Pitts, Jr., *Environ. Sci. Technol.*, 1984, **18**, 365.

[51] A. M. Winer, R. Atkinson and J. N. Pitts, Jr., *Science*, 1984, **224**, 156.

[52] A. M. Winer, J. W. Peters, J. P. Smith and J. N. Pitts, Jr., *Environ. Sci. Technol.*, 1974, **8**, 1118.

[53] T. Y. Chang, *Atmos. Environ.*, 1984, **18**, 191.

[54] R. W. Shaw, in *Meteorological Aspects of Acid Rain*, C. M. Bhumralker, ed., Acid Precipitation Series, Vol. 1., Butterworth, Boston, 1984.

[55] T. A. McMahon and P. J. Denison, *Atmos. Environ.*, 1979, **13**, 571.

[56] World Health Organisation, *Air Quality Criteria and Guides for Urban Air Pollution*, Technical Report Series, No. 5067, Geneva, 1972.

[57] *NIOSH Manual of Analytical Methods*, Volumes 1–7, NTIS.

[58] R. C. Reid, J. M. Pravsnitz and T. K. Sherwood, *The Properties of Gases and Liquids*, McGraw-Hill, New York, 1977.

[59] C. Antoine, *C. R. Acad. Sci.*, *Paris*, 1888, **107**, 681, 836, 1143.
[60] C. B. Willingham, W. J. Taylor, J. M. Pignocco and F. D. Rossini, *J. Res. Natl. Bur. Stand.*, 1945, **35**, 219.
[61] *International Critical Tables*, 1947.
[62] Intersociety Committee, *Methods of Air Sampling and Analysis*, American Public Health Association, 1972.
[63] A. C. Stern, *Air Pollution*, Academic Press, New York, 1976.
[64] A. A. Slowik and E. B. Sansome, *J. Air Pollut. Control Assoc.*, 1974, **24**, 245.
[65] H. C. Wohlers, H. Newstein and D. Daunis, *J. Air Pollut. Control. Assoc.*, 1967, **17**, 753.
[66] R. J. Byers and J. W. Davis, *J. Air Pollut. Control Assoc.*, 1970, **20**, 236.
[67] ASTM, Standard Recommended Practices for Sampling Atmospheres for Analysis of Gases and Vapours, D1605-60, Part 26. *ASTM Annual Book of Standards*, American Society for Testing and Materials, Philadelphia, 1981.
[68] P. D. Crittenden and D. J. Read, *Atmos. Environ.*, 1976, **10**, 897.
[69] Y. Mamane and A. E. Donagi, *J. Air Pollut. Control Assoc.*, 1976, **26**, 991.
[70] R. S. Barrett, *Analyst*, 1981, **106**, 817.
[71] G. O. Nelson, *Controlled Test Atmospheres*, Ann Arbor Science, London, 1971.
[72] T. G. Beckwith and N. L. Buck, *Mechanical Measurements*, Addison Wesley, London, 1961.
[73] W. C. Baker and J. F. Pouchot, *J. Air Pollut. Control Assoc.*, 1983, **33**, 66; 156.
[74] *NIOSH Manual of Analytical Methods Volumes* 1–7. Available from Superintendent of Documents, US Government Printing Office, Washington DC.
[75] US Environmental Protection Agency, *Compendium of Methods for the Determination of Toxic Organic Compounds in Ambient Air*, EPA 600/4-89/017, 1988.
[76] *General Industry Safety and Health Regulations*, *OSHA*, 1974. Available from Superintendent of Documents, US Government Printing Office, Washington DC.

6

Chemical analysis of animal feed and human food

Michael O'Keeffe
Food Analysis Department, The National Food Centre (Teagasc), Dunsinea, Castleknock, Dublin 15

6.1 INTRODUCTION

Analysis of both animal feed and human food is aimed at:

  (i) establishing the nutritional content of the material;
 (ii) determining the commercial value of the material;
(iii) establishing its compliance or otherwise with regulatory requirements;
 (iv) assessing the composition of the product produced in manufacturing (quality assurance, quality control);
  (v) determining the level of anti-nutritional or toxic substances; and
 (vi) establishing compliance with customer requirements.

Many of the analyses undertaken have more than one of the aims outlined above; e.g. compositional analysis of a food product may cover points (iv) and (vi).

Organized analysis of food materials dates back to the 19th century. Methods were developed and encoded in practice by various professsional bodies — e.g. the Association of Official Analytical Chemists (AOAC). These methods then become reference or standard methods, receiving wide applicability and, also, 'acceptability'. Such an approach to uniformity of methodology is important in that it determines comparability between results by different analysts/laboratories. Good examples of these are the proximate analysis of food — crude protein, ether extract (fat), oven moisture, ash, and/or crude fibre. All these determinations are based upon a defined set of procedures to give a proven estimate of the constituent of interest. For example, crude protein is a measure of Kjeldahl nitrogen or the amount of ammonia produced under particular conditions of acid digestion of the sample —

not absolute protein content. Fat determination by ether extraction is a measure of ether-soluble substances under defined conditions of extraction of the sample with that particular solvent. Oven moisture is a measure of the volatile materials removed from the sample on heating under defined conditions of temperature and time. Ash is the residue left from burning the material at a defined temperature. Crude fibre is an estimate of fibre as the residue from digestion of the material with acid and alkali.

In the area of residue analyses in foodstuffs, methods are used which give an accurate measure of the residue level in the food to ensure that the food is safe to eat and/or that regulatory requirements on allowed residue levels are not exceeded. Such methods are validated within a laboratory by recovery studies, i.e. determination of the amount of standard material added to the sample which is recovered by the method procedure, and between laboratories by use of reference materials.

6.2 SAMPLING AND SAMPLE PRE-TREATMENT

6.2.1 Sampling

No matter how good the analysis, or how accurately it can determine the quantity of analyte in a sample, it can at best only give an accurate result for the actual sample which is analysed (often only a few grams). Sampling is probably the most neglected aspect of the procedure for determining the quantity of an analyte. In many cases the analyst has little if any influence on the quality of the sampling or of the sample sent to the laboratory, and bad experience of results not matching the expectations of the client has led to the development of the term 'sample, as received'. Of course, mis-matching of analytical results with client expectations are not due solely to poor sampling; apart from poor quality analyses, they may be due to poor production practices or wishful thinking.

6.2.1.1 *Influence of sample type*

There are what could be termed relatively easy sampling situations and very difficult ones. The aim is to obtain from as few analyses as possible an accurate picture of the 'lot'. The lot can be anything from the national population of beef animals (level of banned hormones being used in beef production), to a ship-load of grain (protein content), to a large clamp of silage (digestibility), or to a vat of minced meat (proximate analysis). The homogeneity/heterogeneity of the lot determines the sampling procedure to be adopted, as does the information required (information on protein level or on absence of drug residues).

6.2.1.2 *Number of samples*

Statistically designed sampling plans can be used to determine sampling procedure; depending on the standard deviation(s) of the measured analyte, the number of samples required to give a defined level of accuracy can be established [1]. It should be noted that if the analytical uncertainty is less than 1/3 of the sampling uncertainty, additional reduction of the analytical uncertainty is of little significance.

The *sampling procedure* depends on the nature of the lot (i.e. the bulk material being studied) — size, sub-lots, etc. — and the degree of homogeneity of the material. The different types of samples that can be taken include random samples (evenly-spaced), representative samples (only if truly homogeneous), and composite samples [especially for heterogeneous materials to obtain average result(s) for a lot]. It should be remembered that a lot may be relatively homogeneous for one analyte (e.g. moisture, protein and grain) but heterogeneous for another (e.g. aflatoxin hot spot).

The following are some of the practical points to be considered in sampling different materials. For powders, mixing is required before sampling to prevent layering. For liquids, it is important to mix thoroughly and/or to sample at different heights (e.g. milk and fat). For solid particulates (e.g. grain), a composite sample may be taken composed of a number of individual samples taken from different parts of the lot. For solid material (e.g. butter/cheese) or for compacted material (e.g. hay/silage) a corer should be used. Suitable sampling procedures are described in the analytical literature [2,3].

It is most important to ensure that the sample does not change in its chemical/physical composition over time, after the sample has been taken. Processes that can occur include gain/loss of moisture, changes to status of fat, deterioration of analytes due to temperature effects, etc.

Normally, animal or human food samples are stored at refrigeration temperature if analysis is to be within a day or so, or in frozen storage for a longer storage period. Frozen storage is considered to be the best condition to maintain a sample in a state similar to when it was taken.

Samples received at the laboratory must be labelled and stored under appropriate conditions. Sample preparation procedures are aimed at producing a representative test-sample from the sample as received [4]. Grinding, milling, mincing and homogenizing are commonly used to produce a highly homogeneous sample which may be sub-sampled, and from which the weighed portion for analysis can be taken in confidence. However, with powders and liquids, thorough mixing before and during weighing of the test portion is extremely important.

Sample preparation techniques and sample storage must be such as not to cause changes to the sample (e.g. temperature increase during grinding can cause loss of moisture or volatiles, freezing/thawing repeatedly causes losses of residues). To cater for unavoidable losses of moisture in grain, for example, some sample preparations are done according to very well defined conditions of mill type, time, speed, etc. This ensures comparability of results between samples within a laboratory and between different laboratories.

6.3 COMPOSITIONAL ANALYSIS

The macro constituents of a food, determined by compositional (proximate) analysis, are moisture (water), protein, fat, carbohydrate and ash (minerals). Analysis for these basic components of food is undertaken to establish the nutritional value of the food which, in the case of animal feed, is required to determine the quantity of the food to be fed to animals to achieve targeted production (growth, milk production). In addition to this basic compositional analysis, analysis of the micro constituents of a food, such as vitamins, and characterization of the types of macro constituents (essential amino acids in the protein, types of lipids in the fat, sugars, starch content and digestibility of the carbohydrate fraction and mineral characterization of the ash) may be required to establish the nutritional quality of the food.

Food analyses are directed also at the eating quality of the food in such areas as the levels of added substances (salts, preservatives, plant proteins to animal food products) and the freshness of the food (rancidity measurements).

6.3.1 Moisture

The moisture content of food, and conversely the total solids or dry matter content, gives useful information for foodstuff specifications, nutrition and processing. The moisture content of foodstuffs varies widely, from <25% for grains, hay, compounded animal feeds and milk powders, to 50–75% for meats, dairy products and silages, and to >80% for milk, fruit and vegetables (Table 6.1).

Table 6.1 — Moisture content (%) of foods and feeds.

Milk powders	3–10
Barley, wheat	7–22
Hay	15–20
Compound feed	10–20
Bread	25–40
Cheese	28–79
Meat	50–75
Grass silage	75–85
Milk	87
Vegetables	12–95
Fruit	70–90

While moisture is determined commonly by drying of the sample under controlled conditions, moisture may instead be determined by direct measurement of water content by distillation, or by chemical or physical techniques.

6.3.1.1 Oven moisture

Drying methods include air-oven drying, vacuum-oven drying and microwave-oven drying. The basic air-oven drying method is carried out by drying an accurately weighed sample in an oven for specified time and temperature or to constant weight; typical time/temperature combinations used are 6 hr or overnight at 100–103°C (meat, animal feed), or 2 hours at 130°C (grain). Typically, about 5 g of sample are weighed (to the nearest mg) in a predried and preweighed dish, the dish with sample is placed in the oven for the specified period, then removed and allowed to cool in a desiccator. When cool, the dish is reweighed and the moisture content is expressed as the loss in weight on drying [5]. Various types of air ovens are used for moisture determination, including convection and forced-draft ovens. Special enclosed systems, such as the Brabender oven which has a built-in balance allowing for weighing of the dried sample *in situ*, are also used.

Because volatile components of the food may be lost in air-oven drying, the determination by this method is referred to as 'moisture and volatile content'. For foods which have a significant content of material which is volatile at 100°C or if decomposition is likely to occur (as with foods with high sugar contents), vacuum-oven drying may be used. In the vacuum-oven method, drying is done at reduced pressure and at a temperature of about 70°C, to constant weight [6].

Microwave ovens are becoming increasingly popular for rapid estimation of moisture content. Such ovens, with a built-in balance coupled with automatic calculation of moisture content, can determine moisture in food such as meat in periods of less than 5 min. These systems are of particular attraction for food-production plants, where rapid determination of the composition of raw ingredients or of the recipe during formulation is needed for process control [7].

6.3.1.2 Distillation and chemical methods

Direct measurement of water content by distillation or chemical techniques is used where more accurate determination of moisture is required. In the distillation method, the sample is placed in a flask with a solvent (such as toluene) which is immiscible with and less dense than water, and which has a boiling point above 100°C. The mixture is heated and the toluene/water distills off and is collected in a receiver under a condenser [8]. The common apparatus for this procedure is the Dean-and-Stark receiver, which is graduated to allow for direct reading of the volume of water in the sample (at the interface between the lower water layer and the upper toluene layer) (Fig. 6.1). This method, although it avoids the problem of over-estimation of moisture content resulting from loss of volatiles in the air-drying methods, may have problems from the limited precision of measuring the water volume in the receiver and the possible distillation of water-soluble components such as acids. Correction for these acids may be made by titration of the distilled water with alkali.

Moisture content may be determined directly by Karl Fischer titration, which is based on a reaction of water with iodine and sulphur dioxide in the presence of methanol and pyridine [9]. This two-stage reaction occurs as follows:

(a) $I_2 + SO_2 + 3C_5H_5N + H_2O \rightarrow 2C_5H_5N.HI + C_5H_5N.SO_3$

(b) $C_5H_5N.SO_3 + CH_3OH \rightarrow C_5H_5NHSO_4CH_3$

Karl Fischer titration for water determination may be carried out by a variety of procedures, but commonly a dedicated apparatus incorporating a system for electro-metric determination of titration end-point is used. The water in the sample is extracted by refluxing with anhydrous methanol and an aliquot of the extract is titrated with the Karl Fischer reagent (sulphur dioxide, pyridine and iodine in anhydrous methanol). To ensure accurate determinations by the Karl Fischer titration procedure, the reagents and apparatus must be kept free from atmospheric moisture, and the ratio of extracting solvent (methanol) to sample quantity must be large to ensure complete water extraction. This technique is particularly suited to determinations on foods with low moisture contents.

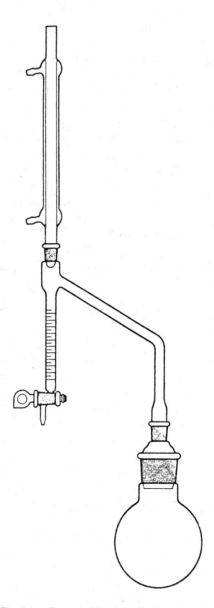

Fig. 6.1 — Dean and Stark distillation apparatus.

6.3.1.3 *Physical methods*

Many varied physical methods for moisture determination in foodstuffs are available, of which nuclear magnetic resonance (NMR) and near infrared (NIR) spectroscopy are the most commonly used. In the case of NMR spectroscopy, the sample is placed in a stationary magnetic field and a variable radio-frequency field is applied which causes oscillation of the hydrogen nuclei. The strength of the signal resulting can be used to determine moisture content from a calibration curve of signal amplitude *vs.* moisture content.

NIR spectroscopy has been applied widely both in the reflectance and transmitance modes for determination of moisture in foodstuffs [10]. This technique had its first general application in the analysis of grain and has been used, also, for dairy products, meat, animal foodstuffs and a variety of other foods. The NIR instrument can be calibrated for a number of food components (e.g. moisture, protein, fat) and results are available simultaneously within a few minutes [11]. Measurements are made in the near infrared region (700–2500 nm) containing overtone and combination bands of the fundamental infrared vibrations. The sample may be presented in a number of modes in the instrument, and moisture (or other components) determined from selected wavelength combinations. In the case of moisture, 1940 nm and 1450 nm are commonly used wavelengths (Fig. 6.2). This technique is

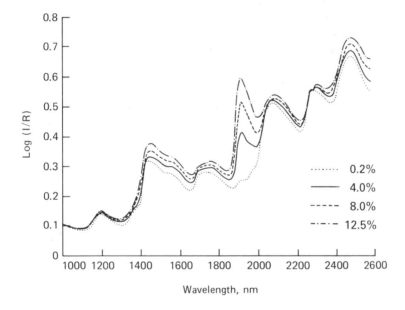

Fig. 6.2 — NIR reflectance spectrum of water in wheat (reproduced from K. H. Norris in *Modern Methods of Food Analysis*, K. K. Stewart and J. R. Whitaker, eds. pp. 167–186. The AVI Publishing Company, Westport, Connecticut, USA, 1984, by permission).

very rapid but depends on calibration of the instrument with a range of samples for which accurate reference analysis data are available. More details are given in Chapter 3 of this volume.

6.3.2 Protein (nitrogen, crude protein)
The protein content of food is calculated from the nitrogen content determined by analysis. Appropriate factors, based on the proportion of nitrogen in the proteins in

the particular food, are used to calculate 'crude' protein from the analytically determined nitrogen content. Nitrogen can occur in food in forms other than protein-nitrogen, but such non-protein nitrogen generally does not contribute significantly to the total nitrogen.

Two chemical procedures are used for the determination of total organic nitrogen in food: the Kjeldahl and Dumas methods. The most commonly used, the Kjeldahl method, involves the digestion of the food sample by heating with sulphuric acid and catalysts to reduce the organic nitrogen to ammonium ions, which can be determined by distillation/titration or by colorimetry. Kjeldahl nitrogen determination has been used for crude protein estimation in most foodstuffs [12]. The method has been widely used since its introduction over 100 years ago, although it has been much modified both in the reagent mixture used for digestion and in the apparatus used, culminating in highly automated systems [13].

A sample of food (commonly 0.5 to 2.0 g) is placed in a digestion flask or tube, a catalyst mixture added (e.g. potassium and/or sodium sulphate to raise the boiling point, and copper as a catalyst), and concentrated sulphuric acid. The digestion mixture is heated to a temperature of about 400°C, until digestion of the sample is complete (at least 2 hr) and the digest is allowed to cool. After dilution with water, an excess of sodium hydroxide is added to the digest to liberate ammonia. The ammonia is distilled into a receiving flask containing a 4% solution of boric acid with indicator (commonly methyl red/methylene blue). When distillation of ammonia is complete, the distillate in the boric acid is titrated to its original pH with dilute hydrochloric acid (0.1M). The quantity of HCl required for titration is a measure of the ammonia distilled into boric acid which, in turn, is a measure of the nitrogen in the sample (Fig. 6.3).

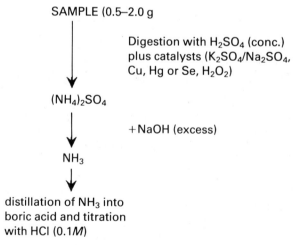

SAMPLE (0.5–2.0 g

Digestion with H_2SO_4 (conc.)
plus catalysts (K_2SO_4/Na_2SO_4,
Cu, Hg or Se, H_2O_2)

$(NH_4)_2SO_4$

+NaOH (excess)

NH_3

distillation of NH_3 into
boric acid and titration
with HCl (0.1M)

Fig. 6.3 — Kjeldahl nitrogen determination.

Alternatively, the ammonium ion content can be determined colorimetrically. Commonly used methods are those based on the reaction of ammonium ions with phenol and sodium hypochlorite in alkaline conditions to give a blue colour which is

LECO nitrogen determinator
– flow diagram –

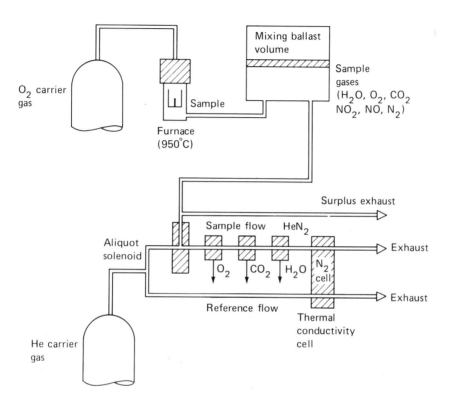

Fig. 6.4 — Crude protein determination by the Dumas method for total nitrogen with the 'LECO' apparatus.

Table 6.2 — Factors used for calculation of protein content from the total nitrogen determined in various foods

Wheat, barley	5.83
Soya	5.71
Gelatin	5.55
Meat	6.25
Milk and milk products	6.38
General factor	6.25

measured at 630 nm [14]. An alternative is the Nessler reaction where ammonium ions react with potassium mercuric iodide, in alkaline conditions, to produce a coloured complex which is measured at 425 nm [15]. The advantage of these colorimetric procedures is that they may be readily automated in a continuous-flow analysis system.

The traditional distillation and titration determination has been automated by a number of apparatus manufacturers. After preparation of sample digests, all additional steps are automated and Kjeldahl nitrogen (or crude protein) results are produced within 5–10 min.

The Dumas method, which is based on the conversion of organic nitrogen into gaseous nitrogen, is becoming increasingly popular with the development of fully-automated systems. This technique, which dates from over 150 years ago, involves combustion of the sample at temperatures in excess of 900°C with copper oxide and platinum catalyst, resulting in the production of a mixture of gases. An aliquot of the gas mixture is introduced automatically into a gas chromatograph which measures the nitrogen content and determines the nitrogen in the sample (Fig. 6.4). Apart from weighing of the sample (0.2–0.5 g) and loading onto a sample carousel, all steps in the procedure are fully automated and the determination is complete within a few minutes.

Because of its history as a reference technique, the Kjeldahl method for total nitrogen is used for calibration of the Dumas procedure, but both methods compare well for crude protein determination. The Dumas method has been applied very successfully for determination of total nitrogen in grains and, more recently, in meats [16].

'Crude protein' is calculated from total nitrogen, determined by either the Kjeldahl or Dumas procedures, by use of factors which relate nitrogen content to protein for particular foodstuffs. Factors of 5.70, 6.25 and 6.38 are used for wheat grains, meat and forages, and dairy products, respectively, on the basis of nitrogen constituting 17.5%, 16% and 15.7% of the protein in these food products (Table 6.2).

6.3.3 Oil and fat

Procedures for determination of oil/fat utilize the particular solubility characteristics of this component, in that it is soluble in non-polar organic solvents (diethyl ether, petroleum ether, hexane, chlorinated hydrocarbons, etc.). Oil/fat is extracted from the sample with an organic solvent and determined by gravimetric or specific gravity techniques.

The degree to which the oil/fat is available to the extracting solvent varies with sample and oil/fat type. In the case of processed foods and in some animal feedstuffs, oil/fat may not be totally extracted owing to binding of oil/fat to proteins and carbohydrates. In these cases the sample is prepared for solvent extraction by acid digestion ($3M$ hydrochloric acid) prior to the extraction.

The classical extraction of oil/fat from food samples is carried out by continuous (or repeated) washing of the sample with diethyl ether or petroleum ether [17]. The weighed sample in an extraction thimble is placed in the sample chamber of a Soxhlet apparatus which is connected between a weighed flask containing solvent and a reflux condenser; the flask is heated and the sample in the extraction thimble is repeatedly washed with condensing solvent, which recyles into the solvent flask by a siphoning mechanism each time the sample chamber fills with solvent (Fig. 6.5).

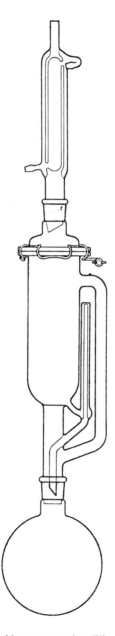

Fig. 6.5 — Soxhlet apparatus for oil/fat determination.

After a defined period of extraction (usually 4 hr), the ether is evaporated from the flask and the last traces of ether removed by heating in an oven. The oil/fat content in the sample is determined by weighing the flask after cooling in a desiccator.

While the Soxhlet method is generally applicable to food samples, some food types have oil/fat in a form which is not readily removed by simple extraction with ether. In the case of oilseeds such as rapeseed, the sample must be ground with sand before Soxhlet extraction to ensure that the oil is available to the extracting solvent. The procedure for oilseed requires an intermediate regrinding of the partially-extracted sample to ensure total recovery of the oil [18]. Pretreatment of the sample by acid hydrolysis to release bound forms of fat has already been mentioned, but some sample types require more exhaustive extraction procedures. Because oil/fat in milk occurs as an oil-in-water emulsion, sulphuric acid is used to break the emulsion and remove protein [4]. The Babcock and Gerber methods are based on the use of sulphuric acid, using special bottles with graduation for direct reading of fat content (Babcock) or a butyrometer for determination from specific gravity readings (Gerber). Other methods which utilize detergents, rather than sulphuric acid, to disperse the emulsions have been developed for oil/fat in milk [19]. For milk products, the sample is digested in an ammonia/alcohol solution to ensure dispersion of the oil/fat prior to extraction with diethyl and petroleum ether — the Rose–Gottlieb method.

Automated systems based on the Soxhlet extraction procedure have been developed by a number of instrument manufacturers. These systems give a relatively rapid (approximately 2-hr) extraction of oil/fat from the sample, with all the steps of extraction, rinsing and removal of solvent being automated [20].

A very popular rapid oil/fat analysis system based on methanolic extraction into perchloroethylene and determination by specific-gravity change is the Foss-let method [21]. In this method, the oil/fat is extracted from the sample into perchloro-ethylene by intimate crushing and mixing of the sample with the solvent in a metal cup with vibrating weight. The resulting homogenate is filtered and the specific gravity of the solvent plus oil/fat solution determined electronically and expressed as a percentage of oil/fat. A complete analysis by this method can take as little as 10 min and it has been widely used for oil/fat determinations in meat, oilseeds and grains. For reliable results, this technique must be standardized against a reference procedure, such as the Soxhlet method [17].

Both NMR and NIR spectroscopy are used for oil/fat deterrmination and have been particularly useful for oilseed analyses [22]. Rapid methods for fat determination in meat have been developed which are based on X-ray absorption by intact cuts of meat (differential absorption between lean and fat) [23]. Other rapid measurements of fat content in meat are based on microwave technology [24].

6.3.4 Carbohydrate and fibre

Carbohydrate levels in foodstuffs range from negligible (meat) to a major constituent (grains, fruit, forages) (Table 6.3). For compositional analysis of a food sample, carbohydrate is often estimated by difference from an analysis of the sample for moisture, protein, fat and ash. Although such an estimation may be sufficient for

Table 6.3 — Carbohydrate and fibre content (%) of foods and feeds

	Carbohydrate	Fibre
Barley, wheat	60–85	6–10
Hay	5–15	30–50
Bread	40–60	3–9
Cheese	<1	0
Meat	0	0
Grass silage	5–25	25–45
Milk	5	0
Vegetables	1–50	1–25
Fruit	3–20	<1–10

compositional analysis of foods such as meat products, where the nature of the added carbohydrate (wheat flour) is known, adequate nutritional information may require characterization of the carbohydrate. This is particularly the case for establishing the digestible and indigestible (fibre) portions of the carbohydrate fraction.

Carbohydrate consists of monosaccharides (e.g. glucose, fructose, galactose), disaccharides (e.g sucrose, lactose, maltose) and oligosaccharides (2 to 8 monosaccharide units), 'available' polysaccharides (starch, dextrins and glycogen) and 'unavailable' polysaccharides or 'fibre'. Procedures for determination of starch and sugars in food are treated in the next section (Nutritional Analysis). In this section (Compositional Analysis) the procedures for determination of fibre are described.

The structural polysaccharides of plant cell walls constitute the fibre component in food. Within this category of structural polysaccharides, there are gradations of digestibility: hemicelluloses>celluloses>lignin. The original estimation of fibre (or indigestible material) was developed in the mid-nineteenth century and has been very widely used since then, particularly in the analysis of animal feedstuffs. Feeding of farm animals is usually done at a rate designed to give particular yields (of milk or meat); in designing the feeding programme for animals, it is important to establish how available the particular feed is to the animals. This estimation of digestibility, i.e., crude fibre, is determined on a sample of dried, ground, defatted feed by boiling sequentially in acid and alkali, washing the digested materials with alcohol and weighing the dried residue, which is then corrected for ash content (Fig. 6.6) [25].

The crude fibre procedure was the only method available for determination of feed digestibility for a century and was, therefore, of considerable value to agriculture. However, the technique has major shortcomings in that it is highly susceptible to changes in procedure and sample characteristics. In addition, for different feeds, the relationship between crude fibre and true digestibility varies. Because of these drawbacks, alternative chemical methods were developed in the 1960s for both animal feed and human food. These techniques are based on use of detergent (with or without acid) digestion of the sample to identify more precisely the structural polysaccharide components. In the case of animal feedstuffs, boiling of the dried, defatted sample with detergent (cetyltrimethylammonium bromide) at neutral pH gives a residue containing hemicellulose, cellulose, lignin and proteins linked to cell walls [26]. Boiling of the sample with 0.5M sulphuric acid plus detergent gives a

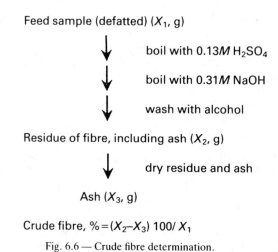

Feed sample (defatted) (X_1, g)

boil with 0.13M H$_2$SO$_4$

boil with 0.31M NaOH

wash with alcohol

Residue of fibre, including ash (X_2, g)

dry residue and ash

Ash (X_3, g)

Crude fibre, % = (X_2–X_3) 100/ X_1

Fig. 6.6 — Crude fibre determination.

residue containing cellulose and lignin, while further digestion of this residue with concentrated sulphuric acid gives a residue containing lignin. In this way, by sequential digestion of the sample of feedstuff (Fig. 6.7), precise identification of the components of the fibre may be obtained. In the case of human food, the important measure of 'dietary fibre' is this neutral detergent fibre, consisting principally of hemicellulose, cellulose and lignin. For human food, dietary fibre is an important nutritional factor because of the importance of fibre for a wholesome and healthy diet.

Automated equipment exists for determination of fibre by methods similar to those described above. With such automated equipment, the addition of reagent and washing of digested material are carried out automatically.

Although the detergent fibre techniques are recognized to be a considerable improvement on the crude fibre technique, they do not measure exactly what occurs under physiological conditions in the digestion of humans or different types of animals (monogastric, ruminants). In agricultural research, true digestibility of feed has been determined by *in vivo* trials in which the contribution of a feed to animals can be estimated from measurements of feed intake and production of waste products [27]. Such trials are expensive and lengthy and are not suitable for routine analysis of feedstuffs. An *in vitro* procedure was developed to mimic the digestion of the feed by the ruminant animal. A dried, ground, defatted feed sample is incubated sequentially with rumen liquor (harvested from experimental animals) and with pepsin–HCl at 38°C for two periods of 48 hr, with continuous mixing of the sample. The residue remaining after washing and drying, corrected for ash content, is called 'dry organic matter digestibility' (DOMD) [28].

In laboratories with facilities for holding experimental animals, this technique offers a measure of digestibility which closely approximates to the true digestibility of the feed. An alternative digestibility measurement has been developed which utilizes microbial cellulase enzymes to mimic *in vitro* digestibility [29]. This method avoids the need for experimental animals as a source of rumen liquor and gives more reproducible results but, again, is only an estimate of true digestibility.

Feed sample

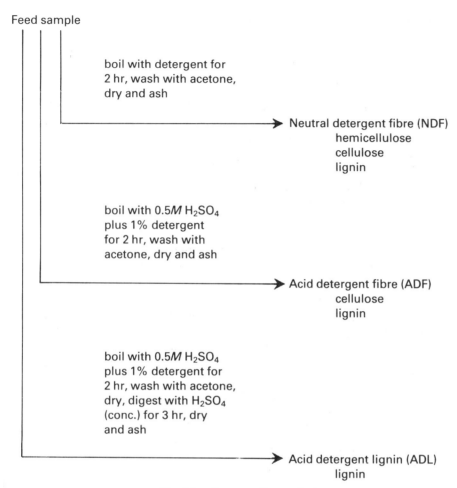

boil with detergent for
2 hr, wash with acetone,
dry and ash

→ Neutral detergent fibre (NDF)
    hemicellulose
    cellulose
    lignin

boil with 0.5M H_2SO_4
plus 1% detergent
for 2 hr, wash with
acetone, dry and ash

→ Acid detergent fibre (ADF)
    cellulose
    lignin

boil with 0.5M H_2SO_4
plus 1% detergent for
2 hr, wash with acetone,
dry, digest with H_2SO_4
(conc.) for 3 hr, dry
and ash

→ Acid detergent lignin (ADL)
    lignin

Fig. 6.7 — Detergent fibre methods.

6.3.5   Ash

Ash is composed of minerals and silica (in the case of animal foodstuffs such as hay
and silage, which may contain soil). Ash is determined by combustion of the organic
matter in the food and measurement of the inorganic residue [30]. The weighed food
samples are combusted at about 550°C in silica crucibles in a muffle furnace. Ashing
is conveniently done overnight, and the weight of ash is determined after cooling in a
desiccator to room temperature. Complete ashing is established when the ash is
white, indicating that all carbon has been removed. Some samples require charring
on a hot-plate prior to ashing in a muffle furnace, to ensure complete combustion of
the food sample. Dry ashing is simple and attractive as a procedure for preparing
samples for mineral determinations. Mineral determinations may be made on the ash
but losses due to the volatilization at high temperatures can be a problem; when the
simple dry ashing procedure is used, determination of inorganic components such as

sodium chloride are better done by extraction of the salt from the sample in a separate analytical procedure [31].

6.4 NUTRITIONAL ANALYSIS

The previous section discussed the procedures used to give a general compositional analysis of food, separating the food into the major components of moisture, protein, oil/fat, carbohydrate (plus fibre) and ash. This section describes the further characterization of the major components, particularly with respect to nutritionally important constituents. Certain amino acids, the fatty acid profile, starch and sugars, and essential minerals are constituents of protein, oil/fat, carbohydrate, and ash, respectively, which, together with vitamins, are of nutritional importance in food.

6.4.1 Amino acids

Whereas 24 amino acids are constituents of proteins, only 8 of these are considered to be 'essential' in the sense that they must be in the diet protein as they cannot be synthesized in the human or animal body (Table 6.4). It is important, therefore, to identify the amino acids in the protein of foods so that diets can be formulated which contain the necessary quantities of these essential amino acids. In the area of animal nutrition, in particular, determination of the amino acid content of the food proteins is important. A common analysis of compounded animal feedstuffs is for lysine, because the usual raw ingredients for such compounded feed (e.g. cereal grains) are deficient in lysine.

Table 6.4 — Essential amino acids in proteins (g/100 g of protein)

Amino acid	Skimmed milk	Soya	Beef	Egg	Fish	Yeast
Lysine	8.6	6.8	8.3	6.3	6.6	6.8
Tryptophan	1.5	1.4	1.0	1.5	1.6	0.8
Phenylalanine	5.5	5.3	3.5	5.7	4.1	4.5
Methionine	3.2	1.7	2.8	3.2	3.0	2.6
Threonine	4.7	3.9	4.5	4.9	4.8	5.0
Leucine	11.0	8.0	7.2	9.0	10.5	8.3
Isoleucine	7.5	6.0	4.7	6.2	7.7	5.5
Valine	7.0	5.3	5.1	7.0	5.3	5.9

From Lichtfield and Sachsel (1965).

Analysis of food proteins for their constituent amino acids, the amino acid profile, involves the isolation of proteins, acid (HCl) hydrolysis of the proteins and determination of the amino acids [32]. The quantitative analysis of amino acids has been carried out by ion-exchange chromatography, using as mobile phase a series of buffers which separate the amino acids on the basis of their pK_a values (Fig. 6.8). More recently, these dedicated amino acid analysers have been replaced by HPLC systems which give faster and more efficient determination of amino acids with more compact equipment.

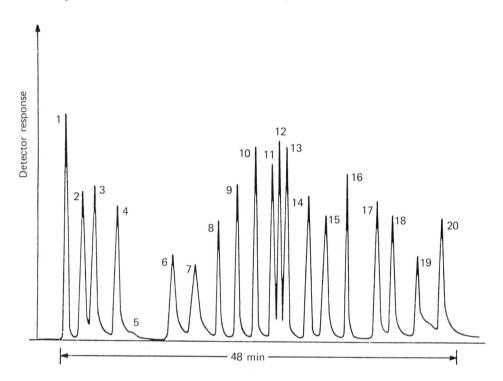

Fig. 6.8 — Chromatogram of amino acids (reproduced from A. Pryde and M. T. Gilbert *Applications of High Performance Liquid Chromatography*, p. 148, Chapman & Hall, London, 1979, by permission).

In the analysis of proteins from compounded animal feeds for lysine, determination of total lysine gives an overestimate of the quantity of this essential amino acid which is nutritionally available. Only lysine molecules present in the protein in positions where the ε-amino group of the amino acid is free are available to the animal (Fig. 6.9). The lysine molecules with a free ε-amino group are derivatized with fluorodinitrobenzene (FDNB) prior to hydrolysis of the protein, then the derivatized lysine (ε-DNP-lysine) is determined as 'available lysine' [33].

6.4.2 Fatty-acid profile

The quantitative determination of the constituent fatty acids in the oil/fat of food is important both for identification of the oil/fat and to provide nutritional information on the types of fatty acids (saturated, monounsaturated and polyunsaturated) in the food.

For fatty-acid profile analysis, the oil/fat is extracted from the food with chloroform/methanol. Water is added to the extract to cause separation of the chloroform layer containing the oil/fat. Alternatively, oil/fat may be extracted from the sample with petroleum ether.

A sample of the oil/fat is hydrolysed by refluxing with methanolic sodium hydroxide, and subsequently methyl esters of the fatty acids are prepared by further

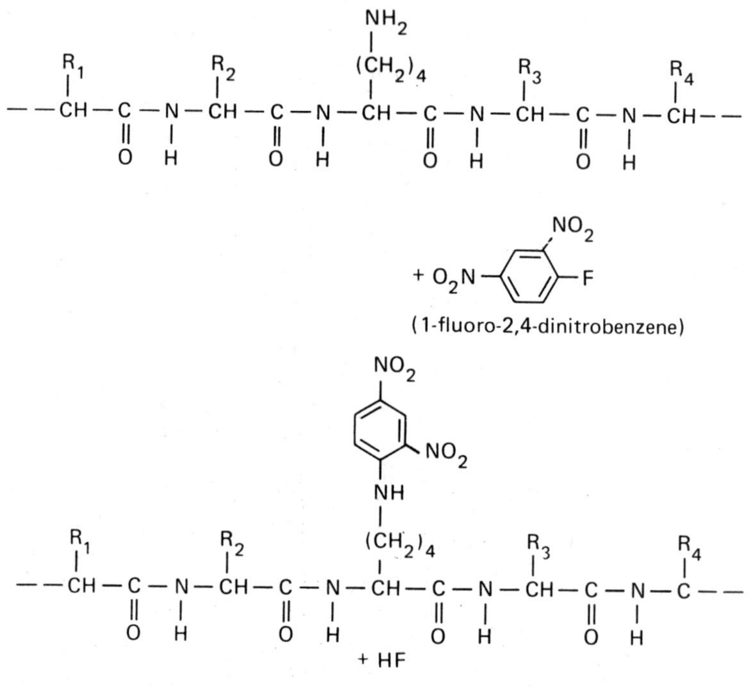

Fig. 6.9 — Available lysine determination (derivatization with FDNB of ε-amino groups).

refluxing with boron trifluoride–methanol. The fatty-acid methyl esters are extracted into heptane for analysis by gas chromatography. Several other procedures, involving base- or acid-catalysed reactions are available for oil/fat hydrolysis and preparation of methyl esters [34]. The fatty acid profile for the oil/fat is determined by gas chromatography on a polar polyester liquid (stationary) phase, with a flame-ionization detector (FID). The peaks on the chromatogram are identified by comparison with the retention times for standard mixtures of fatty-acid methyl esters (Fig. 6.10). Each peak area is expressed as a percentage of the total area of peaks on the chromatogram, to give the composition of fatty acids in the oil/fat.

The identity of a pure oil/fat can be established from tables of fatty-acid composition of oils/fats in the literature (Table 6.5). In the case of mixtures of oils/fats, determination of the constituent oils/fats and the proportions present may be difficult. However, for mixtures of oils/fats and for food product samples, it is often the proportions of different types of fatty acids which are of most interest. Fatty-acid profile analyses may be used also to check on the suitability of oil/fat for use as a food or food ingredient. An example of this is the determination of the content of the antinutritional fatty acid, erucic acid (C22:1), in rapeseed oil [35] (Fig. 6.11).

6.4.3 Starch and sugars

Although it is common in compositional analysis of food to calculate the percentage carbohydrate content as the difference between 100% and the sum of the percentages of moisture, protein, oil/fat and ash, there are many situations where an

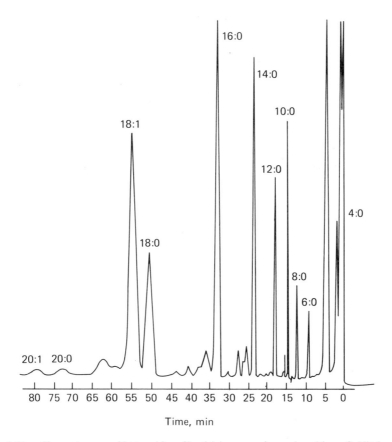

Fig. 6.10 — Chromatogram of fatty-acid profile of dairy cream (reproduced from C. Hitchcock and E. W. Hammond in *Developments in Food Analysis Techniques, 2*, R. D. King, ed. pp. 185–224. Applied Science, Barking, Essex, 1980 with permission).

accurate determination of the nutrient carbohydrate fraction (starch and sugars) is required.

Free sugars (monosaccharides and disaccharides) are determined as the substances extractable from the food with hot, 80% ethanol. After extraction of the sugars, the disaccharides are converted into monosaccharides by acid hydrolysis with concentrated HCl. The sugars are determined as reducing sugars from their oxidation by copper ions. The final determination procedure involves measurement of the amount of reduced copper either directly or indirectly by titration [36]. A common procedure (developed by Schoorl) consists of reaction of the reducing sugar extract with excess of copper sulphate, and determination of the unreduced copper sulphate by iodometric titration with potassium iodide and sodium thiosulphate [37]. (Fig. 6.12). This method gives a composite determination of total reducing sugars.

For more detailed identification of sugars in a food sample, high-performance liquid chromatography (HPLC) may be used [38]. The ethanolic extract of the food,

Table 6.5 — Fatty-acid composition (%) of oils and fats

Fatty acid	Butterfat	Lard	Maize oil	Olive oil	Rapeseed oil	Sunflower oil	Beef tallow
C_4	4						
C_6	2						
C_8	1						
C_{10}	2						
C_{12}	3					0.5	
C_{14}	13	0.5–3				0.2	1–6
C_{15}	1.5						1.5
C_{16}	26	20–32	8–19	10–18	2–5	3–10	20–37
$C_{16:1}$	2	2–5		1			1–9
C_{17}	0.5	0.5					0.5–2
C_{18}	13	5–24	0.5–4	2–3	1–2	1–10	6–40
$C_{18:1}$	28.5	35–62	19–50	57–78	13–30	14–65	26–50
$C_{18:2}$	3	3–16	34–62	6–14	10–25	22–75	0.5–5
$C_{18:3}$	0.5		2	0.5	5–10	0.5	0.5
C_{20}				1		0.3	
$C_{20:1}$		1.5			5–15		
$C_{22:1}$					20–50		
C_{24}					1		

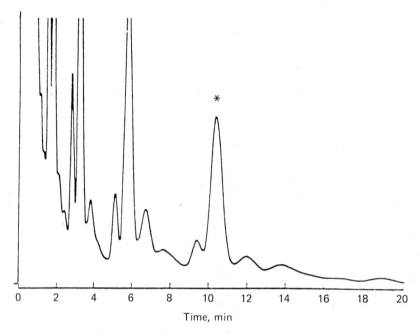

Fig. 6.11 — Chromatogram of rapeseed oil containing erucic acid (reproduced, with permission, from Ref. [49]).

(a) Cu^{2+} + reducing sugar $\rightarrow$ Cu^+ + oxidized sugar + Cu^{2+}
    (excess)

(b) $2Cu^{2+} + 2I^- \rightarrow 2Cu^+ + I_2$

(c) $I_2 + 2S_2O_3^{2-} \rightarrow 2I^- + S_4O_6^{2-}$

Fig. 6.12 — Determination of reducing sugars (after Schoorl).

with or without subsequent hydrolysis of disaccharides, is chromatographed on an ion-exchange column, with an acetonitrile/water mobile phase. The separated sugars are detected by a refractive index detector, and the quantities of the individual sugars in the food sample are calculated by comparison of sample peak responses with peak responses obtained from injected sugar standards (Fig. 6.13).

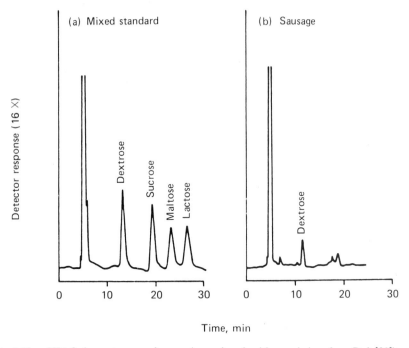

Fig. 6.13 — HPLC chromatogram of sugars (reproduced, with permission, from Ref. [38]).

The residue of the food sample after extraction of sugars with hot ethanol is used for determination of the starch content. Methods based on polarimetric, titrimetric, colorimetric, enzymatic and chromatographic techniques are available [39]. Starch may be extracted with hot, dilute hydrochloric acid and the starch content determined on a polarimeter from the degree of optical rotation caused by the starch extract. An alternative approach (based on the Schoorl method for sugars) involves extraction of the sugar-free residue with hot ethanolic potassium hydroxide, hydrolysis of the extracted starch with hydrochloric acid and determination of the resulting glucose concentration in the hydrolysed extract as reducing sugar by iodometric titration (as described for free sugars above). The starch content is calculated from the measured glucose content by multiplying by a factor of 0.9.

Apart from the methods described for starch and sugar here based on polarimetry, iodometric titration and high-performance liquid chromatography, a number of colorimetric and enzymatic methods are commonly used. Commercial kits are available for the determination of specific sugar(s) in food samples. These systems are based on a dehydrogenase enzyme specific to a particular sugar which utilizes the coenzyme nicotinamide adenine dinucleotide phosphate (NADP). The reaction of the enzyme with the sugar involves reduction of NADP to NADPH, which is measured spectrometrically at 340 nm. Where disaccharides are to be assayed, the disaccharide is split by a specific enzyme to produce the constituent sugars, then one of the constituent sugars is determined by the appropriate dehydrogenase reaction and NADP/NADPH measurement [40]. Examples of these reactions are as follows:

$$\text{sucrose} + H_2O \xrightarrow{\text{β-fructosidase}} \text{glucose} + \text{fructose}$$

$$\text{fructose} + \text{ATP} \xrightarrow{\text{hexokinase}} \text{fructose-6-phosphate} + \text{ADP}$$

$$\text{glucose} + \text{ATP} \xrightarrow{\text{hexokinase}} \text{glucose-6-phosphate} + \text{ADP}$$

$$\text{fructose-6-phosphate} \xrightarrow{\text{phosphoglucose isomerase}} \text{glucose-6-phosphate} + \text{ADP}$$

$$\text{glucose-6-phosphate} + \text{NADP}^+ \xrightarrow{\substack{\text{glucose-6-phosphatede} \\ \text{dehydrogenase}}} \text{gluconate-6-phosphate} + \text{NADPH} + H^+$$

6.4.4 Trace elements

Determination of trace elements in food samples is of importance because of the nutritional requirement by animals and humans for trace elements. Elemental analysis may be carried out on the ash resulting from the dry ashing procedure described in Section 5.3, but ashing may result in losses of some elements from absorption or volatilization [41]. The procedure of wet ashing (or digestion) is preferred for most elemental analyses.

Complete digestion of the food sample can be achieved by use of a single acid (nitric acid) or by use of di-acid or tri-acid mixtures, with or without an oxidizing agent such as hydrogen peroxide, at high temperatures. The most common acid mixture for such digestions is a mixture of nitric and sulphuric acids [42], but nitric–sulphuric–perchloric mixtures are also used [43]. Because of the very low levels of trace elements in food samples, special precautions must be taken in the laboratory to prevent losses or contamination of samples, glassware or apparatus. Such precautions include rigorous cleaning of glassware and utensils, use of high-purity grades of water and reagents, and environmental cleanliness (e.g. dust-free air). For trace-element analyses, the use of reference samples or materials is particularly important.

The principal trace elements of interest in animal and human food are calcium, magnesium, iron, phosphorus, sodium and potassium. After sample preparation by dry ashing and solubilizing of the ash, or wet digestion with acids, trace elements are determined by instrumental methods. Sodium and potassium are commonly determined by flame photometry, where the sample digest solution is aspirated into a flame, the elements are atomized and the atoms absorb energy and move to an excited state. The excited atoms return to the ground (unexcited) state by releasing radiation of a specific wavelength (emission). The intensity of this emission is proportional to the concentration of the element in the sample digest solution. The emission wavelengths for sodium and potassium are 589 nm and 767 nm, respectively [44].

Calcium, magnesium and iron are determined by atomic-absorption spectrophotometry. The sample digest is aspirated into an elongated flame (air/acetylene or other combinations) where the metals are atomized. For each element, a specific hollow-cathode lamp produces radiation of the wavelength that will excite the atoms in the flame (calcium 422.7 nm, magnesium 285.2 nm, iron 248.3 nm). The concentration of each element in the same digest is related to the decrease in the intensity of radiation from the hollow cathode lamp caused by this absorption/excitation [45]. Atomic-absorption spectrophotometry, therefore, functions in a directly opposite way to flame photometry and is much more sensitive and selective.

Phosphorus is determined as phosphate by reaction with ammonium molybdate to form phosphomolybdic acid. The phosphomolybdic acid is reduced to give a blue colour which is measured colorimetrically at a wavelength of 880 nm [46].

6.4.5   Vitamins
Vitamins are essential nutrients in that they perform catalytic functions for metabolism in humans and animals. Because they occur in different food types, vitamin deficiency can be avoided only by including a variety of ingredients in the diet. For both human and animal foods, vitamins are commonly added to ensure that dietary requirements are met. Vitamins are classified in terms of their solubility characteristics as fat-soluble (A, D, E) and water-soluble (B group, C). Fat-soluble vitamins are not readily excreted and are stored in the body so that the possibility of deficiency arises mainly in the young human or animal. In the case of water-soluble vitamins, a regular supply (to counteract excretion) is required from the diet.

Analytical methods for vitamins are numerous; some examples are listed in Table 6.6.

6.5   RESIDUE ANALYSES FOR FOOD SAFETY
The task of ensuring the safety of the food supply for humans is extensive and complicated, involving considerable resources in inspection, sampling and analysis. The most important aspect of food safety is the food product manufacturing system used, and its control through Good Manufacturing Practice (GMP) and quality assurance procedures such as Hazard Analysis at Critical Control Points (HACCP), particularly to ensure microbiological safety of food [47]. However, the food safety

Table 6.6 — Selected methods for vitamin analyses in foods.

Vitamin	Extraction	Separation	Determination
1. *Fat-soluble vitamins*			
A	Saponification and extraction of the unsaponifiable portion with ether	alumina column	spectrophotometry, or fluorimetry
D	(as for vitamin A, above)	alumina column	GC/ECD of halogenated esters of HPLC with UV detection (254 nm)
E	(as for vitamin A, above)	thin-layer chromatography	spectrophotometry
A,E	Saponification and extraction of the unsaponifiable portion with hexane	HPLC on alumina or C_{18} column	vit. A by fluorimetry (330 nm/514 nm), vit. E by spectrophotometry (460 nm)
2. *Water-soluble vitamins*			
B_1	digestion with acid, and with enzymes (protease and phosphatase)	HPLC on silica column	fluorimetry, by post-column derivatization with alkaline ferricyanide (366 nm/464 nm)
B_2	(as for vitamin B_1, above)		fluorimetry (435 nm/545 nm)
Niacin	(as for vitamin B_1, above)		fluorimetry by post-column derivatization with cyanogen bromide and aminoacetophenone (435 nm/545 nm)
C	extraction with metaphosphoric acid solution	filtration	fluorimetry of phenylenediamine derivative (350 nm/430 nm)

Reprinted from G. Brubacher, W. Muller-Mulot and D. A. T. Southgate, eds., *Methods for the Determination of Vitamins in Food: recommended by COST 91*, Elsevier Applied Science, London, 1985, by permission of the copyright holders.

chain begins much earlier, right at the production systems used for foods of plant and animal origin. In the case of foods of animal origin, the question of food safety begins with the safety of the animal feedstuff ingredients used. The chain of safety, therefore, runs through the feedstuff ingredients, the compounded feed, the animal husbandry system (including veterinary drug and other treatments or the treatment of crops with pesticides), the 'harvesting' of the raw food ingredients (e.g. cereals, fruit, vegetables, milk, meat, eggs), the storage treatments, the processing, retailing and domestic storage and cooking of food. At all of these stages, food safety considerations arise.

The scope of this treatment of the subject of food safety must be limited; it will not cover microbiological food safety or the whole area of food additives and possible toxic chemicals arising during food processing. It will be limited to the question of potentially toxic chemicals in food (or feed) arising either naturally or during husbandry practices (agrochemicals, veterinary drugs). It is important to emphasize that the following discussion refers only to undesirable chemicals in food, either because of their proven toxicity or potential toxicity or because of regulatory requirements which preclude their occurrence in food, or set limits which are accceptable. In practice, chemical residues may occur in foods at levels which are not harmful to the consumer.

6.5.1 Natural toxic chemicals in food

Natural toxic chemicals in foods (or feeds) may be divided conveniently into two categories:

(a) normal components of natural food products
(b) natural contaminants of natural food products

6.5.1.1 Normal toxic components of food

In this category are those chemicals produced by plants which have toxic effects for humans and/or animals. Some examples of these are fatty acids (such as erucic acid in oil seed rape), glucosinolates in oil seeds, saponins in plants and stimulants (such as theobromine in cocoa) [48]. The latter example cannot be regarded as a toxic chemical *per se*, no more than caffeine in coffee, but is a good example of an undesirable substance from a regulatory viewpoint, in this case the anti-doping regulations applying to horse-racing.

The determination of residues of **erucic acid** in oil seed demonstrates some interesting analytical procedures. As with fatty-acid profile analysis (see Section 6.4.2), the content of erucic acid is determined by gas-liquid chromatography with a flame-ionization detector. The oil is extracted from the seed with hexane or petroleum ether, the fatty acids in the oil are released by alkaline hydrolysis of the oil, then methyl ester derivatives of these are produced by reaction with boron trihydride–methanol solution. The gas-chromatographic conditions are set such that erucic acid (a 22-carbon fatty acid with a single double bond, $C_{22:1}$) is separated on the chromatogram. Because accurate quantification of the erucic acid content is required (as distinct from the relative values required in the fatty-acid profile analyses) an internal standard is used to monitor the completeness of the methylation procedure and the quantification by gas chromatography [49]. Tetracosanoic acid ($C_{24:0}$) was used as internal standard originally because it was well separated from erucic acid in the chromatogram. However, it was established subsequently that tetracosanoic acid may occur as a minor component in rapeseed oil and tricosanoic acid ($C_{23:0}$) was selected as being more suitable for use as an internal standard (Fig. 6.14)

This analysis demonstrates both the importance of an internal standard for accurate determinations, particularly when procedures such as derivative formation are involved; and the importance of selecting an internal standard which will not occur as a natural component of the sample. Accurate determination of erucic acid is important because regulations require that foods must not contain erucic acid at a level greater than 5% of the fatty acids in the food [50]. The erucic acid content is expressed as a percentage of the total fatty acid content.

Glucosinolates are a group of substituted glucoses which are hydrolysed during their metabolism by the animal or human (Fig. 6.15). The products of the metabolism of glucosinolates are isothiocyanates and/or nitriles, which cause depression in

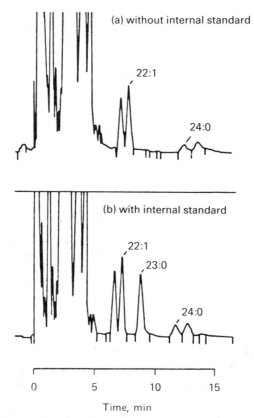

Fig. 6.14 — GC chromatograms of rapeseed oil, showing use of tricosanoic acid as an internal standard for erucic acid analysis (reproduced from N. L. Wilson *J. Sci. Food Agric.* 1981, **32**, 1103, by permission).

the synthesis of thyroid hormones. Because of the toxic effects associated with glucosinolates, new low-glucosinolate varieties of rapeseed have been bred, and rapeseed is routinely assayed for glucosinolate content to determine its suitability for incorporation into animal feed or human food.

The assay for glucosinolates involves extraction of the ground seed with metha-nol–water (70:30) and clean-up of the filtered extract on an ion-exchange column, such as Amberlite IR-120. The glucosinolates are desulphated on a DEAE-Sepha-dex A-25 column and determined by GLC (with FID detection) of trimethylsilyl derivatives [51] or by HPLC (with UV detection) of the desulphoglucosinolates [52] (Fig. 6.16).

Natural toxic substances which are a normal component of the plant, such as glucosinolates, are thought to have evolved as a protective mechanism, either to prevent consumption of the plant by grazing animals, or as a form of natural pesticide [53]. Other examples of such substances are the glycoalkaloids and saponins (Fig. 6.15). Many analytical methods are used for determination of the content of natural toxins in plants, particularly GLC and HPLC, TLC and immunoassays.

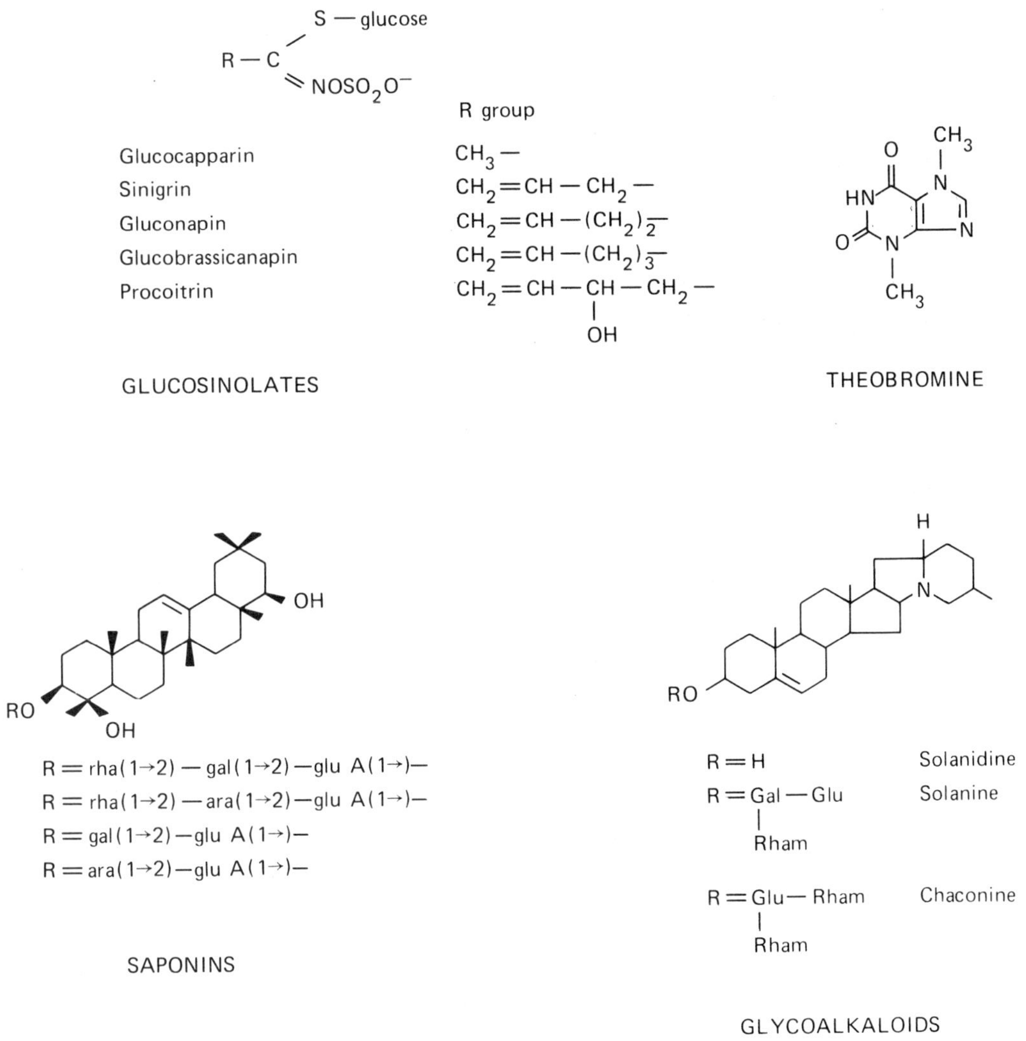

Fig. 6.15 — Chemical structures for natural toxins.

The analysis for **theobromine** in feed ingredients is of interest (a) because of the largely regulatory basis for it being described as an undesirable substance and (b) because it demonstrates how undesirable substances can be unwittingly incorporated into food or feed. Theobromine is an alkaloid with cardiac stimulant properties and is one of a list of banned drugs which is screened for in the urine of race-horses (Fig. 6.15).

A wide variety of raw ingredients are used in the formulation of a compound feed for animals, and plant fractions from food production are commonly used where they can contribute to the target values for composition of the feed, i.e. specified levels of

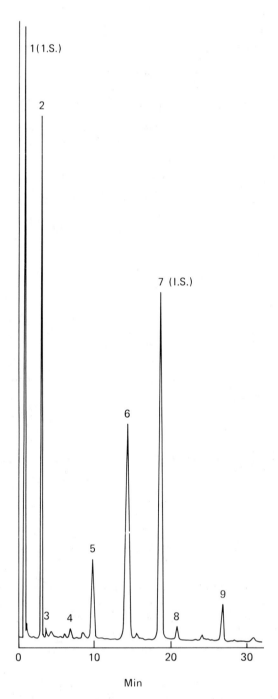

Fig. 6.16 — Chromatogram of glucosinolates in rapeseed (reproduced, with permission, from Ref. [52]).

protein, fat, carbohydrate, etc. Cocoa husks are a by-product of the manufacture of chocolate and have been used as a raw ingredient for compound feed for horses. The low levels of theobromine arising from the cocoa husks were determined subsequently in the analysis of horse urine, leading to disqualification of the horses.

Theobromine may be determined in feed by hot-water extraction of the feed, clarification of the extract with Carrez reagents, and determination by HPLC on a reversed-phase octadecylsilane (C_{18}) column and measurement by UV detection at 270 nm [55].

A problem with this relatively simple analysis for theobromine in feed was the ocurrence of 'false positive' results for some feeds, owing to co-elution on the HPLC system of an interfering substance at the same retention time as theobromine. To prevent this, an additional step was introduced into the analytical procedure: a silica column clean-up of the clarified feed extract. This solid-phase extraction (SPE) procedure, achieved with a simple disposable cartridge-type column, for example 'Sep-pak' (Waters) or 'BondElut' (Analytichem), is sufficient to provide a sample extract in which theobromine can be determined unequivocally.

6.5.1.2 Natural toxic contaminants of food
The major natural contaminants of food are the mycotoxins, toxic chemicals produced by fungal organisms which grow on plant foods either during plant growth or, more often, during post-harvest storage. The mycotoxins are considered to be one of the most toxic groups of chemicals occurring in food. They arise wherever fungal growth occurs. and this is normally associated with less than ideal storage conditions (e..g. damp, warm conditions) for grains and fruit. Table 6.7 lists some of the main categories of mycotoxins, the organisms which produce them, and the toxic effects associated with each category of mycotoxin. Structures for some of the principal mycotoxins are shown in Fig. 6.17.

Table 6.7 — Mycotoxins in food and feed

	Produced by	Effects
Aflatoxins	Aspergillus	Carcinogens
Ochratoxin A	Aspergillus+penicillum	Affect kidney function
Zearalenone	Fusarium	Oestrogenic, affect fertiliity
Tricothecenes	Fusarium	Feed refusal, vomiting

Analysis for residues of mycotoxins in food and feed is important because of their relatively high toxicity. Maximum limits for residues of the various mycotoxins are specified for foodstuffs and these limits are generally close to the limits of detection for the analytical methods. A particular problem with analysis of plant foods for mycotoxins is that the toxins are not uniformly distributed; fungal growth is often localized to suitable microclimate situations within the store, giving rise to 'hot-spots' of mycotoxin occurrence. Because of this problem, sampling for mycotoxin analyses

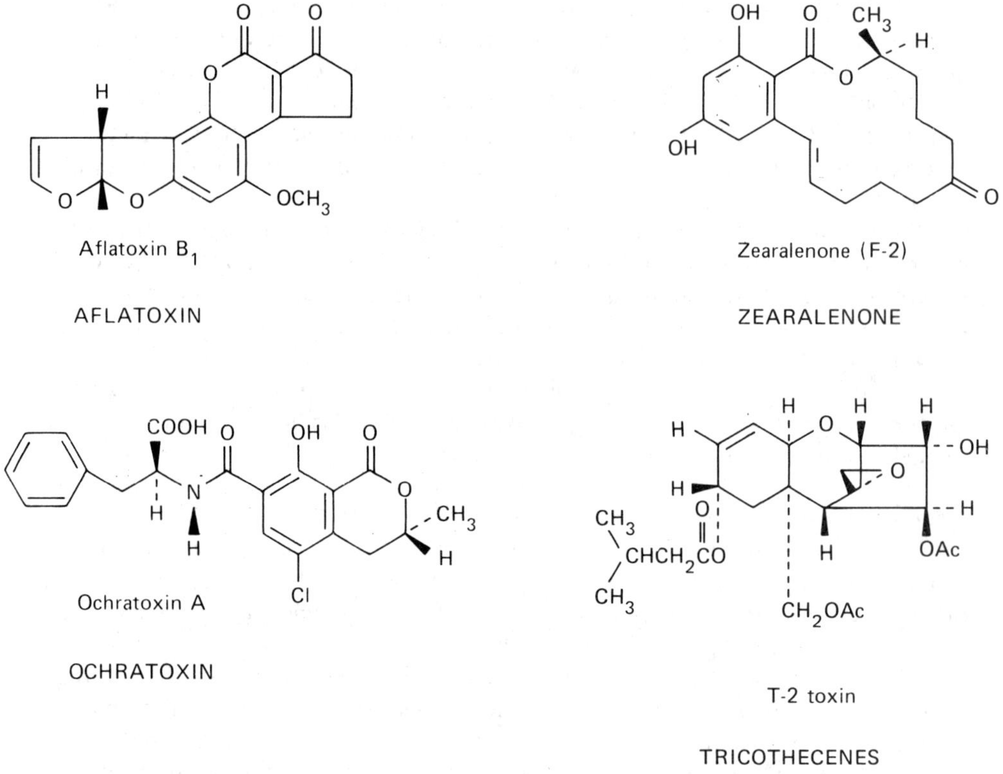

Fig. 6.17 — Chemical structures for mycotoxins.

requires very thorough procedures (see Section 6.2 [56]). A large number of samples must be taken and composited, with complete mixing and comminution to ensure that the analytical sample is representative. Specific regulations on sampling procedures are laid down for mycotoxin analyses [57].

The aflatoxins are of particular concern because of their carcinogenic properties. Aflatoxins are a group of similar compounds of which the major one, both in terms of occurrence and toxicity, is aflatoxin B_1. Analyses for aflatoxin B_1 are carried out both in human food (e.g. grain, dried fruits) and also in plant materials, such as grain, intended for animal feedstuffs. If contaminated feed is consumed by animals, residues of aflatoxins or of toxic metabolites (such as aflatoxin M_1 in milk) can occur in foods derived from the animals (e.g. meat, milk, eggs). Considerable attention is given to ensuring that milk-based human foods, particularly baby foods, are free of aflatoxin M_1 [58].

A similar situation arises with ochratoxin in pig meat. Pigs fed intensively on grain-based diets can accumulate residues of ochratoxin in the tissues. Countries with a large pig-meat industry are developing systems for routine screening of carcasses for residues of ochratoxin [59].

Analyses of foods for residues of mycotoxins are carried out by a wide variety of analytical techniques [60]. A review of the literature reveals that thin-layer chromatography (TLC), high-performance liquid chromatography (HPLC) and enzyme-linked immunosorbent assays (ELISA) are the major techniques used. As for other residue analyses, gas chromatography/mass spectrometry (GC/MS) methods increasingly are being reported for mycotoxin analyses. A general scheme of a reference method for the analysis of aflatoxins in plant foods is shown in Fig. 6.18.

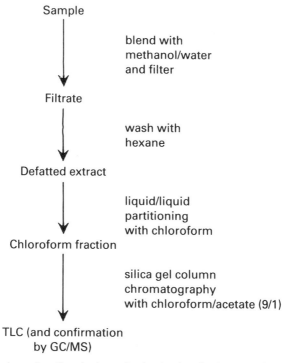

Sample

blend with
methanol/water
and filter

Filtrate

wash with
hexane

Defatted extract

liquid/liquid
partitioning
with chloroform

Chloroform fraction

silica gel column
chromatography
with chloroform/acetate (9/1)

TLC (and confirmation
by GC/MS)

Fig. 6.18 — General scheme for aflatoxin determination in plant food samples (reproduced from M. W. Trucksess, W. C. Brumley and S. Nesheim, *J. Assoc. Off. Anal. Chem.*, 1986, **67**, 973, by permission).

This procedure involves solvent extraction of the residue from the food, and clean-up of the sample extract by liquid/liquid partitioning and column chromatographic procedures, prior to determination of residue levels. The fluorescent property of aflatoxins allows for highly specific and sensitive determination of aflatoxin residues by using an HPLC with a fluorescence detector [61].

Immunological methods have found widespread use in analyses for mycotoxins (and other food residues). They have been applied both at the level of sample clean-up/concentration through immunoaffinity columns (IAC), and at the level of residue determination through immunoassays (radioimmunoassay, RIA or enzyme immunoassay, EIA) [62]. The production of antibodies which bind specifically the analyte of interest are the basis for immunological methods. Such antibodies, produced by animals immunized with the analyte of interest, or through additional tissue-culture techniques (monoclonal antibodies), are covalently bound to a gel to provide the immunoaffinity column. The sample extract is passed through the column and the analyte is selectively bound by the antibodies. The bound analyte is released from the column in a concentrated and purified state by elution with alcohol or buffer (Fig. 6.19).

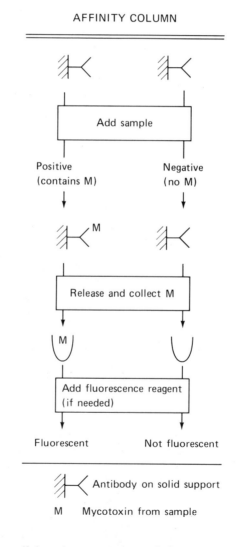

Fig. 6.19 — Immunoaffinity column extraction and clean-up procedure for mycotoxins (reproduced, with permission, from Ref. [62]).

Antibodies also form the basis for the determination technique of immunoassay. Competitive immunoassays depend on competition between a known quantity of the substance of interest and the unknown residue amount of the same substance in the sample extract for binding to a limited quantity of antibody. Markers, normally either radio-labelled material (RIA) or enzyme activity (EIA), are used to measure indirectly the amount of antibody-bound analyte. Figure 6.20 gives examples of different formats for immunoassays.

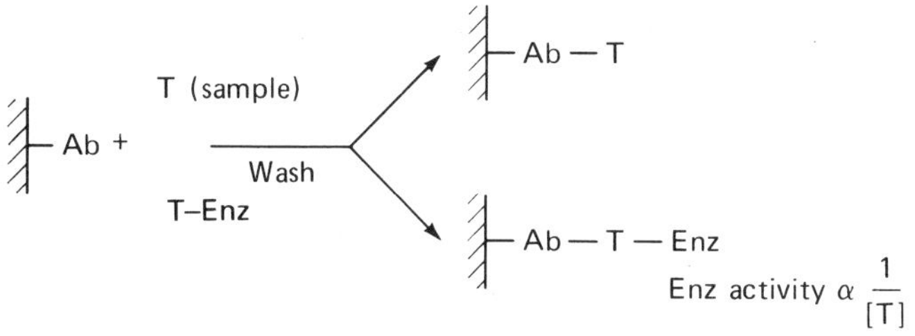

Competitive ELISA technique: Ab, antibody; T, toxin; T-Enz, enzyme-labelled toxin

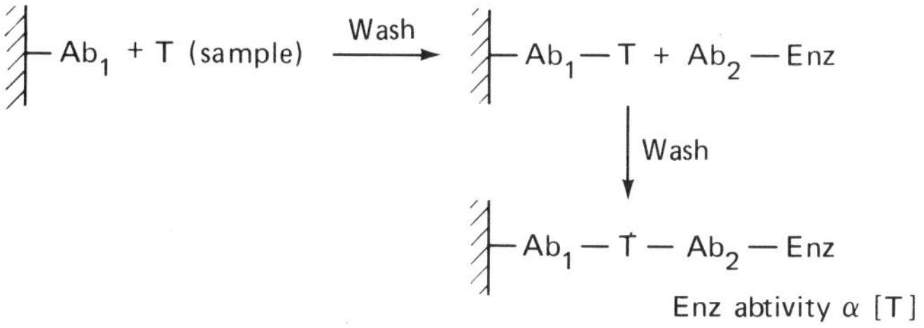

Non-competitive direct sandwich ELISA technique: T, toxin; Ab_1, antibody to first antigenic determinant on toxin; Ab_2-Enz, enzyme-labelled antibody to second antigenic determinant on toxin

Fig. 6.20 — Formats for immunoassays (reproduced from P. D. Patel in *Immunoassays in Food Analysis*, B. A. Morris and M. N. Clifford, eds., pp. 141–155, Elsevier Applied Science, London, 1985, by permission).

Immunoassays have been developed for all areas of residue analyses and are particularly useful as relatively rapid screening techniques for checking on food safety. Immunoassay kits are available for a broad range of residues in food. In addition, commercial test systems, in the form of cards, tubes and other simple systems, have been developed based on immunoassay, which can give a very rapid yes/no result on the presence or absence of a particular residue in a foodstuff (Fig. 6.21).

6.5.2 Agrochemicals and food safety

Modern, intensive agricultural peroduction systems utilize an array of agrochemicals. In animal production, veterinary drugs, such as antibiotics and hormones,

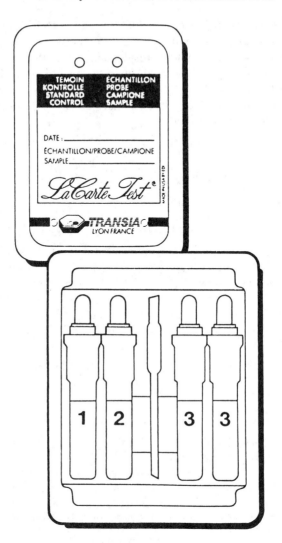

Fig. 6.21 — Card test systems for residue analyses.

insecticides, anthelmintics and growth-promoting agents are widely used. In plant production, pesticides, including fungicides, herbicides and insecticides, and fertilizers are used. These agrochemicals allow for very intensive production systems, but they are the cause of much consumer concern on food safety, so their use must be controlled and food must be tested for residues [63].

6.5.2.1 *Veterinary drugs*
The major veterinary drugs used in animal production are the antibiotics. Antibiotics are required to treat illness in individual animals (therapeutic use), but are also used to control or prevent disease outbreaks in intensive production systems (prophylactic

use), such as in poultry and pig production. In this context, medicated feed (i.e. animal feed containing levels of antibiotics lower than the therapeutic dose) is commonly used, not only for its disease prevention benefits but also for the improved performance and growth of animals which results. The potential food safety problems arising from the use of antibiotics in animal production are the occurrence of toxic residues in food and the induction of resistant strains of bacteria in animals and humans [64,65].

The very effective antibiotic, chloramphenicol, can cause a severe toxic reaction in some humans, even when occurring at residue levels in food. For this reason, chloramphenicol is banned for most uses in food-producing animals.

The problem of bacterial resistance is a complex and much more widespread one. The data in Fig. 6.22 illustrate the similarity of classes of antibiotics used in human medicine and animal husbandry.

The uses of antibiotics in animal husbandry are controlled mainly through requirements for a veterinarian's prescription and a record of use by the farmer. Foods of animal origin are tested for residues of antibiotics to ensure that these are absent or occur at levels below the maximum residue limit (MRL) set for each antibiotic. Apart from the potential human safety problem arising from residues of antibiotics in food, they are a food processing concern. Residues of antibiotics (such as penicillins) in milk may prevent the growth of bacterial cultures (e.g. *Streptococcus thermophilus*) used for cheese and yoghurt manufacture. While antibiotics are an important tool in animal husbandry, many of the problems associated with their use can be avoided by observance of suitable withholding periods between treatment of animals with antibiotics and use of the animal products (milk, meat, eggs) for human food [66].

Testing of foods for residues of antibiotics is carried out by a wide range of analytical techniques, but these can be divided into the two broad categories of microbiological assays and chemical assays. Microbiological assays are based on interference by the antibiotic residue in the food with the growth of a test bacterial organism. The advantage of microbiological assays is that they require little if any sample pretreatment. The disadvantages are the non-specific nature of these assays and the limited quantification possible. Microbiological assays for antibiotics utilize the measurement of interference in bacterial growth of various forms such as

(a) dimensions of zones of inhibition around samples of tissue,
(b) turbidity measurements of growth media as indicative of the degree of bacterial inhibition, and
(c) colorimetric determination of the inhibition of acid production in milk due to penicillin residues.

Chemical assays for antibiotic residues are specific and quantitative. Methods based on TLC, HPLC and GC/MS are common [67–69]. Determination of sulpha drugs in pork is an important example of such assays. Pork is required to contain sulphamethazine residues at levels below 0.1 ppm. A screening procedure widely used in pork slaughter plants is based on the established relationship between urine and muscle sulphamethazine residue concentrations [70]. A TLC kit method for

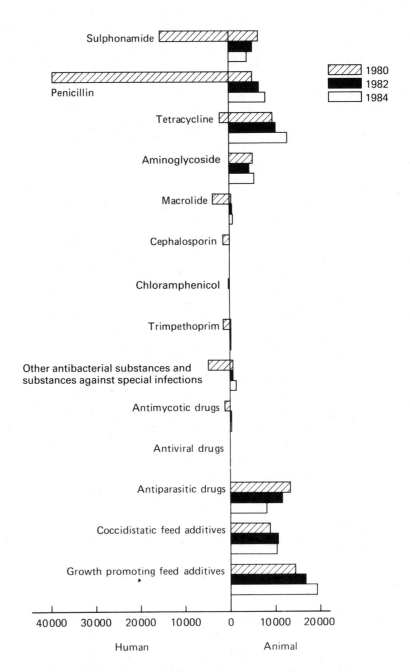

Fig. 6.22 — Consumption of antibiotics by humans and animals in Sweden during 1984 (reproduced from M. Wierup, C. Lowenhielm, M. Wold-Troell and I. Igenas *Vet. Res. Commun.*, 1987, **11**, 397, by permission).

screening pork urine samples, which has cut-off points of 0.4 and 1.3 ppm sulphamethazine in urine, allows for identification of carcasses containing sulphamethazine
residues at levels in excess of 0.1 ppm in kidney and muscle respectively. The
presence of sulphamethazine residues in tissue can be confirmed by an HPLC
method, as outlined in Fig. 6.23. This procedure involves extraction of residue

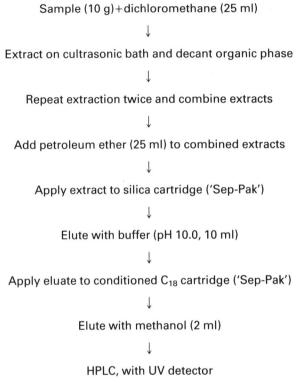

Sample (10 g)+dichloromethane (25 ml)

↓

Extract on cultrasonic bath and decant organic phase

↓

Repeat extraction twice and combine extracts

↓

Add petroleum ether (25 ml) to combined extracts

↓

Apply extract to silica cartridge ('Sep-Pak')

↓

Elute with buffer (pH 10.0, 10 ml)

↓

Apply eluate to conditioned C_{18} cartridge ('Sep-Pak')

↓

Elute with methanol (2 ml)

↓

HPLC, with UV detector

Fig. 6.23 — HPLC determination of sulphamethazine residues in meat (reproduced, with permission, from Ref. [71]).

from homogenized tissue with dichloromethane, by using sonication, and a clean-up/
concentration of extract with a dual solid-phase extraction procedure on silica and
C_{18} cartridges. The sulphamethazine residues are determined by reversed-phase
HPLC on a C_{18} column with a buffer–acetonitrile mobile phase and UV detection at
254 nm [71]. Residue levels of 0.01 ppm may be determined easily with this
procedure.

6.5.2.2 Growth-promoting agents
Improvements in meat production have always been directed at increasing the size
and quantity of lean meat from the individual animal. This has been done by
improved animal husbandry and feeding, by selection through breeding and, in the
last twenty years, by the use of growth-promoting agents. Although the gains to meat
production from better husbandry, feeding and breeding have been considerable,

the return from the use of growth-promoting agents has been the most significcant. Particularly for beef animals, production increases of up to 10% have been found to result [72].

The main classes of growth-promoting agents are the anabolic agents and the partitioning agents. Anabolic agents are either the naturally-occurring steroid hormones, oestradiol, progesterone and testosterone, or synthetic steroids or substances with steroid-like activity. The natural steroid hormones have both oestrogenic, gestagenic or androgenic effects and anabolic (i.e. muscle growth) effects when used in the correct combinations for particular types of animals. For example, combination treatments with oestradiol and testosterone give maximal growth increase for heifers, while combination treatments with oestradiol and progesterone give maximal growth increase for steers. The synthetic (or exogenous) substances, such as diethylstilboestrol, trenbolone or zeranol have very potent anabolic effects. Zeranol has the additional advantage of reduced oestrogenic side-effects on treated animals compared with oestradiol [73]. The anabolic agents are administered to animals by injection or as a feed additive (e.g. diethylstilboestrol), or as implanted pellets in a non-edible part of the animal, such as the ear (e.g. the natural steroids, trenbolone and zeranol).

The partitioning agents are β-adrenergic agonists, such as clenbuterol, salbutamol and cimaterol, which cause a selective development of muscular tissue at the expense of fat tissue. This is achieved by the action of these agents in reducing protein breakdown and/or enhancing protein formation. The partitioning agents are normally administered to animals as feed additives. Partitioning agents, unlike the anabolic agents, have been associated with the development of meat of poorer eating quality in terms of greater toughness [74].

Although some growth promoting agents are permitted to be used in many countries throughout the world, within Europe there is a total ban on the use of anabolic or partitioning agents for meat-production purposes [75]. Analyses for growth-promoting agents are undertaken in many laboratories to control the use of illicit substances and to guarantee the safety of meat for the consumer. The analytical approach within a regulatory situation normally is based on a dual system: rapid screening analyses on a random selection of animals or carcasses, with confirmatory analyses on carcasses identified as apparent positives in the screening assay. Where substances are banned for use, screening assays may be performed on the edible tissues [76] or on body fluids such as urine or bile [77]. Many screening assays are undertaken on urine because of the faster pre-treatment required relative to tissue samples. Immunoassays are commonly used for screening purposes, as they allow for large numbers of samples to be processed together [78]. Immunoassays used for screening purposes must be sufficiently specific so that very few false negatives result and the number of false positives is low. Where a significant number of false positives are identified by the screening assay, it is not effective in reducing the number of samples for the slower and more expensive confirmatory assay.

Flow diagrams for a typical screening immunoassay and for a GC/MS confirmatory assay are shown in Figs. 6.24 and 6.25. For banned substances, the screening and confirmatory assays may be undertaken on edible tissues or on body fluids or excreta. It is important that the sensitivities of the screening and confirmatory techniques are

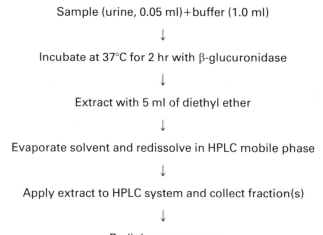

Sample (urine, 0.05 ml)+buffer (1.0 ml)

↓

Incubate at 37°C for 2 hr with β-glucuronidase

↓

Extract with 5 ml of diethyl ether

↓

Evaporate solvent and redissolve in HPLC mobile phase

↓

Apply extract to HPLC system and collect fraction(s)

↓

Radioimmunoassay

Fig. 6.24 — Screening assay for residues of growth-promoting agents by radioimmunoassay (reproduced from R. J. Heitzman, ed., *Residues in food producing animals and their products: reference materials and methods* (in press, 1991), by permission).

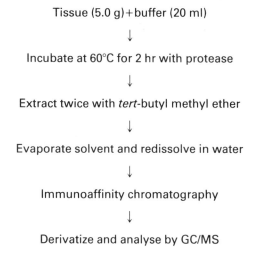

Tissue (5.0 g)+buffer (20 ml)

↓

Incubate at 60°C for 2 hr with protease

↓

Extract twice with *tert*-butyl methyl ether

↓

Evaporate solvent and redissolve in water

↓

Immunoaffinity chromatography

↓

Derivatize and analyse by GC/MS

Fig. 6.25 — Confirmatory assay for residues of growth-promoting agents by gas chromatography–mass spectrometry (reproduced from R. J. Heitzman, ed., *Residues in food producing animals and their products: reference materials and methods* (in press, 1991), by permission).

similar; if the screening assay is much more sensitive than the confirmatory assay, many of the samples found to be positive initially will not be confirmed in the second analysis. To deal with this problem, the analytical sensitivity of the screening method may not be used, but rather a practical 'action limit' is set for the screening procedure which is related to the sensitivity of the confirmatory technique. Only samples which are determined in the screening procedure as having residue levels above the 'action limit' are declared as apparent positives for re-analysis by the confirmatory procedure.

Apart from the regulatory analyses for banned substances, analyses for growth promoting agents in meat are undertaken to certify that food is safe for the consumer. In the case of many of the growth promoting agents, while measurable residue levels may be present in excreta or in organs such as the liver and kidney, residue levels in meat may be very low or non-detectable [79]. For this reason, residue determinations on meat are often insufficient to guarantee that the food is from animals untreated with growth promoting agents. The β-agonist, clenbuterol, is an example of such an agent where residue levels are detectable in liver or urine for periods after no residues are detectable in muscle [80]. To provide the consumer with confidence in meat, a multisystem approach is required including regulations, protocols for approved production practices, monitoring on farms and at slaughter plants, and analyses of the meat. Such an approach is the basis for a quality-assurance programme for meat production.

6.5.2.3 Pesticides

Pesticides are used in all types of food production and storage, and analyses for pesticide residues in food are a major analytical requirement. In plant production, typical uses of pesticides are as insecticides to prevent attack by insects on the seed or the growing plant, as herbicides to control the growth of weeds, and as fungicides to prevent damage to the plants from fungal growth. Stored grain, vegetables, and fruit are treated with fungicides to reduce losses during storage. In animal production, insecticides are commonly used to prevent infestation of animals with insect pests. Pesticides have been used in agriculture for well over a century and are considered now to be an essential (or at least a very necessary) element for intensive agricultural production systems. It has been estimated that pesticides prevent production losses of at least 30% and post-harvest losses (during storage) of at least 20%. In addition, they contribute to the quality of food, in terms of shape, colour, size, etc. [81].

There are many hundreds of pesticides, grouped according to basic molecular structure classes (Table 6.8). The organochlorine compounds (such as DDT,

Table 6.8 — Main classes of pesticides

Class	Examples
Chlorinated hydrocarbons	Lindane, Dieldrin
Organophosphorus compounds	Malathion, Phosmet
Carbamates	Methiocarb, Carbaryl
Triazines	Atrazine
Benzimidazoles	Benomyl
Nitriles	Bromoxynil
Pyrethroids	Permethrin

dieldrin, lindane, etc.) are highly persistent and are no longer approved in many countries for food uses because of the residue levels resulting in the food. Pesticides have been developed which combine both high efficacy and ready degradability, such as the organophosphorus and carbamate compounds, and a wide range of commercial products are available. Because of the widespread use of pesticides in agriculture it is to be expected that residues will occur in food. The regulatory approach for pesticides has been to register each compound for uses on particular crops/animals with dosage levels and stage of growth/withholding periods specified. In addition, a comprehensive list of Maximum Residue Limits (MRLs) have been specified for a wide range of plant and animal foodstuffs. The setting of MRLs (as mg of pesticide residue per kg of food) has been done mainly by the CODEX Committee of the FAO/WHO and these levels have been adopted into national regulations. MRLs are determined based on various factors, such as

(a) the toxicity of the particular pesticide,
(b) the residue levels likely to arise from good agricultural practice, and
(c) the importance of the particular foodstuff in the normal diet.

An example of a CODEX list of MRLs is shown in Table 6.9.

Table 6.9 — CODEX maximum residue limits (MRLs) for Heptachlor (organochlorine pesticide)

Commodity	Maximum residue limit (MRL) (mg/kg)
Carcase meat	0.2 (in the carcase fat)
Carrots	0.2
Citrus fruit	0.01
Cottonseed	0.02
Eggs	0.05 (on a shell-free basis)
Milk	0.006
Pineapples	0.01 (in the edible portion)
Poultry	0.2 (in the carcase fat)
Raw cereals	0.02
Soya bean oil (crude)	0.5
Soya bean oil (edible)	0.02
Tomatoes	0.02
Vegetables (except carrot, soyabean, sugar beets and tomatoes)	0.05

Analyses for pesticide residues in food depend heavily on GLC procedures, with several detectors. Many pesticide residue analyses are multi-residue procedures, such as those for organochlorine pesticides or organophosphorus pesticides. These procedures are designed to extract pesticides of a particular class from the food sample, involving clean-up steps which will result in 1–3 fractions for analysis by GLC, giving high recovery (>80%) of a wide range of pesticides of that class. A typical multi-residue method for organochlorine pesticides in fatty foods (Fig. 6.26)

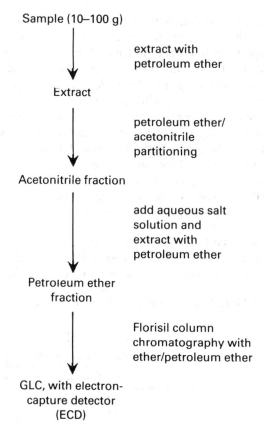

Fig. 6.26 — Multi-residue method for organochlorine pesticide residues in fatty foods (reproduced from J. N. Seiber in *Pesticide Residues in Food*, pp. 142–152, Technomic Publishing, Lancaster, Pennsylvania, 1988, by permission).

allows for the determination of over 15 compounds and metabolites in two fractions from a Florisil column, using an electron-capture detector (ECD) [82]. The fat is extracted from the ground/homogenized sample with petroleum ether and the pesticide residues are transferred by liquid/liquid partitioning to acetonitrile. After back-extraction from the acetonitrile phase into petroleum ether, the sample extract is cleaned up on a Florisil column and residues are eluted in two fractions of 6% and 15% diethyl ether in petroleum ether, respectively. By using the more selective flame-photometric (FPD) or thermionic specific (TSD) detectors, over 20 organophosphorus compounds can be determined in non-fatty foods by using a less extensive sample clean-up procedure (Fig. 6.27). The pesticides are extracted from the food with acetone and cleaned up by partitioning between dichloromethane and petroleum ether [83]. The organophosphorus pesticide residues can be determined on the concentrated extract by using the selective detectors.

GLC or gradient HPLC systems are used to obtain good resolution of the pesticides (Fig. 6.28). Much advance in pesticide residue analysis has been made

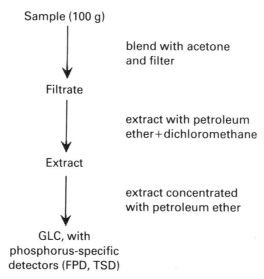

Fig. 6.27 — Multi-residue method for organophosphorus pesticide residues in non-fatty foods (reproduced from J. N. Seiber in *Pesticide Residues in Food*, pp. 142–152, Technomic Publishing, Lancaster, Pennsylvania, 1988, by permission).

through the development of faster, more efficient and automated sample extraction/ clean-up procedures. Automated gel-permeation chromatography (GPC) [84] and solid-phase extraction (SPE) [85] systems have also been developed for pesticide residue analyses.

Under European Community and national regulations, MRLs have been set for over 100 pesticides in fruit and vegetables, and for approximately 20 and 10 pesticides in cereals and in meat/milk/eggs, respectively [86]. The food supply is monitored for residues of these pesticides [87], and actions are taken to ensure that the food supply does not contain residue levels in excess of the MRLs. Detectable residues are found in a significant number of the food samples analysed, but levels in excess of the MRLs are found only in less than 1% of samples. Monitoring for pesticide residues in food is carried out on the foodstuff as it is traded; in the case of citrus fruit this means on the whole fruit including skin, and in the case of meat it means in fresh, uncooked meat. A number of factors can reduce the actual exposure of the consumer to pesticide residues in food, including commercial processing procedures, peeling and/or washing of fruit and vegetables and heating of food during cooking [88]. The less persistent types of pesticides are most subject to reduction during heating, such as the organophosphorus and carbamate pesticides. To get a more accurate measure of human exposure to pesticide residues in food, total diet studies are undertaken [89]. For these studies, the cooked meal is taken as a representative sample of the diet and analysed for residue content.

Monitoring programmes such as those described above suggest that the human exposure to pesticides from the diet is, in general, not a serious food safety issue. Most problems of human safety result from gross contamination (accidental or

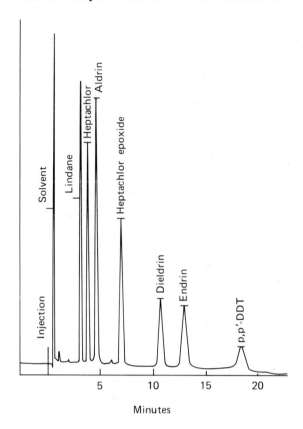

Fig. 6.28 — Gas–liquid chromatogram of organochlorine pesticide residues.

deliberate) of food with pesticides. Operatives involved in the handling, preparation or application of pesticides to crops are also at risk.

6.6 CONCLUSION

Chemical analysis of animal feed and human food is a complex and challenging area of scientific endeavour. It occupies vast resources worldwide in terms of laboratory space, personnel and analytical equipment. It utilizes a comprehensive range of analytical techniques and instrumentation. It covers all areas of the food production system, from a service to basic research in animal and plant husbandry through to nutritional labelling of food products; along the way it covers research and development, monitoring of production, quality assurance and regulatory testing.

    This chapter gives only a short sketch of some of the major aspects of compositional, nutritional and residue analyses. There are many excellent text-books and journals covering the principles and practice of food and feed analysis which will give more information about the topics covered here.

In summary, food and feed analyses are directed to enhancing the quality and safety of human food and offer an interesting and stimulating (if at times frustrating) activity to the analyst.

REFERENCES

[1] W. G. Cochran, *Sampling Techniques*, 2nd Ed. Wiley, New York, 1963.
[2] Association of Official Analytical Chemists, *Official Method of Analysis*, 15th Ed., K. Helrich, ed., Washington, D.C., 1990.
[3] International Standards Organisation, *Meat and meat products — Sampling — Part 1: Taking primary samples*, ISO 3100-1: 1975.
[4] Y. Pomeranz and C. E. Meloan, *Food Analysis: Theory and Practice*, 2nd Ed., Van Nostrand Reinhold, New York, 1987.
[5] Ref. [2], p. 931.
[6] Ref. [2], p. 912.
[7] J. D. Pettinati, *J. Assoc. Off. Anal. Chem.*, 1975, **58**, 1188.
[8] W. A. Dewar and P. McDonald, *J. Sci. Food Agric.*, 1961, **12**, 790.
[9] R. Bernetti, S. J. Kochan and J. J. Pienkowski, *J. Assoc. Off. Anal. Chem.*, 1984, **67**, 299.
[10] B. G. Osborne and T. Fearn, *Infrared Spectroscopy in Food Analysis*, Longman, London, 1986.
[11] M. O'Keeffe, G. Downey and J. C. Brogan, *J. Sci. Food Agric.*, 1987, **38**, 209.
[12] A. L. Lakin, 'Determination of nitrogen and estimation of protein in foods' in *Developments in Food Analysis Techniques*, Vol. 1, R. D. King, ed., Applied Science Publishers, London, 1978.
[13] P. Morries, *J. Assoc. Publ. Analysts.*, 1983, **21**, 53.
[14] L. T. Mann, *Anal. Chem.*, 1963, **35**, 2179.
[15] P. C. Williams, *Analyst*, 1964, **89**, 276.
[16] W. Schmuetz, *Monatsschr. Brauwiss.*, 1989 **42** (7), 297.
[17] Ref. [2], p. 79.
[18] International Standards Organisation, *Oilseeds — determination of hexane extract (or light petroleum extract), called "oil content"*. ISO 659: 1988.
[19] International Dairy Federation, *Dried milk, dried whey, dried buttermilk and dried butter serum. Determination of fat content (Rose Gottlieb reference method)*. IDF Standard No. 9C: 1987.
[20] C. Strugnell, *Irish J. Fd. Sci. Technol.*, 1989, **13** (1), 71.
[21] J. D. Pettinati and C. E. Swift, *J. Assoc. Off. Anal. Chem.*, 1975, **58** (6), 1182.
[22] R. Tkachuk, *J. Amer. Oil Chem. Soc.*, 1981, **57**, 819.
[23] D. H. Kropf, *J. Food Quality.*, 1984, **6**, 199.
[24] A. R. Crosland and N. Bratchell, *J. Assoc. Publ. Analysts*, 1988, **26**, 89.
[25] N. T. Crosby, R. F. Eaton, D. M. Groffman and G. H. Merson, *J. Assoc. Publ. Analysts*, 1989, **27**, 55.
[26] P. J. Van Soest and R. H. Wine, *J. Assoc. Off. Agric. Chem.*, 1967, **50**, 50.
[27] D. J. Minson, 'The measurement of digestibility and voluntary intake of forages with confined animals', in *Forage Evaluation: concepts and techniques,* J. L. Wheeler and R. D. Mochrie, eds., p. 159. CSIRO/American Forage and Grassland Council, Victoria, 1981.
[28] J. M. A. Tilley and R. A. Terry, *J. Br. Grassl. Soc.*, 1963, **18**, 104.
[29] P. Narasimhalu, *Anim. Feed Sci. Technol.*, 1985, **12**, 101.
[30] Ref. [2], p. 70.
[31] International Standards Organisation, *Meat and meat products — determination of chloride content (Reference method)*, ISO 1841: 1981.
[32] D. H. Spackman, W. H. Stein and S. Moore, *Anal. Chem.*, 1958, **34**, 544.
[33] W. R. Peterson and J. J. Warthesen, *J. Food Sci.*, 1979, **44**, 994.
[34] D. Firestone and W. Horwitz, *J. Assoc. Off. Anal. Chem.*, 1979, **62** (4), 709.
[35] R. S. Kirk, R. E. Mortlock, W. D. Pocklington and P. Roper, *J. Sci. Fd. Agric.*, 1978, **29**, 880.
[36] J. H. Lane and L. Eynon, *J. Soc. Chem. Ind.*, 1923, **42**, 32T.
[37] Ref. [2], p. 1023.
[38] M. Sher Ali, *J. Assoc. Off. Anal. Chem.*, 1988, **71** (6), 1097.
[39] M. S. Hui, *Analyst*, 1984, **109**, 1503.
[40] H. U. Bergmeyer, E. Bernt, F. Schmidt and H. Stork, in *Methods of Enzymatic Analysis*, H. U. Bergmeyer, ed., 2nd Ed., Vol. III, Academic Press, New York, 1974.
[41] N. T. Crosby, *Analyst*, 1977, **102**, 225.

[42] G. Tolg, 'Wet oxidation procedures', in *Methodicum Clinicum, Vol. I — Analytical Methods*, F. Korte, ed., Part B, 698, Academic Press, New York, 1974.
[43] T. T. Gorsuch, *The Destruction of Organic Matter*, Pergamon Press, Oxford, 1975.
[44] Ref. [2], p. 870.
[45] Ref. [4], p. 139.
[46] Ref. [2], p. 87.
[47] D. A. Corlett, *Food Technol.*, 1989, **43** (2), 91.
[48] P. R. Cheeke (ed.), *Toxicants of Plant Origin,*, CRC Press, Boca Raton, Florida, 1989.
[49] H. B. S. Conacher and R. K. Chadha, *J. Assoc. Off Anal. Chem.*, 1974, **57** (5), 1161.
[50] Commision of the European Communities, *Directive relating to the fixing of the maximum level of erucic acid in oils and fats intended for human consumption and in food stuffs containing added oil or fat. O.J.* No. L 202, 35, 1976.
[51] B. A. Slominski and L. D. Campbell, *J. Sci. Fd. Agric.*, 1987, **40**, 131.
[52] J. P. Sang and R. J. W. Truscott, *J. Assoc. Off. Anal. Chem.*, 1984, **67** (4), 829.
[53] G. R. Fenwick, I. T. Johnson and C. L. Hedley, *Trends Fd. Sci. Technol.*, 1990, **1**, 23.
[54] M. R. A. Morgan, and H. A. Lee, 'Mycotoxins and natural food toxicants', in *Development and Application of Immunoassay for Food Analysis*, J. H. Rittenburg, ed., 143, Elsevier, London, 1990.
[55] R. S. Hatfull, I. Milner and V. Stanway, *J. Assoc. Publ. Analysts*, 1980, **18**, 19.
[56] D. L. Park and A. E. Pohland, *J. Assoc. Off. Anal. Chem.*, 1989, **72**, 399.
[57] Commission of the European Communities, *Directive establishing community methods of sampling for the official control of feeding stuffs (76/371/EEC). O.J.* No. L 102, 1.
[58] J. Ferguson-Foos and J. D. Warren, *J. Assoc. Off. Anal. Chem.*, 1984, **67**, 1111.
[59] D. M. Rousseau, G. A. Slegers and C. H. Van Peteghem, *J. Agric. Fd. Chem.*, 1986, **34**, 862.
[60] Y. Ueno, 'Mycotoxins', in *Toxicological Aspects of Food*, K. Miller, ed., p. 139, Elsevier, London, 1987.
[61] R. D. Coker, 'High performance liquid chromatography and other chemical quantification methods used in the analysis of mycotoxins in foods', in *Analysis of Food Contaminants*, J. Gilbert, ed., p. 207, Elsevier, London, 1984.
[62] T. J. Hansen, *Tr. Fd. Sci. Technol.*, 1990, **1**, 83.
[63] Commission of the European Communities, *Directive concerning the examination of animals and fresh meat for the presence of residues (86/469/EEC) O.J.* No. L 275, 36.
[64] S. D. Homburg, M. T. Osterholm, K. A. Senger and M. L. Cohen, *N. Engl. J. Med.*, 1984, **311**, 617.
[65] J. R. Walton, *Vet. Rec.*, 1988, **122**, 249.
[66] J. Egan, *Irish Vet. J.*, 1985, **39**, 56.
[67] M. H. Thomas, K. E. Soroka and S. H. Thomas, *J. Assoc. Off. Anal. Chem.*, 1983, **66**, 881.
[68] D. J. Fletouris, J. E. Psomas and N. A. Botsoglou, *J. Agric. Food Chem.*, 1990, **38**, 1913.
[69] J. E. Matusik, R. S. Sternal, C. J. Barnes annd J. A. Sphon, *J. Assoc. Off. Anal. Chem.*, 1990, **73**, 529.
[70] V. W. Randecker, J. A. Reagan, R. E. Engel, D. L. Soderberg and J. E. McNeal, *J. Food Prot.*, 1987, **50**, 115.
[71] N. Haagsma, R. J. Nooteboom, B. G. M. Gortemaker and M. J. Maas, *Z. Lebensm. Unters. Forsch.*, 1985 **181**, 194.
[72] M. G. Keane, *Ir. J. Agric. Res.*, 1987, **26**, 165.
[73] C. K. Parekh and F. Coulston 'Determination of the hormonal no-effect level of zeranol in non-human primates', in *Anabolics in Animal Production — Public Health Aspects, Analytical Methods and Regulation*, E. Meissonnier, ed., p. 353, Office International des Epizooties, Paris, 1983.
[74] D. H. Kretchmar, M. R. Hathaway, R. J. Epley and W. R. Dayton, *J. Anim. Sci.*, 1990, **68**, 1760.
[75] Commission of the European Communities, *Directive prohibiting the use in livestock farming of certain substances having a hormonal action (85/649/EEC) O.J.* No. L 382, 228.
[76] M. O'Keeffe and J. P. Hopkins, *J. Chromatogr. (Biomed. Applic.)*, 1989, **489**, 199.
[77] M. J. Warwick, M. L. Bates and G. Shearer, 'The use of immunoassay for monitoring anabolic hormones in meat', in *Immunoassays in Food Analysis*, B. A. Morris and M. N. Clifford, eds., p. 169, Elsevier, London, 1985.
[78] R. J. Heitzman, 'Immunoassay techniques for measuring veterinary drug residues in farm animals, meat and meat products', in *Analysis of Food Contaminants*, J. Gilbert, ed., p. 73, Elsevier, London, 1984.
[79] M. O'Keeffe 'Tissue levels of the anabolic agents, Trenbolone and Zeranol, determined by radioimmunoassay', in *Advances in Steroid Analysis '84*, p. 225, Elsevier, Amsterdam, 1985.
[80] H. H. D. Meyer and L. Rinke, 'Residues after anabolic dosage of clenbuterol', in *The Control of Fat and Lean Deposition*, University of Nottingham 51st Easter School in Agricultural Science, April, 1991.

[81] C. Ahrens and H. H. Cramer, 'Improvement of agricultural production by pesticides', in *Environment and Chemicals in Agriculture*, F. P. W. Winteringham, ed., p. 151, Elsevier, London, 1984.

[82] International Dairy Federation, *Milk and milk products: recommended methods for determination of organochlorine pesticide residues*, IDF Standard 75B: 1983.

[83] M. A. Luke, J. E. Froberg, G. M. Doose and H. T. Masumoto, *J. Assoc. Off. Anal. Chem.*, 1981, **64**, 1187.

[84] M. L. Hopper and K. R. Griffitt, *J. Assoc. Off. Anal. Chem.*, 1987, **70**, 724.

[85 J. Manes Vinuesa, J. C. Molto Cortes, C. Igualada Canas and G. Font Perez, *J. Chromatogr.*, 1989, **472**, 365.

[86] Commission of the European Communities, *Directive on the fixing of maximum limits for pesticide residues in and on foodstuffs of animal origin* (86/363/EEC). *O.J.* No. L 221, 43.

[87] Ministry of Agriculture, Fisheries and Food, *Report of the Working Party on Pesticide Residues: 1985–88*, HMSO, 1989.

[88] M. E. Zabik, 'Pesticides and other industrial chemicals', in *Toxicological Aspects of Food*, K. Miller, ed., p. 73, Elsevier, London, 1987.

[89] M. J. Gartrell, J. C. Craun, D. S. Podrebarac and E. L. Gunderson, *J. Assoc. Off. Anal. Chem.*, 1985, **68**, 1184.

Index